S0-EFM-528

Advances in Physical Organic Chemistry

Advances in Physical Organic Chemistry

Volume 23

Edited by

D. BETHELL

The Robert Robinson Laboratories
University of Liverpool
P.O. Box 147, Liverpool L69 3BX

ACADEMIC PRESS 1987
Harcourt Brace Jovanovich, Publishers

London Orlando
San Diego New York Austin
Boston Sydney Tokyo Toronto

ACADEMIC PRESS INC. (LONDON) LTD
24/28 Oval Road
London NW1 7DX

United States Edition published by
ACADEMIC PRESS INC.
Orlando, Florida 32887

ISBN 0-12-033523-9
ISSN 0065-3160

TYPESET BY BATH TYPESETTING LTD., BATH, U.K.
AND PRINTED IN GREAT BRITAIN BY ST. EDMUNDSBURY PRESS,
BURY ST. EDMUNDS.

Contents

Contributors to Volume 23

Richard A. Henderson AFRC Unit of Nitrogen Fixation, The University of Sussex, Brighton, BN1 9RQ, U.K.

Sundus Henderson AFRC Unit of Nitrogen Fixation, The University of Sussex, Brighton, BN1 9RQ, U.K.

Michael I. Page Department of Chemical and Physical Sciences, The Polytechnic, Queensgate, Huddersfield, HD1 3DH, U.K.

Hans-Ullrich Siehl Institute of Organic Chemistry, The University of Tübingen, D-7400 Tübingen 1, Auf der Morgenstelle 18, Germany

Glen A. Russell Department of Chemistry, Iowa State University, Ames, Iowa 50011, U.S.A.

The Nucleophilicity of Metal Complexes Towards Organic Molecules

SUNDUS HENDERSON and RICHARD A. HENDERSON

A.F.R.C. Unit of Nitrogen Fixation, University of Sussex, Brighton BN1 9RQ, U.K.

ADVANCES IN PHYSICAL ORGANIC CHEMISTRY
VOLUME 23 ISBN 0 12 033523 9

1 The scope of the review

Nucleophilic aliphatic substitution, as defined by (1), and typified in organic chemistry by the Finkelstein and Menschutkin reactions, can extend into inorganic chemistry with the inclusion of metal complexes which are sufficiently nucleophilic. The aim of this review is to collect the somewhat

$$Y^- + R{-}X \longrightarrow R{-}Y + X^- \tag{1}$$

diverse information which exists on the nucleophilicity of certain metal complexes towards organic molecules. Although this class of reaction has often been included in previous reviews, the main emphasis of these articles has been to compare the nucleophilic and redox pathways of these complexes, rather than to give a detailed discussion of the nucleophilicity.

This review will concentrate on the detailed mechanistic study of the reactions between metal-complex nucleophiles and organic molecules including the stereochemical consequences of the reactions at both the carbon centre (Sections 6 and 7) and the metal (Sections 8 and 9). Furthermore, the influence of substituents on the carbon centre (Sections 11 and 12) and the nature of the solvent (Section 5) will be discussed. However, neither the nucleophilicity of ligands towards organic substrates nor the synthesis of nucleophilic complexes will be covered. The reader interested in the latter is recommended to the reviews by King (1975) and Vaska (1968).

2 Introduction to metal complexes as nucleophiles

The idea of a simple metal complex acting as a nucleophile is conceptually relatively simple. All that is required is a metal centre (the metal and its coligands) which is relatively electron-rich, and which has a vacant site, or a potentially vacant site (via rapid dissociation of a ligand). This is typified by the reaction shown in (2) between the five-coordinate, square-based pyramidal $[Co(dmgH)_2py]^-$ (py = pyridine, dmgH = dimethylglyoxime) and RX.

$$[Co(dmgH)_2py]^- + R{-}X \longrightarrow [Co(R)py(dmgH)_2] + X^- \tag{2}$$

The analogy between (1) and (2) is obvious. However, if the metal nucleophile contains more than one vacant site the liberated group X^- can bind to the metal, as shown in the archetypal reaction of this type shown in (3), involving the square-planar *trans*-$[IrCl(CO)(PPh_3)_2]$.

$$trans\text{-}[IrCl(CO)(PPh_3)_2] + MeI \longrightarrow [Ir(Cl)(I)Me(CO)(PPh_3)_2] \tag{3}$$

It is worthwhile discussing the salient features of reaction (3) at this stage, since they will encompass many of the problems which will recur throughout this chapter. Close inspection of (3) shows that the coordinatively-unsaturated, square-planar iridium complex reacts with methyl iodide to give the coordinatively-saturated, octahedral complex in which both the methyl- and iodo-groups are separately bound to the metal. In accomplishing this reaction the formal $Ir^{(I)}$ site has been oxidised to a formal $Ir^{(III)}$ complex, but only because both the methyl- and iodo-groups are considered as anions when bound to the metal. It is because of this formalism that this type of reaction has become known as "oxidative addition", in which both the formal oxidation-state and coordination number have increased by two units. Ambiguities clearly arise from this formalism primarily because of the inability to define unambiguously oxidation-states, and because although some such reactions are oxidative (e.g. the reaction of Cl_2 with $[Pt^{(II)}L_2Cl_2]$ to give $[Pt^{(IV)}L_2Cl_4]$), others are not, as for instance in the reactions of hydrogen with *trans*-$[IrCl(CO)(PPh_3)_2]$ as shown in (4), in which the classic reducing agent apparently oxidises the metal. Furthermore, in contrast to

$$trans\text{-}[IrCl(CO)(PPh_3)_2] + H_2 \longrightarrow [IrCl(CO)(H)_2(PPh_3)_2] \quad (4)$$

what is expected for an oxidising reaction, the addition of hydrogen to certain iridium complexes is favoured by electron-accepting ligands (Crabtree and Hlatky, 1980). Because of these conceptual problems a new notation for classifying these reactions has been proposed (Crabtree and Hlatky, 1980) in which the three-centre, oxidative additions of (3) and (4) are designated {3,2} reactions. The first figure in parentheses designates the number of centres involved in the reaction and the second figure gives the number of electrons transferred to the metal. The simple ligand addition shown in (5) is designated {2,2}.

$$[FeH(Ph_2PCH_2CH_2PPh_2)_2]^+ + py \longrightarrow [FeH(py)(Ph_2PCH_2CH_2PPh_2)_2]^+ \quad (5)$$

Although this system has the advantage of describing what is happening to the metal during the reaction its use is not essential provided one is conscious of the underlying assumptions in the formalism "oxidative addition". Thus in both reactions (3) and (4) the metal's d-orbital occupancy changes from d^{n-2} to d^n, but never fully becomes d^n since there is (as a molecular orbital description reveals) a two-way flow of electron density. There is no requirement that the flow is balanced at any stage of the reaction (Saillard and Hoffmann, 1984). Both the positions of the coordinated carbon monoxide stretching frequencies in the infrared spectrum of the oxidative addition products of *trans*-$[IrCl(CO)(PPh_3)_2]$ (Vaska, 1968), and X-ray photoelectron spectroscopy of the adducts of $[Ir(phen)(cod)]^+$ (phen = 1,10-phenanthroline; cod = 1,4-cycloctadiene) (Louw *et al.*, 1982) have been used to probe the metal-to-ligand and ligand-to-metal charge transfer. In both studies the MeI adduct has considerable metal-to-ligand charge transfer, and the "oxidation numbers" of the metal (2.58 and 2.48) are in good agreement with the picture described above.

Having clarified what is meant by the term oxidative addition, it is clear that the description is purely stoichiometric and can cover a range of mechanistic pathways. The factors which increase the nucleophilicity of the site (i.e. a relatively low oxidation state and electron-releasing ligands) are the same factors which favour its behaviour as a reductant. As shown by the example in Scheme 1, although a reaction may stoichiometrically resemble a nucleophilic pathway of the form of (1), this cannot be assumed and the simple second-order rate law (first-order in both reactants) cannot distinguish between the two mechanisms.

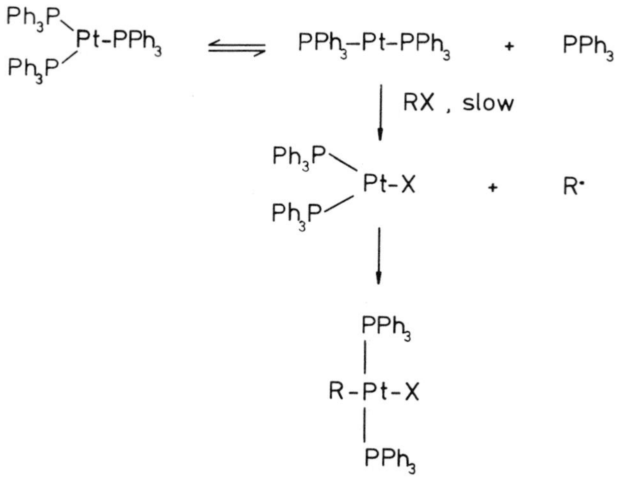

$$(Ph_3P)_2Pt\text{-}PPh_3 \rightleftharpoons PPh_3\text{-}Pt\text{-}PPh_3 + PPh_3$$

$$PPh_3\text{-}Pt\text{-}PPh_3 \xrightarrow{RX,\ \text{slow}} (Ph_3P)_2Pt\text{-}X + R^{\cdot}$$

$$(Ph_3P)_2Pt\text{-}X + R^{\cdot} \longrightarrow R\text{-}Pt(PPh_3)_2\text{-}X$$

Scheme 1

Obviously a discussion of the oxidative addition reactions which proceed by an electron-transfer mechanism is outside the scope of this review; for a thorough appraisal of this area the reviews by Halpern (1970), Lappert and Lednor (1976), Stille and Lau (1977), Deeming (1972, 1974), Kochi (1978), together with the recent article by Hill and Puddephatt (1985) and references therein, should be consulted. For the remainder of this chapter we will be concerned entirely with systems in which the nucleophilicity of the metal complex has been either unambiguously demonstrated, or at least strongly implicated. It is important to be conscious of the spectrum of transition-state structures which could exist in the interaction between a metal centre and a molecule of RX as shown in Fig. 1.

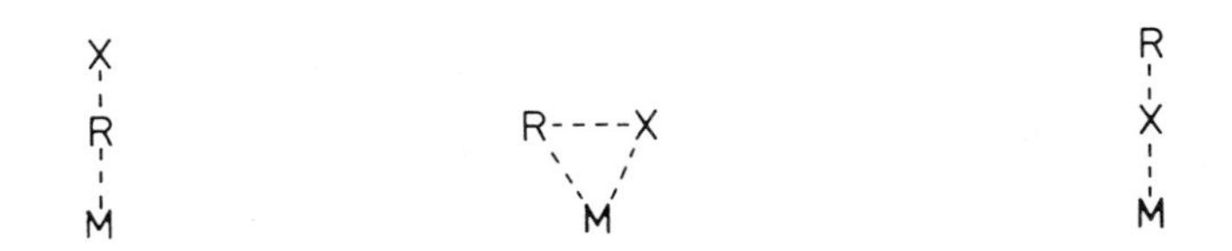

FIG. 1 Spectrum of transition states for the reaction between a metal, M and alkyl halide, RX

At the extremes of this series the metal interacts with the carbon centre and the group X exclusively, whereas an intermediate form involves a three-centred transition state where equal interaction between the metal and both the carbon centre and X occurs. Strictly, only the transition states involving the interaction between the metal centre and the carbon centre are of relevance to the discussion herein. For this reason we have included briefly the mechanistic studies which are beginning to appear in the literature on the homogeneous activation of hydrocarbons.

3 Types of metal-based nucleophiles

Before discussing in detail the nucleophilicity of metal complexes towards organic molecules it is important to appreciate the depth of reactivity types that is being considered. Predominantly the organic substrates are alkyl halides and tosylates, acyl halides, aryl halides and allyl halides together with a limited amount of information of the interactions with propargyl halides, and epoxides.

It is no trivial task to classify the types of metal nucleophiles, but at present it is more sensible to categorise them in terms of their structure than their reactivity. In Table 1 the nucleophiles have been grouped according to their coordination number; this refers, however, to the "reactive" coordination number. Thus although the parent molecule is $[Pd(PPh_3)_3]$ or $[Pd(PPh_3)_2(C_2H_4)]$ the reactive species is the two coordinate $[Pd(PPh_3)_2]$. In some of the complexes the π-bonded η^5-cyclopentadienyl-ligand is present, and is considered to occupy three coordination sites. For an introduction to the nomenclature and bonding of these π-type ligands the reader should consult Cotton and Wilkinson (1980). Two obvious points should be made about this table. The first is that derivatives of some of the complexes have not been included, such as the relatively trivial perturbation of using substituted cyclopentadienyl-ligands. Secondly, the inclusion of a particular complex in this table does not imply that *all* its reactions with organic molecules proceed by a nucleophilic pathway, only that there is good evidence that it does behave as a nucleophile under certain circumstances.

TABLE 1

Types of metal nucleophiles

Coordination number	Complex	Comment
2	$[M(PR_3)_2]$	M = Pd or Pt, R = alkyl or aryl
3	$[MR_3]^-$	M = Sn, Ge or Pb
4	*trans*-$[MX(L)(PR_3)_2]$	M = Rh or Ir, L = CO or N_2, X = halide
	$[M(CO)_4]^{n-}$	M = Co, $n = 1$; M = Fe, $n = 2$; also protonated iron complex, $[FeH(CO)_4]^-$
	$[Rh(CN)_4]^{3-}$	
	$[(\eta^5\text{-}C_5H_5)Ni(CO)]^-$	
5	$[M(dmgH)_2L]^-$	M = Co or Rh, L = a range of neutral two-electron donor ligands
	$[M(CO)_5]^-$	M = Mn or Re
	$[(\eta^5\text{-}C_5H_5)M(CO)L]^{n-}$	$n = 1$, M = Fe or Ru, L = CO; $n = 0$, M = Co, Rh or Ir, L = CO, PPh_3 or $\eta^2\text{-}C_2H_4$
6	$[(\eta^5\text{-}C_5H_5)M(CO)_3]^-$	M = Cr, Mo or W

The reactivity of these complexes will not be discussed individually, but rather the various features of the nucleophilic substitution reaction will be outlined and the complexes' behaviour will be compared and contrasted in this respect.

4 The scale of nucleophilicity

As will be seen in this section, metal complexes can be some of the strongest nucleophiles known, and they have been referred to as supernucleophiles. Furthermore, by subtle changes in the nature of the ligands, the nucleophilicity of the complexes can be varied. Thus although $[RhCl(PPh_3)_3]$ reacts readily with MeI as shown in (6), substitution of one of the phosphine ligands by carbon monoxide to give *trans*-$[RhCl(CO)(PPh_3)_2]$ results in a molecule with a much diminished reactivity towards MeI. This difference in

$$[RhCl(L)(PPh_3)_2] + MeI \longrightarrow [Rh(Me)Cl(I)L(PPh_3)_2] \quad (L = CO \text{ or } PPh_3) \qquad (6)$$

reactivity is a consequence of the greater electron-withdrawing capability of carbon monoxide over the phosphine ligands, thus decreasing the electron density and hence the nucleophilicity of the metal centre (Collman and Roper, 1968 and references therein).

THE INFLUENCE OF LIGANDS ON THE NUCLEOPHILICITY

One of the most common ligands, particularly in this area of chemistry, is the monotertiary phosphine. This ligand is amenable to variation by simple perturbation of the alkyl- and aryl-substituents. Several studies have been reported in which the relative nucleophilicities of the metal complexes have been correlated, on a semi-quantitative basis, with the electron-releasing capability of the phosphine ligand.

The rate of the reaction between MeI and *trans*-$[RhCl(CO)(PR_3)_2]$ decreases with the nature of R in the order: n-$C_8H_{17} \geqslant$ n-$C_4H_9 >$ *p*-$C_4H_9C_6H_4 >$ *p*-$C_2H_5C_6H_4 >$ *p*-$C_6H_{13}C_6H_4 > C_6H_5$. The analysis of the kinetic data for this reaction is complicated by phosphine dissociation from the complex and its subsequent quarternisation by MeI (Franks *et al.*, 1981). The faster reactions with the alkyl substituents are a consequence of the greater electron-releasing capability of these groups. It has been proposed that as the number of carbon atoms increases beyond four, the basicity of the phosphine ligand is not markedly affected and so the trend observed for the first three phosphines reflects the steric influence of these groups. The influence of the *p*-substituents on the reaction rate is essentially that which would be expected on the basis of the electron-releasing effect of the substituent, except that the position of *p*-$C_4H_9C_6H_4$ is slightly anomalous and this may represent a specific solvation effect.

Similarly, studies on the reaction between MeI and $[Pt(PR_3)_3]$ show an increase in rate in the order: R = *p*-$C_2H_5C_6H_4 <$ *p*-$MeC_6H_4 <$ n-$C_{16}H_{33} <$ n-C_8H_{17}, consistent with the various electron-releasing capabilities of the *p*-alkyl groups. The faster reaction with the alkylphosphines again reflects the greater basicity of these phosphines. The increase in reaction rate with decreasing alkyl-chain length possibly has its origins in a steric effect, but more likely this is an electronic or specific solvation effect (Franks and Hartley, 1981).

The kinetics of the reaction shown in Scheme 2 have been interpreted in terms of rate-limiting nucleophilic attack of $[(\eta^5\text{-}C_5H_5)M(CO)L]$ (M = Co, Rh or Ir, L = phosphine) on MeI. Rapid, iodide-induced intramolecular migration of the alkyl-group to the carbonyl-carbon atom yields the acyl-product (Hart-Davis and Graham, 1970). Variation of the phosphine ligand from PPh_3, PPh_2Me to $PPhMe_2$ leads to an increase in the reaction rate. The

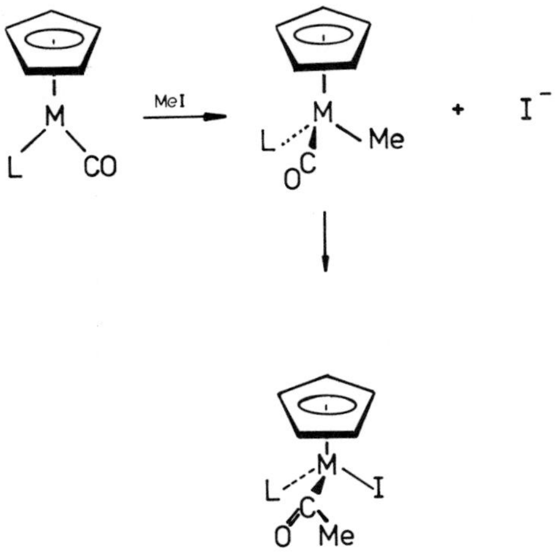

Scheme 2

substituents on the phosphine are some three atoms away from the nucleophilic centre and thus the observed reactivity trend should reflect only electronic effects. However, despite $P(C_6H_{11})_3$ being one of the most basic phosphines used in these studies, the reaction with MeI is one of the slowest, indicating that steric effects can play an important role, which even at three atoms removed from the reaction centre cannot be dissected from the electronic effects. This is further illustrated by comparison of the reactivities of $[(\eta^5\text{-}C_5H_5)Co(CO)L]$ (L = monotertiary phosphine) towards MeI and the analogous nucleophilicities of the substituted pyridines as shown in Table 2.

The nucleophilicities of the complexes $[Co(L'H)_2L]^-$ (L'H = substituted glyoximate ligand) have been studied as a function of varying the glyoximate ligand and the ligand L (Schrauzer and Deutsch, 1969). The nucleophilicities of these complexes towards MeI depends on the energy of the $3d_z^2$-orbital, which can be influenced by the nature of the coligands. Studies on $[Co(dmgH)_2L]^-$ showed that the relative nucleophilicities decrease with L along the sequence: 2,6-lutidine > 2-picoline > $C_6H_{11}NH_2$ > py > $PhNH_2$ > 4CNpy > Me_2S > Ph_3Sb > Bu_3P > Ph_3As > $C_6H_{11}NC$ > Ph_3P. This series demonstrates the effect of both electron-donating substituents increasing the nucleophilicities and substituents with vacant orbitals decreasing the nucleophilicity. The decreasing nucleophilicity along the series corresponds to the decreasing donor, and increasing acceptor, strength of L. The "in plane" ligands (L'H) affect the nucleophilicity in the order shown below:

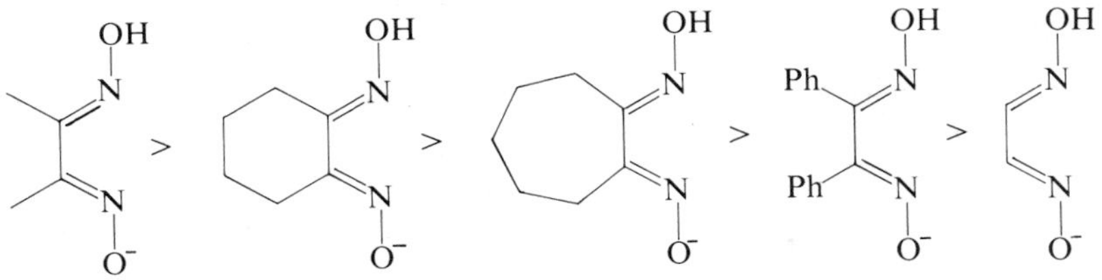

TABLE 2

Comparison of the relative nucleophilicities of substituted pyridines and $[(\eta^5\text{-}C_5H_5)Co(CO)L]$ (L = phosphine) towards MeI[a]

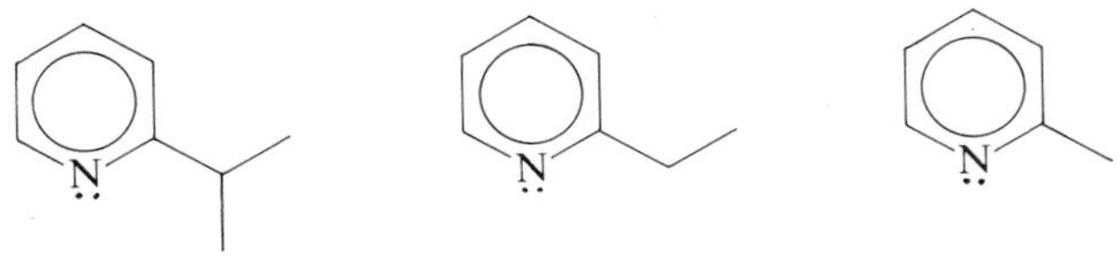

	2-isopropylpyridine	2-ethylpyridine	2-methylpyridine
k_{rel}	1.0	3.2	6.7

	$(\eta^5\text{-}C_5H_5)Co(CO)PPh_3$	$(\eta^5\text{-}C_5H_5)Co(CO)PPh_2Me$	$(\eta^5\text{-}C_5H_5)Co(CO)PPhMe_2$
k_{rel}	1.0	6.1	10.7

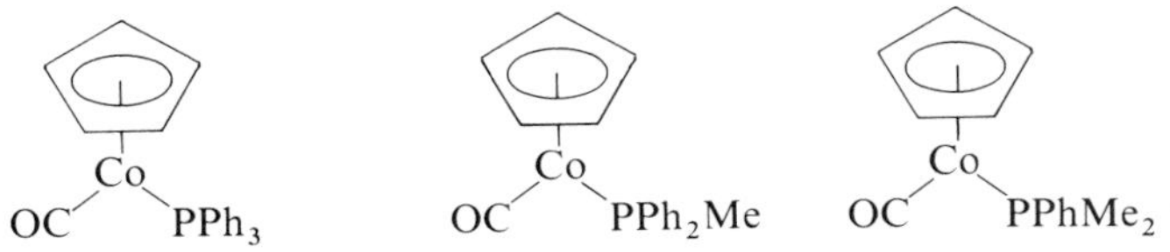

[a] Data of Hart-Davis and Graham (1970)

The strong inductive effect of the methyl-substituents renders dimethylglyoximate the most electron-releasing. The difference in reactivity between the cyclohexyl- and cycloheptyl-ligand has been attributed to the angular dependence of the orbital overlap between the nitrogen donor and the cobalt atoms, which is perturbed by the conformational preferences of the two rings. As would be expected, the formally analogous complexes containing the ligands [1] and [2] are weaker nucleophiles because of the decreased negative charge on the complex.

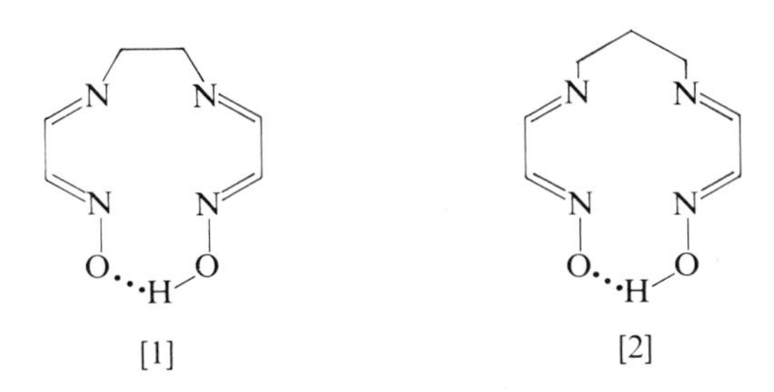

The effect of changing either the axial ligand L or the "in-plane" ligand (L′H) is relatively small, but the latter has the greater effect. For instance $[Co(dmgH)_2L]$ shows a change in rate of *ca* 50 in going from L = H_2O to PR_3. In contrast, the analogous species containing ligand [1] is somewhat anomalous exhibiting a factor of 500 for the same change in L (Toscano and Marzilli, 1984).

In an extensive study of the reactions of MeI with *trans*-$[IrCl(CO)L_2]$ (L = $AsPh_3$, $P(OPh)_3$, PPh_3, $P(p\text{-}MeC_6H_4)_3$, $PMePh_2$, PEt_3 or PMe_2Ph) the relative nucleophilicity of the complexes as a consequence of the phosphine was found to exhibit the same pattern of reactivity as already discussed. The arsine complex reacts only slightly faster than its phosphine analogue (Kubota *et al.*, 1973). The rates for the reactions of MeI with $[IrYCl_3]$ increase in the order Y = $N_2 > PPh_3 > CO$ which reflects the electron-density at the metal; from N_2 to CO the π-accepting capabilities of these ligands progressively increase.

Changing the anion in the complex *trans*-$[IrX(CO)(PPh_3)_2]$ augments the rate of reaction with MeI in the order $F \gg N_3 > Cl > Br > NCO > I > NCS$; however, the variation of reaction rate for the whole series only covers two orders of magnitude (Kubota *et al.*, 1973). It has been pointed out that this order is that expected on the basis of the Hard Soft Acid Base (HSAB) postulate (Pearson, 1985). An interesting aspect of this series is the order of reactivity with respect to halide, or pseudohalide, which is the opposite of that found for the oxidative addition reactions of the same substrates with dioxygen, dihydrogen (Vaska, 1968; Chock and Halpern, 1966), benzenethiols (Gaylor and Senoff, 1972) and organic azides (Collman *et al.*, 1968). The more polarisable iodine atom renders the iridium metal the most basic, as determined by the protonation of *trans*-$[IrX(CO)\ (PPhMe_2)_2]$ with benzoic acid (Deeming and Shaw, 1971), and hence it would be expected that the iodo-complex would be the most nucleophilic. It has been proposed that the contrasting influence of the halido-groups on the reactivity of the complexes towards the various substrates is attributable to the different geometries (and hence different electronic requirements) of the transition states as shown in Fig. 2, where the oxygen molecule is bound to the metal in a three-centre transition state, giving a pseudo-six-coordinate species, whereas with MeI the transition state is only five-coordinate.

FIG. 2 The different transition state structures for the reactions between *trans*-$[IrX(CO)(PPh_3)_2]$ and oxygen or methyl halide

In a further study, the reactions of *trans*-$[IrCl(CO)L_2]$ (L = $PEtPh_2$ or PEt_2Ph) and *trans*-$[IrCl(CO)\{P(C_6H_4Z\text{-}p)_3\}_2]$ (Z = Cl, F, H, Me or MeO) with alkyl halides have been investigated (Ugo *et al.*, 1972). The reactions with both methyl and benzyl halides were affected by the phosphines in the order: $PEtPh_2 > PEt_2Ph \gg PPh_3$. The anomalous position of PEt_2Ph has been ascribed to steric effects, and this is consistent with the lower entropies of activation for reaction (7) when L = ethyl-substituted phosphines.

$$trans\text{-}[IrCl(CO)L_2] + RX \xrightarrow{k_2} [IrCl(X)(R)(CO)L_2] \quad (7)$$

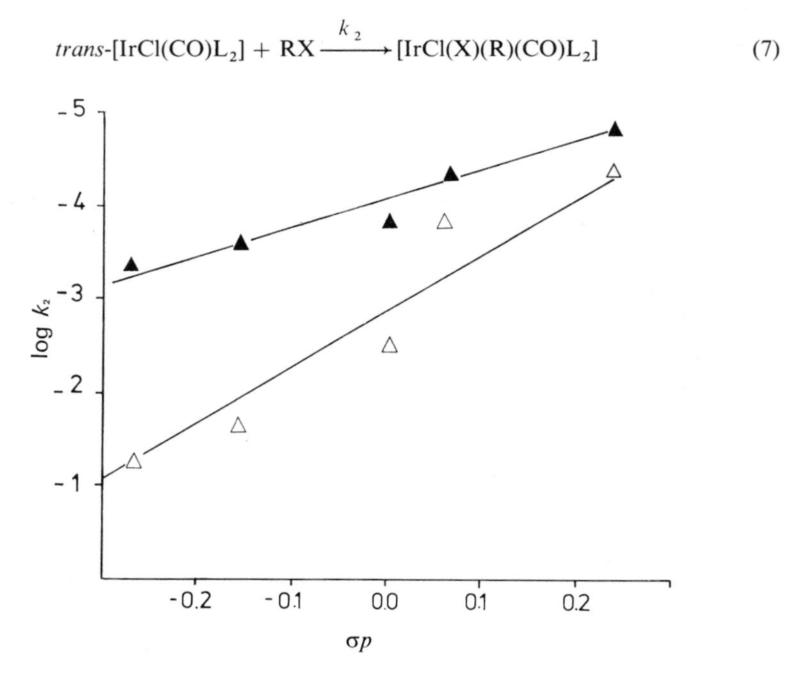

FIG. 3 Hammett plots for the reactions of *trans*-$[IrCl(CO)(PAr_3)_2]$ with $PhCH_2Cl$ (▲) and MeI (△)

The *p*-substituted arylphosphine ligands represent an ideal system to study the electronic effects of reaction (7) since they are uncomplicated by steric effects. Marked differences in the reaction rate with both MeI and $PhCH_2Cl$ are observed with a change of the *p*-substituent. A graph of $\log_{10} k_2$ where k_2 is the second-order rate constant for reaction (7) against the Hammett σ_p constant yields two lines as shown in Fig. 3, for the two alkyl halides. The reaction constants (ρ) are -6.4 (MeI) and -2.6 (Ph_2CH_2Cl). The large negative value with MeI is indicative of a positive charge formed at the iridium atom in the transition state. The less negative value for $PhCH_2Cl$ demonstrates that the binding of the benzyl group to the metal in the transition state is weaker than is the binding of the methyl group.

THE NUCLEOPHILICITY OF METAL COMPLEXES

Relatively little systematic work on the nucleophilicities of metal complexes has been performed. Of particular interest would be the definition of a scale of nucleophilicity based on isoelectronic metals with the same coligands. The variety of different types of nucleophiles and the dichotomy of electron transfer *vs* nucleophilic mechanisms precludes any extensive comparison. However, it is possible to compare the relative nucleophilicities within a given group.

The relative rates of reaction between MeI and $[(\eta^5\text{-}C_5H_5)\,M(CO)(PPh_3)]$ (M = Co, Rh or Ir) are in the order: Co(1.0), Rh(1.4), Ir(8.0) (Hart-Davis and Graham, 1970), whereas for the reaction between pentafluoropyridine and $[M(CO)_2(PPh_3)_2]$ the order is : Co(1.0), Rh(857), Ir(357) (Booth *et al.*, 1975). Although both studies demonstrate the greater nucleophilicity of the heavier members of the triad, it is clear that the relative reactivities of the members can vary. Such behaviour is also manifest in the acid strengths of transition metal hydrides (Pearson, 1985; Shriver, 1970; Jordan and Norton, 1982 and references therein; Ziegler, 1985).

Comparison of $[M(dmgH)_2OH]^-$ (M = Co or Rh) in its reaction with MeI has a Pearson nucleophilicity of Co(14.3) and Rh(13.7) (Schrauzer *et al.*, 1968; Weber and Schrauzer, 1970).

In an early study (Dessy *et al.*, 1966), reduction of homobimetallic complexes at a mercury electrode was used to generate the corresponding mononuclear anions *in situ* as exemplified in (8). Although the reaction

$$[(\eta^5\text{-}C_5H_5)Mo(CO)_3]_2 \xrightarrow[0.1M\ Bu_4N^+\ ClO_4^-]{Hg,\ 2e,\ MeOCH_2CH_2OMe} 2\,[(\eta^5\text{-}C_5H_5)Mo(CO)_3]^- \qquad (8)$$

products were not isolated, studies of the reaction between a wide variety of nucleophiles and several alkyl halides were determined semi-quantitatively. The span of reactivity observed covers more than 12 orders of magnitude, with the nucleophilicity of PhS^- lying in the centre of the scale. A linear relationship between $\log_{10} k_2$ and $-E_{\frac{1}{2}}^{\text{ox}}$ (measured at a Pt electrode) is observed as defined by (9). The parallelism between nucleophilicity (and basicity) and the oxidation potential of a complex is a consequence of a nucleophilic displacement (or protonation) at the metal formally resulting in an oxidation of the metal in the transition state (or product).

$$\log_{10} k_2 = 6.6E_{\frac{1}{2}}^{\text{ox}} - 3.0 \tag{9}$$

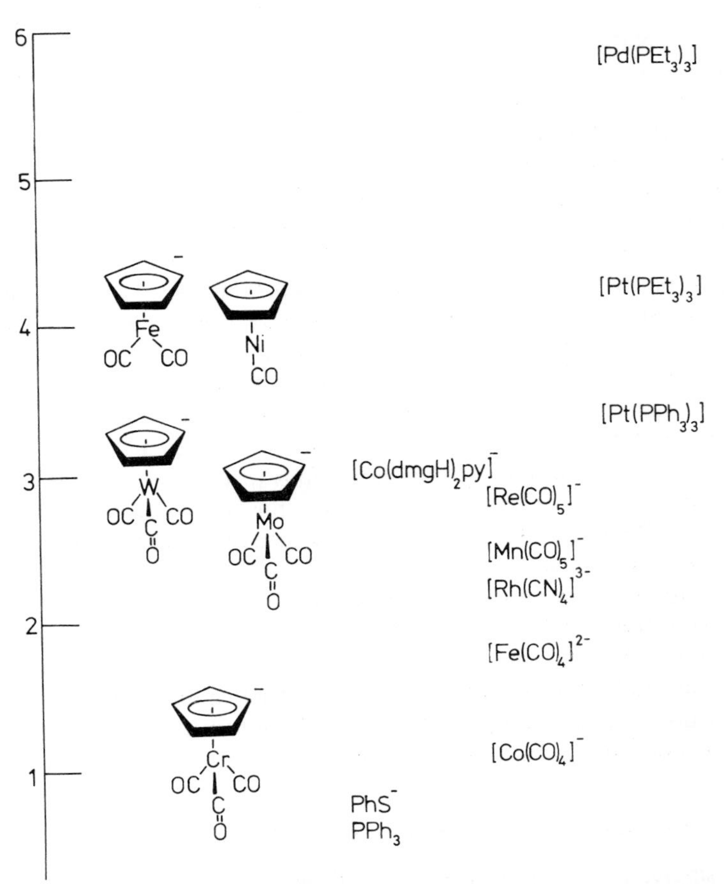

FIG. 4 Pictorial representation of the nucleophilicities of various transition metal complexes. The scale is $\log_{10} (k_I/k_{OTs})$, where k_I is the second order rate constant for the reaction of the nucleophile with MeI, and k_{OTs} is the corresponding value for MeOTs

More recently an extensive study on the nucleophilicities of several transition metal complexes has been reported (Pearson and Figdore, 1980). Experimental difficulties over the range of nucleophiles studied precluded the use of a single solvent. Because of the range of solvents employed, and the various influences on the reaction rates with variously charged nucleophiles (Section 5), it is not easy to establish an absolute scale of nucleophilicity. However the accumulation of data presented by Pearson and Figdore together with other studies has led to the nucleophilicity scale shown in Fig. 4 (Collman *et al.*, 1977; Hart-Davis and Graham, 1970; Douek and Wilkinson, 1969; Pearson and Gregory, 1976; Collman and MacLaury, 1974 and Dessy *et al.*, 1966).

The collected data give the order of transition metal nucleophiles as follows: Ni(0) $\gg$ Pd(0) $\gg$ Pt(0), (for the species $[M(PR_3)_2]$) and Fe(0) > Ru(0), (for the species $[(\eta^5\text{-}C_5H_5)M(CO)_2]^-$), Re($-1$) > Mn($-1$), (for the species $[M(CO)_5]^-$), Ir(1) > Rh(1) > Co(1), (for the species $[(\eta^5\text{-}C_5H_5)M(CO)(PPh_3)]$) and W(0) > Mo(0) > Cr(0), (for the species $[(\eta^5\text{-}C_5H_5)M(CO)_3]^-$). Clearly there is no simple criterion for establishing the reactivity of a given "complex-type" even within the same group.

The reluctance of alkyl sulphonates to react by a free-radical pathway has led to a proposed kinetic distinction between nucleophilic and electron-transfer pathways (Pearson and Figdore, 1980). Failure to react with methyl tosylate (MeOTs), but rapid reaction with MeI could be a criterion of the electron-transfer pathway. For the reactions of a large range of metal complexes there exists a linear correlation between $\log_{10} k_2^{\mathrm{MeOTs}}$ and $\log_{10} k_2^{\mathrm{MeI}}$ indicative of an S_N2 mechanism common to all the reactions. However, there are notable exceptions to this graph, such as $Li_2[PtMe_4]$, which does not lie on the line despite strong indications of its nucleophilicity. Further corroboration of a common mechanism for all these complexes is the observed isokinetic plot. However the closeness of the isokinetic temperature to room temperature ($T = 282 \pm 22$ K) and the variety of solvents used in these studies make any such conclusions from these correlations highly dubious. Indeed $[Co(CN)_5]^{3-}$ also lies on this isokinetic plot, but is known to react with MeI by halogen-atom abstraction (Chock and Halpern, 1969).

As indicated throughout this section, the nucleophilicity of the complexes is critically dependent upon the nature of the solvent, and this aspect will be discussed in more detail in the next section.

5 The influence of the solvent on the nucleophilicity

Surprisingly, relatively little systematic work has been performed aimed specifically at understanding the influence that the solvent has on the nucleophilicity of the complexes. An early study (Schrauzer and Deutsch,

1969) on the reactivity of vitamin B_{12} with MeCl, and of $[Co(dmgH)_2(PBu_3)]^-$ with EtBr, concluded there was no large solvent effect on these reactions in methanol, ethanol, propan-1-ol and H_2O, together with some mixtures. However, the range of solvent types covered is clearly very narrow, and subsequent, more extensive, studies have demonstrated that the reactions of many transition-metal nucleophiles are facilitated by polar solvents.

The rate of reaction of $[(\eta^5\text{-}C_5H_5)M(CO)(PPh_3)]$ (M = Co or Rh) with MeI is very dependent on the solvent as shown in Table 3 (Hart-Davis and Graham, 1970). Studies on the reaction of $[PtPh_2(bipy)]$ (bipy = 2,2′-bipyridine) with MeI or $PhCH_2Br$ in a range of solvents (benzene, ethyl acetate, dichloromethane, acetone or nitromethane) showed, as with $[(\eta^5\text{-}C_5H_5)Co(CO)(PPh_3)]$ and *trans*-$[IrCl(CO)(PPh_3)_2]$, that a factor of 10–15 in the reaction rate was observed in going from the least to the most polar. This implies a common nucleophilic reaction for all the complexes (Jawad and Puddephatt, 1976).

TABLE 3

The influence of the solvent on the rate of the reaction between $[(\eta^5\text{-}C_5H_5)M(CO)(PPh_3)]$ (M = Co or Rh) and MeI

Solvent	THF[a]	Me_2CO	CH_2Cl_2	CH_3CN
Dielectric constant	7.4	20.7	9.08	36.0
M = Co, k_{rel}	0.2	0.9	1.0	*ca* 2.5
M = Rh, k_{rel}			49.3	1.0

[a] THF = tetrahydrofuran

Early work comparing the reactivity of *trans*-$[IrCl(CO)(PPh_3)_2]$ towards the substrates O_2, H_2 and MeI (Chock and Halpern, 1966) demonstrated that the greater sensitivity of the reaction with MeI to a change in solvent was a consequence of a less negative $\Delta S^{\ddagger}$, compensated only in part by an increase in $\Delta H^{\ddagger}$. Thus, in the strongly ionising solvent N,N-dimethylformamide (DMF), the reaction with MeI is seventeen times faster than that in benzene at 30°C. This is similar to the characteristics of the Menschutkin reaction, and thus the reaction of *trans*-$[IrCl(CO)(PPh_3)_2]$ with MeI was concluded to proceed via the five-coordinate transition state shown in Fig. 1. Further work on the reactions of *trans*-$[IrCl(CO)(PPh_3)_2]$ with MeI in a variety of solvents corroborated these conclusions (Ugo *et al.*, 1972). The qualitative observations that the reactions of *trans*-$[IrCl(CO)(PMe_3)_2]$ with $CH_3CHBrCO_2Et$ and $PhCHFCHBrCO_2Et$ are faster in the more polar solvents such as N-methylpyrrolidone (NMP) and DMF than in benzene raises doubts on the use of such a probe as a mechanistic criterion, since these reactions proceed by an electron-transfer mechanism (Labinger and Osborn, 1980) (see Section 6).

The problems in defining a nucleophilicity scale using a variety of solvents have been outlined (Pearson and Figdore, 1980). The influence of the solvent on the nucleophilicity of neutral complexes such as *trans*-$[IrCl(CO)(PPh_3)_2]$ with MeI is relatively simple. For instance, the reaction in tetrahydrofuran (THF) is three times faster than it is in benzene (Ugo *et al.*, 1972). However, when the nucleophile is anionic, then comparison of the reactivities in two different solvents is not simple. Electrostatic arguments would indicate that a free anion would be more reactive in a medium of lower dielectric constant, but extensive ion-pairing with counter cations greatly augments the reactivity. Several studies on the nature of some transition-metal nucleophiles in solution have been reported, and these give important insight into the reactivities of these species.

Detailed studies of the solution infrared spectrum (carbonyl stretching region) have established the sites and extent of alkali-cation interaction with $[(\eta^5\text{-}C_5H_5)Fe(CO)_2]^-$ (Pannell and Jackson, 1976; Nitay and Rosenblum, 1977); $[Mn(CO)_5]^-$ (Darensbourg *et al.*, 1976; Pribula and Brown, 1974); $[Co(CO)_4]^-$ (Edgell *et al.*, 1978; Edgell and Chanjamsri, 1980); $[V(CO)_5L]^-$ (L = PPh_3, $P(OPh)_3$, $P(Bu^n)_3$ and CNMe) (Darensbourg and Hanckel, 1981, 1982) and $[(\eta^5\text{-}C_5H_5)M(CO)_3]^-$ (M = Cr, Mo or W) (Darensbourg *et al.*, 1982). In all cases the cations are closely associated with the carbonyl oxygen atoms.

Simplistically it might be considered that ion-pairs would always be poorer nucleophiles than the "bare" anions. However, quite the opposite has been observed in the reactions of epoxides with $[(\eta^5\text{-}C_5H_5)Fe(CO)_2]^-$ (Nitay and Rosenblum, 1977), the reactions of benzyl halides with $[Co(CO)_4]^-$ (Moro *et al.*, 1980) and $[Mn(CO)_4L]^-$ (L = CO, PR_3 or $P(OR)_3$) (Darensbourg *et al.*, 1976) and $[(\eta^5\text{-}C_5H_5)Mo(CO)_3]^-$ (Darensbourg *et al.*, 1980). These results have been interpreted in terms of a highly-ordered transition state in which the cation assists carbon-halogen cleavage as shown in Fig. 5.

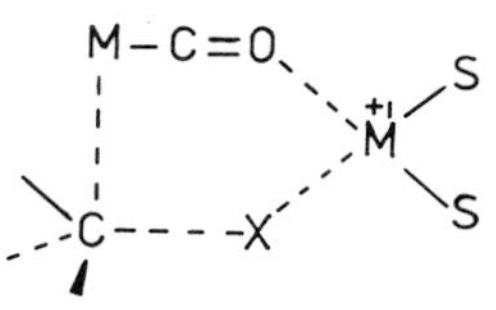

FIG. 5 Highly ordered transition state for the reaction between a metal carbonyl complex and alkyl halides in the presence of alkali metal cations

The reaction of $[(\eta^5\text{-}C_5H_5)Mo(CO)_3]^-$ with $PhCH_2Cl$ in THF is more rapid in the presence of sodium ion, and the addition of hexamethylphosphoramide (HMPA) (which coordinates to the sodium) inhibits the reaction. However, the analogous reactions of Bu^nI show the opposite effect. These effects are a consequence of the local environment of the nucleophile, rather than a bulk medium effect. Reaction (10) between $[(\eta^5\text{-}C_5H_5)W(CO)_3]^-$ and $CH_3C{\equiv}C(CH_2)_4I$ is faster, however, when the tungsten complex is present as the (Bu_4N^+) salt rather than as the Na^+ salt, presumably because of the greater nucleophilicity of the complex in the presence of the more charge-dispersed cation (Watson and Bergman, 1979).

$$[(\eta^5\text{-}C_5H_5)W(CO)_3]^- + CH_3C{\equiv}C(CH_2)_4I \longrightarrow [(\eta^5\text{-}C_5H_5)W\{(CH_2)_4C{\equiv}CCH_3\}(CO)_3] + I^- \quad (10)$$

It is possible to investigate much purer solvent effects using the large bis(triphenylphosphine)iminium cation (PPN^+), which, although associated with $[(\eta^5\text{-}C_5H_5)Mo(CO)_3]^-$ in THF, does not electronically perturb the carbonyl groups. The second-order rate constants for the reactions between $[(\eta^5\text{-}C_5H_5)Mo(CO)_3]^-$ and Bu^nI or $PhCH_2Cl$ in acetonitrile and THF are shown in Table 4. These values give further credence to the proposed S_N2 mechanism. The charge separation present in a transition state of S_N2 character would be less stabilised in a polar solvent than the ground state of an anionic nucleophile and neutral alkyl halide, giving rise to a correspondingly higher activation energy. The data observed for the reactions with $PhCH_2Cl$ indicate a greater degree of charge separation in the transition state with this substrate.

TABLE 4

Kinetic data for the reactions of (PPN) $[(\eta^5\text{-}C_5H_5)Mo(CO)_3]$ with alkyl halides[a]

Solvent	$10^4k_2/dm^3mol^{-1}s^{-1}$	
	Bu^nI	$PhCH_2Cl$
MeCN	1.64	1.60
THF	26.0	0.63

[a] Darensbourg *et al.*, 1982

A detailed study on the influence of the solvent on the nucleophilic reactivity of $Na_2[Fe(CO)_4]$ towards alkyl halides and sulphonates, and the dramatic effect that ion-pairing has on the reactivity, have been reported (Collman *et al.*, 1977). The reactions studied are typified by (11).

$$[Fe(CO)_4{}^{2-} + RX \longrightarrow OC{-}Fe(R)(CO)(CO)_2^- + X^- \quad (11$$

In NMP or THF the values of the two successive ion-pair dissociation constants of $[Fe(CO)_4]^{2-}$ shown in (12) were estimated from a combination of conductometric titrations in the presence and absence of the cryptand,

$$Na_2[Fe(CO)_4] \overset{K_{1D}}{\rightleftharpoons} Na^+ + Na[Fe(CO)_4]^- \overset{K_{2D}}{\rightleftharpoons} 2Na^+ + [Fe(CO)_4]^{2-} \quad (12)$$

Kryptofix 2,2,2. The corresponding values and limits estimated from this study were: in NMP, $K_{1D} = 0.28$, $1 \times 10^{-3} > K_{2D} > 1 \times 10^{-5}$ mol.dm^{-3}; in THF, $K_{1D} \sim 5 \times 10^{-6}$, $K_{2D} \ll 5 \times 10^{-6}$ mol.dm^{-3}.

The mechanism of the reaction with various alkyl halides in NMP is believed to involve nucleophilic attack by both the solvent-separated ion-pairs $Na_2[Fe(CO)_4]$ (k_2) and $[Na(Fe(CO)_4]^-$ (k'_2) (where $k'_2 = 0.18\,dm^3\,mol^{-1}\,s^{-1}$ and $k_2 < \frac{1}{3}\,k'_2$) as evidenced by the non-linear dependence of the observed rate constant on the concentration of complex. In THF the dissociation of the ions is less extensive, and the rate of reaction with alkyl halides is correspondingly slower. However, the addition of polar solvents or crown ethers increases the rate of the reaction, and this has been ascribed to the kinetically more reactive, more dissociated species.

6 Stereochemical changes at a saturated carbon centre

In general, the most convincing evidence that the reaction of an alkyl halide or tosylate with a metal complex adopts a nucleophilic displacement (S_N2) mechanism is the demonstration that inversion of the configuration of the saturated carbon centre has occurred as a consequence of this process. However, as will be seen, inversion of the configuration of the carbon atom is not an essential consequence in the reactions of coordinatively-unsaturated, transition-metal nucleophiles.

REACTIONS OF COORDINATIVELY-SATURATED COMPLEXES

In practice, it is not easy to establish the stereochemical consequence of the reaction between a chiral carbon centre and a metal complex because of the lack of crystallographic information establishing the absolute configurations of metal-alkyl complexes. In order to determine the stereochemistry of the addition, one strategy is subsequently to cleave the metal—carbon bond with a reagent so as to regenerate the original organic compound. Provided that the stereochemistry of the cleavage reaction is known, the stereochemistry of the nucleophilic displacement reaction can be established. As an example, the sequence of reactions shown in Scheme 3 demonstrates the inversion of configuration of the nucleophilic displacement between $[(\eta^5\text{-}C_5H_5)Fe(CO)_2]^-$ and $MeCHBrC_2H_5$, but only because it is implicitly assumed that the cleavage of the iron—carbon bond occurs without affecting the configuration of the chiral carbon centre (Johnson and Pearson, 1970).

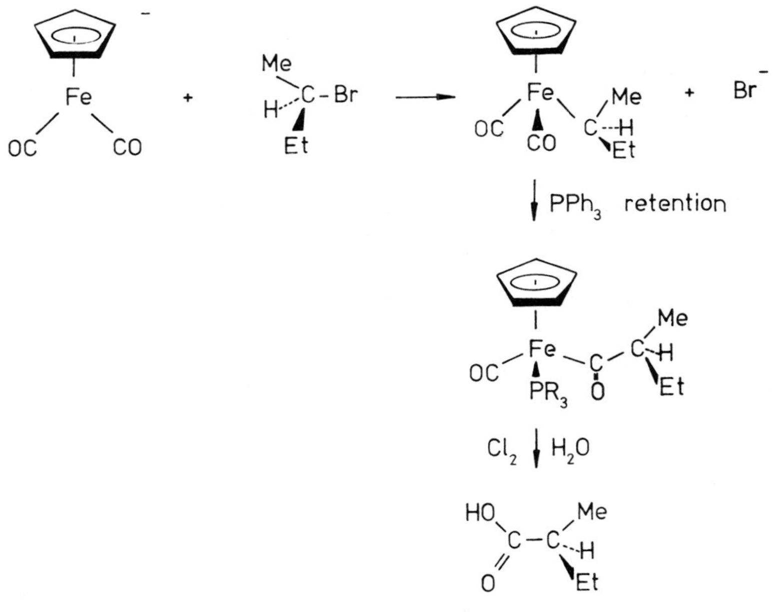

Scheme 3

Examples are known, however, in which halogen cleavage of metal-alkyl bonds occurs with retention (Whitesides and Boschetto, 1969; Calderazzo and Noack, 1966; Johnson, 1970; Slack and Baird, 1974, 1976) and others with inversion (Whitesides and Boschetto, 1971; Jensen and Davis, 1971; Jensen *et al.*, 1971; Anderson *et al.*, 1972). Without firm evidence about the stereochemistry of the cleavage of metal-alkyl bonds, little can be said about the stereochemistry of the reaction between metal nucleophiles with alkyl halides.

A further complication that can occur in these stereochemical studies was demonstrated in the reaction between $[Mn(CO)_5]^-$ and optically active $MeCHBrCO_2Et$, where subsequent bromine cleavage was used to establish the stereochemistry of the reaction. The stereospecificity observed in this reaction is low (*ca* 60%) and this is a consequence of racemisation of the alkyl-complex in the presence of an excess of $[Mn(CO)_5]^-$ as shown in (13) (Johnson and Pearson, 1970).

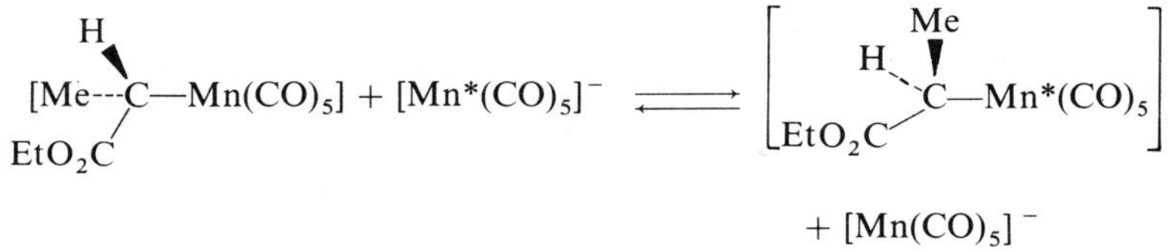

$$[Me\text{---}C(H)(EtO_2C)\text{—}Mn(CO)_5] + [Mn^*(CO)_5]^- \rightleftharpoons [H(Me)(EtO_2C)C\text{—}Mn^*(CO)_5] + [Mn(CO)_5]^- \quad (13)$$

The elegant work of Whitesides and coworkers represents an unambiguous means of determining the stereochemistry of the reaction between nucleophiles and saturated carbon centres. The work centres around the stereospecific preparations of *erythro-* and *threo*-3,3-dimethylbutan-1-ol-1,2-d_2[$Bu^tCH(D)CH(D)OH$]. The two isomers are readily distinguished by their characteristic vicinal coupling constants {J(erythro) = 9.5 Hz; J(threo) = 5.8 Hz} (Whitesides and Boschetto, 1969; Block *et al.*, 1974). In the original studies, conversion of the *erythro*-alcohol to the bromobenzenesulphonate (with retention of configuration) and subsequent reaction with $[(\eta^5\text{-}C_5H_5)Fe(CO)_2]^-$ produced a product whose ^1H nmr spectrum consists of a single AB quartet δ1.38, $J = 4.4$ Hz. Independent analysis of the vicinal coupling constants for both the complexes containing the *threo-* ($J = 4.6$ Hz) and *erythro*-ligand ($J = 13.1$ Hz) showed that the reaction between *erythro*-$Bu^tCH(D)CH(D)SO_3C_6H_4Br$ and $[(\eta^5\text{-}C_5H_5)\text{-}Fe(CO)_2]^-$ occurs with inversion of configuration at C-1 as shown in Scheme 4.

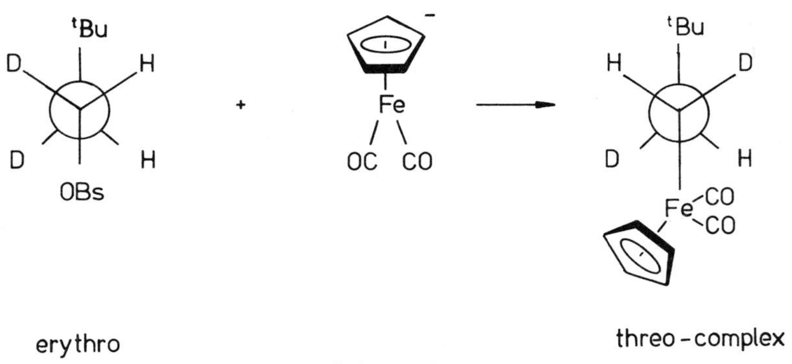

Scheme 4

Subsequent studies, occasionally using the trifluoromethane sulphonate derivative, have extended this work to less potent nucleophiles such as

$[(\eta^5\text{-}C_5H_5)Mo(CO)_3]^-$, $[Co(dmgH)_2py]^-$, $C_6H_5Se^-$ and Me_3Sn^- (Block and Whitesides, 1974). In all the studies, except those with the tin compound, inversion of configuration at C-1 was found with a stereoselectivity of greater than 95%. However, with Me_3Sn^- only 80% inversion was observed, a point to which we shall return later.

Another area where ^1H nmr spectroscopy can be used to determine the stereochemistry at the carbon atom is demonstrated in the reactions of substituted cyclohexanes with $[Co(dmgH)_2py]^-$, as shown in the selective examples (14) and (15). The stereochemistry of the displacement (inversion of the configuration of the carbon centre) and the strong preference for the metal residue to occupy the less congested equatorial site, defines the orientation of the substituent Y at the 4-position.

$[Co(dmgH)_2py]^-$ + Y–(cyclohexyl)–X ⟶ [Y–(cyclohexyl)–$Co(dmgH)_2py$] + X^- (14)

$[Co(dmgH)_2py]^-$ + Y–(cyclohexyl)–X ⟶ [Y–(cyclohexyl)–$Co(dmgH)_2py$] + X^- (15)

(X = Y = Br or X = OTs, Y = OH)

Demonstration of the configurations of the products shown in (14) and (15) comes from comparison of the ^1H nmr spectra of the products with those of the formally analogous t-butyl-substituted cyclohexyl bromides, where the t-butyl-group, like the metal residue, is forced into an equatorial position (Jensen *et al.*, 1970). Alternatively assignments can be made on the basis of the magnitude of the coupling constants: $J(\text{ax–ax}) = 8\text{–}14$ Hz > $J(\text{eq–eq})$ or $J(\text{ax–eq}) = 1.7$ Hz. However, the assignments based on these coupling constants are less compelling, amounting to a distinction based on the resolution of a signal into a doublet or the observation of a broad singlet.

Finally the reaction of α-bromocamphor with isotopically-labelled $[^2HFe(CO)_4]^-$ gives rise to camphor-3-exo-d as shown in (16). This result

(α-bromocamphor: =O, H, Br) $\xrightarrow{[^2HFe(CO)_4]^-}$ (camphor-3-exo-d: =O, ^2H, H) (16)

has been interpreted (Alper, 1975) in terms of an S_N2 attack at the carbon atom, with inversion of configuration. As discussed before, however, in the absence of any evidence concerning the stereochemistry of the subsequent cleavage, this conclusion must remain tentative.

REACTIONS OF COORDINATIVELY-UNSATURATED COMPLEXES

For the reactions of coordinatively-saturated nucleophiles with alkyl halides, the overwhelming evidence is that they proceed with inversion of configuration at the carbon atom as expected for an S_N2-type reaction. With the square-planar, coordinatively-unsaturated reactants such as *trans*-$[IrCl(CO)(PPh_3)_2]$, however, the possibility exists that either inversion or retention of configuration at carbon can occur, depending upon whether the transition state is five-coordinate or six-coordinate with a "side-bonded" approach of the alkyl halide to the metal centre (Fig. 1). Indeed this area has been one of some controversy over the years.

In 1970 two conflicting reports on the stereochemistry of the addition of chiral alkyl halides to square-planar iridium(I) complexes appeared. In one report it was claimed that the reaction of *trans*-$[IrCl(CO)(PPh_2Me)_2]$ with optically active $CH_3CHBrCO_2Et$ occurred with retention of configuration as shown in Scheme 5 (Pearson and Muir, 1970). This result is consistent with a "six-coordinate intermediate", as is the lack of incorporation of any free halide into the product (Section 8). However, the conclusions should again be treated with care since the study employs the cleavage of the iridium–carbon bond by halogen, and without knowing the stereochemistry of this reaction little can be said about the stereochemistry of the displacement.

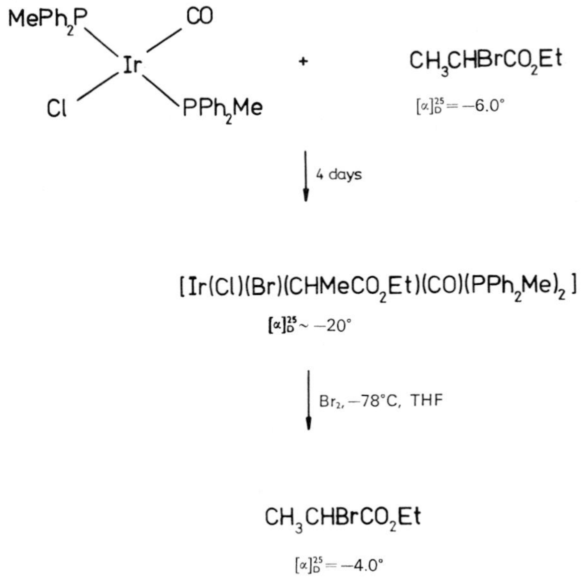

MePh$_2$P, CO, Ir, Cl, PPh$_2$Me + $CH_3CHBrCO_2Et$ $[\alpha]_D^{25} = -6.0°$

↓ 4 days

$[Ir(Cl)(Br)(CHMeCO_2Et)(CO)(PPh_2Me)_2]$ $[\alpha]_D^{25} \sim -20°$

↓ Br_2, −78°C, THF

$CH_3CHBrCO_2Et$ $[\alpha]_D^{25} = -4.0°$

Scheme 5

In contrast, the reaction of *trans*-$[IrCl(CO)(PMe_3)_2]$ with *trans*-2-fluorocyclohexyl bromide was claimed to involve inversion of the configuration at the carbon atom as shown in (17) on the basis of ^{19}F nmr spectroscopy of the product (Labinger *et al.*, 1970).

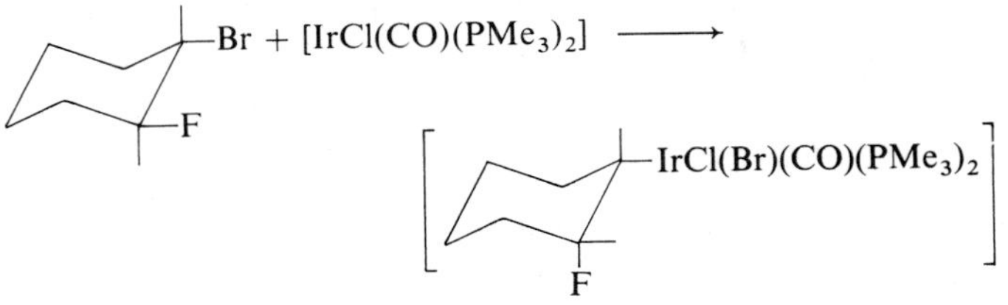

—Br + $[IrCl(CO)(PMe_3)_2]$ ⟶ $[$—$IrCl(Br)(CO)(PMe_3)_2]$ (17)

A subsequent study of reaction (17) claimed that under the conditions reported by Labinger *et al.* no reaction occurred (Jensen and Knickel, 1971) and that the infrared spectrum of the reaction mixture was sensitive to traces of radical initiators (e.g. dioxygen). However, more detailed studies of the reactions between a variety of alkyl halides and *trans*-$[IrCl(CO)(PR_3)_2]$ has clarified this confused literature. In these studies the alkyl group contains a fluorine atom which can be used as the nmr probe (Bradley *et al.*, 1972; Labinger *et al.*, 1973; Labinger and Osborn, 1980).

The contentious studies with *trans*-2-bromofluorocyclohexane were repeated with *trans*-$[IrCl(CO)(PMe_3)_2[$ and the products isolated and purified fied. The correspondingly much improved ^{19}F nmr spectrum (compared to the previous study) now demonstrated that two products were formed. Furthermore, the reaction of *cis*-2-bromofluorocyclohexane with *trans*-$[IrCl(CO)(PMe_3)_2]$ yielded an identical ^{19}F spectrum clearly indicating that complete loss of stereochemistry at the carbon atom had occurred. The two most likely products are [3] and [4].

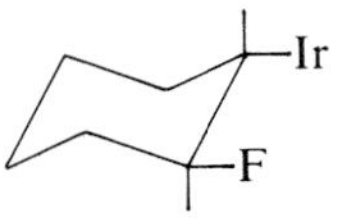

[3] 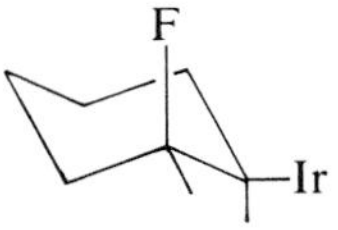 [4]

Loss of stereochemical integrity has also been observed in the reactions of *trans*-$[IrCl(CO)(PMe_3)_2]$ with 1-bromo-2-fluoro-2-phenylethane-1-d_1. The reaction of the (*RR*,*SS*)- and (*RS*,*SR*)-diastereoisomers produces the same product mixture, demonstrating the complete racemisation at the carbon centre.

Studies on (*RR*,*SS*)- and (*RS*,*SR*)-ethyl-3-bromo-3-fluoro-3-phenylpropionate with *trans*-$[IrCl(CO)(PMe_3)_2]$ also demonstrated that the same products were formed from both isomers in a non-stereospecific process. This result brings into question the proposed retention of configuration observed in the reactions of $MeCHBrCO_2Et$ and *trans*-$[IrCl(CO)(PPh_3)_2]$ (Pearson and Muir, 1970). Using a much improved preparation of ethyl-(*R*)-(+)-α-bromopropionate, the reaction with *trans*-$[IrCl(CO)(PMe_3)_2]$ yielded a product in which complete racemisation occurred at the carbon centre. These reactions were all repeated with a series of complexes of formula, *trans*-$[IrCl(CO)L_2]$ ($L = PMe_2Ph$, PPh_2Me or $AsMe_2Ph$) and in all cases complete loss of stereochemical integrity at the chiral centre was observed.

One possible means of racemisation of the alkyl complex containing an α-hydrogen, is rapid proton exchange as shown in (18). However, this

$$\mathrm{Ir{-}C(Me)(H)(CO_2Et)} \underset{H^+}{\rightleftharpoons} \mathrm{Ir{=}C(Me)(CO_2Et)} \overset{H^+}{\rightleftharpoons} \mathrm{Ir{-}C(Me)(CO_2Et)(H)} \quad (18)$$

possibility was eliminated since the reaction of ethyl (R)-(+)-α-bromopropionate with *trans*-$[IrCl(CO)(PMe_3)_2]$ in the presence of MeO^2H shows no incorporation of deuterium into the product. Furthermore the isolation of one diastereoisomer of $[Ir(PhCHFCHCO_2Et)Cl(Br)(CO)(PMe_3)_2]$ excludes the possibility of rapid epimerisation. Changing the solvent from benzene to the more polar NMP and DMF does not affect these stereochemical results.

These stereochemical results, together with some further mechanistic studies on the same reactions (Labinger *et al.*, 1980) have led to the conclusion that the reactions of saturated alkyl halides (except methyl derivatives), vinyl and aryl halides and α-haloesters with *trans*-$[IrCl(CO)(PR_3)_2]$ proceed by a radical chain pathway, whereas methyl, benzyl and allyl halides and α-haloethers probably react by a nucleophilic displacement process.

One final type of coordinatively-unsaturated complex where stereochemical studies have demonstrated the nucleophilic character of the metal towards alkyl halides is $[Pd(PR_3)_3]$. These reactions contrast with those of the analogous platinum complexes where the participation of free radicals has been demonstrated (Stille and Lau, 1977 and references therein).

The reaction of *cis*-3-acetoxy-5-carbomethoxycyclohexene with $[Pd(PPh_3)_3]$, and then cleavage of the product with methylphenylsulphonylacetate gives the *cis*-cyclohexene product as shown in (19) (Trost and Strege, 1977). The formation of this isomer is a consequence of two inversions, since the reaction of the intermediate π-bonded palladium

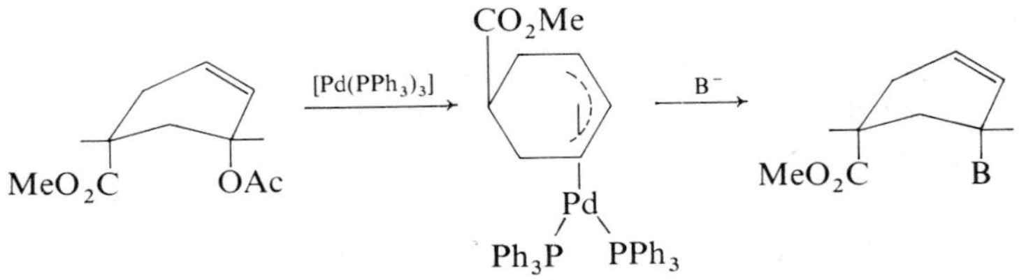

CO_2Me

MeO_2C OAc $\xrightarrow{[Pd(PPh_3)_3]}$ Pd, Ph_3P PPh_3 $\xrightarrow{B^-}$ MeO_2C B (19)

complex with the acetate has been shown to occur with inversion of configuration, thus defining the S_N2-type attack of $[Pd(PPh_3)_3]$ on the cyclohexene derivative.

The reactions of some optically active benzyl halides have provided further evidence for the nucleophilicities of the $[Pd(PR_3)_3]$ species (Lau *et al.*, 1974, 1976; Wong *et al.*, 1974; Stille and Lau, 1976). The salient features of these studies are summarised in Scheme 6. The stereochemistry of the addition is established by reaction of the palladium-alkyl complex with carbon monoxide (this "insertion" is known to take place by an intramolecular migration process, with retention of configuration in the migrating alkyl group), and subsequent formation of an ester from this acyl complex.

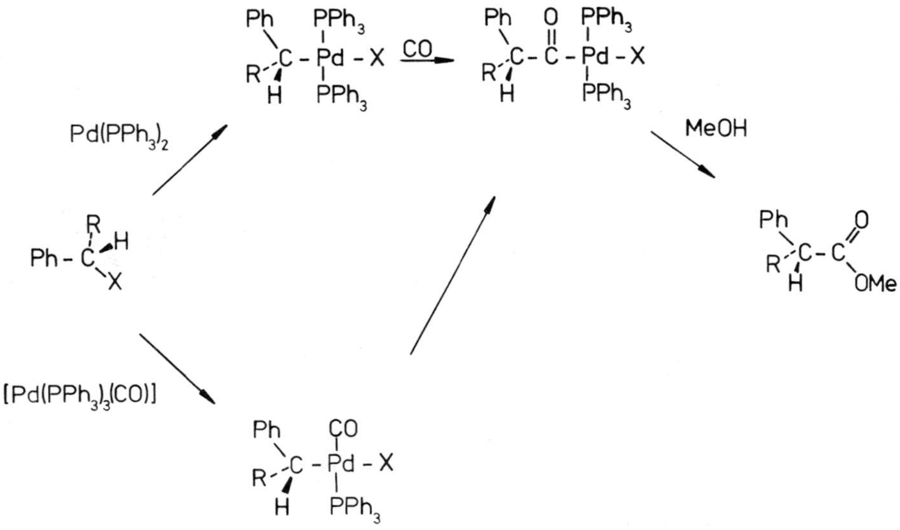

Scheme 6

These subsequent reactions occur at a position remote from the chiral centre. In this way the additions were shown to proceed quantitatively (or essentially so) with inversion of the configuration of the carbon centre.

However, in the analogous reactions of $[Pd(PEt_3)_3]$ and PhCHDX (X = Cl or Br), some loss of stereochemical integrity (*ca* 30%) is observed (Becker and Stille, 1978). This has been ascribed to rapid nucleophilic exchange after the formation of $[Pd(CHDPh)Cl(PEt_3)_2]$ which inverts the configuration of the alkyl-group. When the substrate is PhCHDBr only 19% net inversion is observed and the isotopically labelled bibenzyl (PhCHDCHDPh) is an isolable product. This result has been attributed to a competitive, one-electron transfer reaction leading to the radical pair shown in [5].

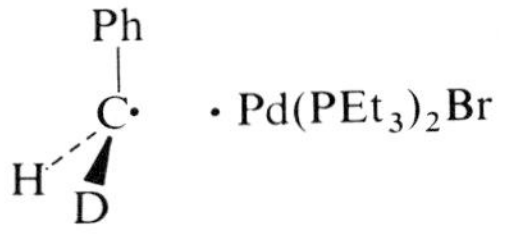

Ph
C· · $Pd(PEt_3)_2Br$
H D

[5]

A less clear-cut demonstration of the nucleophilicity of $[Pt(PPh_3)_2]$ comes from its reaction with optically active 8-(α-bromoethyl)quinoline, as shown in (20). The configuration of the product was deduced by Brewster's rules (Sokolov, 1976). However, prior coordination of the nitrogen atom to the metal could predefine the pathway.

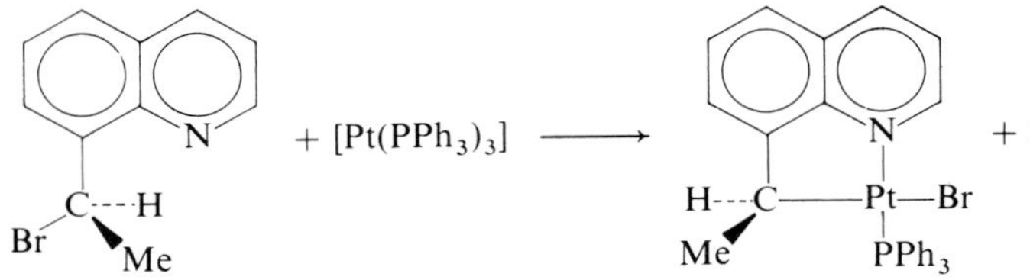

$$\text{8-(CHBrMe)quinoline} + [Pt(PPh_3)_3] \longrightarrow [Pt(C,N\text{-CHMe-quinolinyl})Br(PPh_3)] + 2PPh_3 \quad (20)$$

Although it has been deduced from several studies that the reaction between $[Pd(PR_3)_3]$ and allyl acetates occur with inversion of the configuration of the carbon atom (Trost and Verhoeven, 1978, 1980), it is only recently that direct proof has been obtained in the reaction of $[Pd(PPh_3)(Ph_2PCH_2CH_2PPh_2)]$ with (*S*)-(*E*)-3-acetoxy-1-phenyl-1-butene as shown in (21). The gross stereochemistry of the product was established by ^{1}H nmr spectroscopy, and measurement of its optical rotation, and comparison with that of the optically pure isomers prepared by an alternative route, demonstrated that the reaction occurred with 81% inversion of the carbon's configuration (Hayashi *et al.*, 1983).

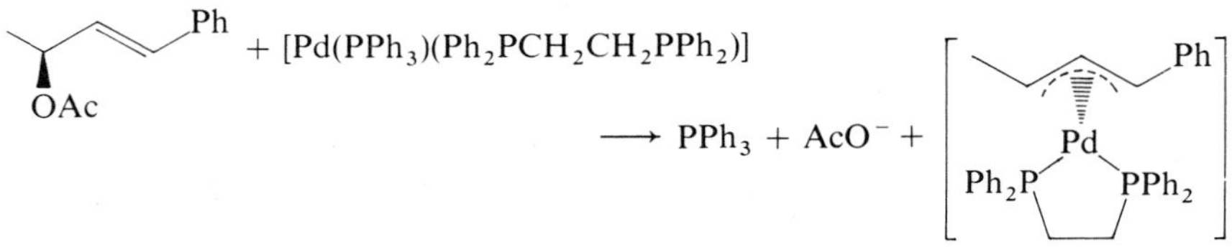

$$\text{MeCH(OAc)CH=CHPh} + [Pd(PPh_3)(Ph_2PCH_2CH_2PPh_2)] \longrightarrow PPh_3 + AcO^- + [Pd(\eta^3\text{-MeCHCHCHPh})(Ph_2PCH_2CH_2PPh_2)]^+ \quad (21)$$

REACTIONS OF Me_3Sn^-

The reactivity of Me_3Sn^- has been reserved for a separate section, since with this species particularly the electron-transfer and nucleophilic pathways are energetically very similar. In this section we will not review in detail the somewhat controversial literature concerning the electron-transfer mechanisms; for this the reader should consult Kitching *et al.* (1978); and references therein. Instead, only those studies relevant to the compound's nucleophilicity will be discussed.

As indicated earlier, the 80% inversion of configuration at the carbon atom observed in the reactions of Bu^{t}CHDCHDOTs with Me_3Sn^- (Block and Whitesides, 1974) should only be considered as a minimum value. However the stereochemistry of the reactions of alkyl compounds with Me_3Sn^- has generally been thought to be a straightforward indicator of the mechanism; the nucleophilic pathway gives a product of inverted carbon configuration whereas the electron-transfer pathway yields a racemic product. The demonstration of free-radical reactions which proceed with inversion of the configuration at carbon (Kuivila and Alnajjar, 1982; Ashby and DePriest, 1982) must cast some doubt on this stereochemical distinction between the two pathways.

The reaction of Me_3Sn^- with 2-butyl bromide has been shown to proceed with inversion of the carbon's configuration (Jensen and Davis, 1971), but many subsequent stereochemical studies have shown more complicated behaviour.

The reactions of Me_3Sn^- with *cis*-4-t-butylcyclohexyl bromide and with *trans*-4-t-butylcyclohexyl tosylate have been shown by ^1H nmr spectroscopy to give the same product, and on the basis of independent evidence this product has been assigned the *cis*-configuration as shown in Scheme 7. Thus the reaction of the tosylate occurs by a nucleophilic displacement reaction with inversion of configuration at the carbon centre, and the bromide reaction apparently proceeds with retention (Koermer *et al.*, 1972).

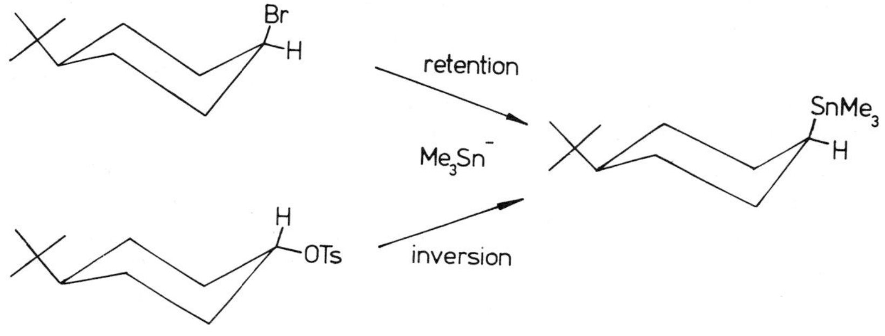

Scheme 7

Furthermore, the reaction of Me_3Sn^- with 1-bromo-1-methyl-2,2-diphenylcyclopropane is said to occur with retention of configuration at the carbon atom, based on the net retention upon HCl quench of the stannyl-complex (Sisido *et al.*, 1967) as shown in (22). The reactions occurring with retention presumably proceed via a four-centre transition state.

Br Ph Me Ph $\xrightarrow{Me_3Sn^-}$ Me_3Sn Ph Me Ph $\xrightarrow{HCl}$ H Ph Me Ph (22)

The stereochemistry of the reaction between 7-bromonorbornene and Me_3Sn^- is strongly dependent upon the solvent (Kuivila *et al.*, 1972). In particular it is observed that, as the coordinating capability of the solvent towards the cation (associated with Me_3Sn^-) increases, the proportion of inversion at the carbon centre increases.

In a detailed study of the reactions of Me_3Sn^- with *cis*- or *trans*-4-t-butylcyclohexyl bromides in THF, mixtures of products were obtained with complete loss of stereochemical integrity at the carbon atom, thus implicating radical reactions (San Filippo Jr *et al.*, 1978). Similar studies of Me_3M^- (M = Ge or Sn) with 4-alkylcyclohexyl bromides and the corresponding tosylates showed that the latter react with inversion of configuration, consistent with an S_N2 mechanism, whereas the bromides react via a non-stereospecific process for which several possible pathways have been discussed (Kitching *et al.*, 1978).

The reactions of cyclopropylmethyl bromide proceed via free, non-caged, radicals. The stereochemistry of the reaction is also sensitive to the counter-cation of Me_3Sn^-. Thus the lithium salt at $-70°C$ gives a higher proportion of the product with inverted configuration than the sodium salt. This indicates that at low temperatures there is a nucleophilic pathway operating. The low-energy radical pathway for the reactions of Me_3Sn^- is consistent with the observations (Koermer *et al.*, 1972) that olefins are sometimes observed as minor products of the reactions shown in Scheme 7.

More recently, studies of the reactions of (R)-2-$C_8H_{17}X$ (X = Cl, Br or OTs) with Me_3SnLi have shown that the stereoselective inversion of configuration decreases along the series OTs > Cl > Br and, furthermore, the temperature and mode of addition affects the selectivity (San Filippo Jr and Silbermann, 1981). There has also been criticism of earlier work in which trapping agents were used to detect radical reactions; it has been claimed that the trapping agents perturb the reaction and give rise to radical mechanisms.

REACTIONS OF BINUCLEAR COMPLEXES

An interesting, and to date unique, study of the stereochemical consequences of the reaction between an alkyl halide and a binuclear nucleophile comes from the study of the reactions of $[\{(\eta^5\text{-}C_5H_5)Co\}_2(\mu\text{-}CO)_2]^-$. This species reacts with methyl iodide to give the binuclear product shown in (23).

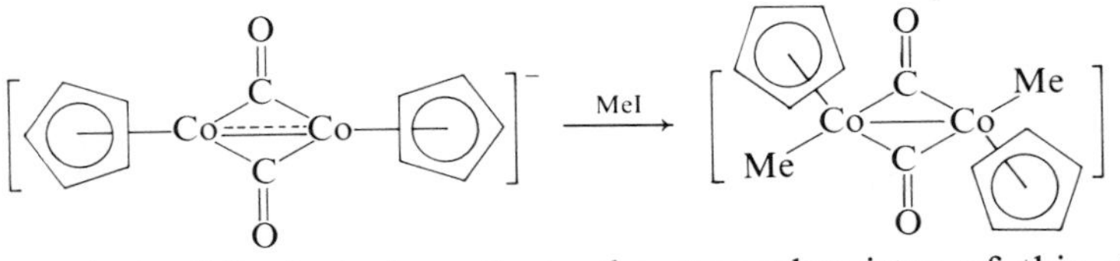

(23)

However, it is difficult to investigate the stereochemistry of this simple reaction because reactions with alkyl halides involving alkyl groups larger than ethyl do not lead to stable products.

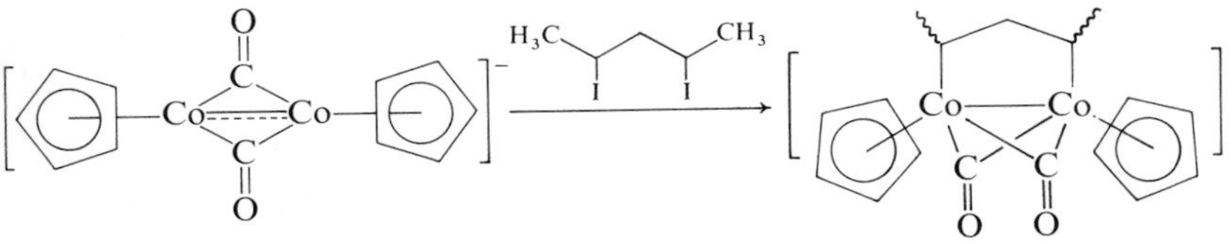

(24)

The stereochemistry of the metallacycle formation was determined as shown in (24). The alkylation of the dicobalt anion with α,γ-diiodides occurs with a complicated stereochemistry as evidenced by the studies with *meso*- and (±)-2,4-dimethylpentane. The mechanism of the reaction is shown in Scheme 8. The initial alkylation occurs with complete loss of stereochemical

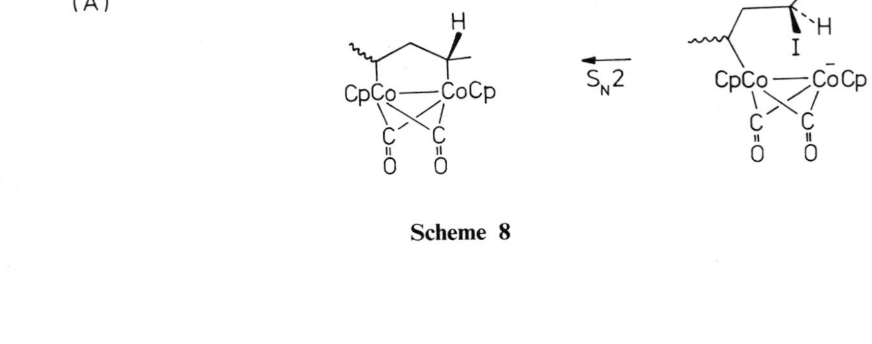

Scheme 8

integrity as a consequence of an electron-transfer process. Subsequent reduction of the monoalkylated product is followed by ring-closure with predominant inversion of configuration, presumably because of an S_N2 mechanism (Yang and Bergman, 1983).

7 Stereochemical changes at an unsaturated carbon centre

REACTIONS OF COORDINATIVELY-SATURATED COMPLEXES

A detailed study of the reactions between alkynyl or vinyl halides and $[Co(dmgH)_2py]^-$ has demonstrated a complex pattern of behaviour (Cooksey *et al.*, 1972). At low concentrations of nucleophile over short reaction times, the reaction with prop-2-ynyl halides yields the prop-2-ynylcobaloxime, presumably by a nucleophilic displacement pathway. However, at longer reaction times in the presence of an excess of cobaloxime, allenylcobaloxime is observed. This is formed as a consequence of an S_N2' displacement of one cobaloxime by another attacking the initially formed prop-2-ynylcobaloxime as shown in Scheme 9. The reactions of substituted

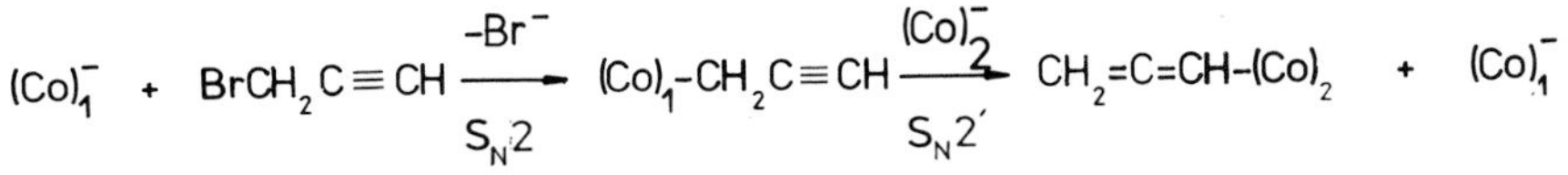

$$(Co)_1^- + BrCH_2C{\equiv}CH \xrightarrow[S_N2]{-Br^-} (Co)_1\text{-}CH_2C{\equiv}CH \xrightarrow[S_N2']{(Co)_2^-} CH_2{=}C{=}CH\text{-}(Co)_2 + (Co)_1^-$$

Scheme 9

prop-2-ynyl halides results only in the substituted allenyl products as shown in (25). This is a consequence of the hindered access to the saturated carbon centre making the direct S_N2' mechanism the one of lowest energy. Similar conclusions have been reached for the reactions of allyl halides with the cobalt nucleophile. Thus the reaction of either α- or γ-methylallyl halides with $[Co(dmgH)_2py]^-$ gives the γ-methylallylcobaloxime, the former reactions occurring by an S_N2' pathway, and the latter by an S_N2 route. These studies demonstrate that the cobaloxime nucleophile is more sensitive to steric effects than more conventional nucleophiles such as the alkoxide ion.

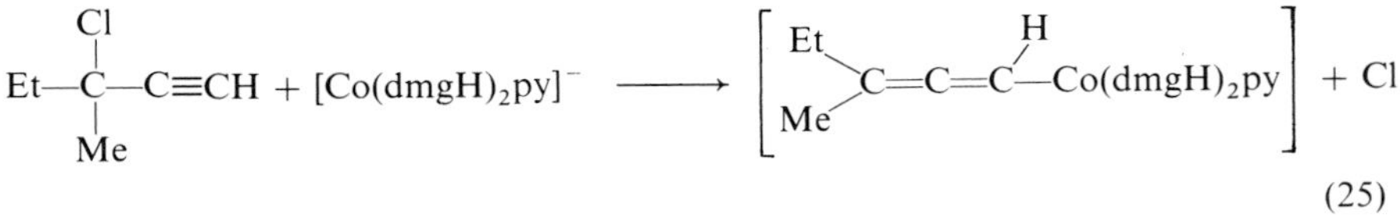

$$\text{Et—}\underset{\text{Me}}{\overset{\text{Cl}}{\text{C}}}\text{—C}{\equiv}\text{CH} + [\text{Co(dmgH)}_2\text{py}]^- \longrightarrow \left[\begin{matrix}\text{Et}\\ \text{Me}\end{matrix}\!\!>\!\text{C}{=}\text{C}{=}\underset{}{\overset{\text{H}}{\text{C}}}\text{—Co(dmgH)}_2\text{py} \right] + \text{Cl}^- \qquad (25)$$

Stereochemical investigation of the reaction between vinyl halides and $[Co(dmgH)_2L]$ (L = py, H_2O or $PhNH_2$) indicate an addition-elimination mechanism (Van Duong and Gandemer, 1970). The reactions of the cobaloximes with *cis*- or *trans*-2-phenylethenylbromide give the *cis*- and *trans*-vinylcobaloximes respectively.

REACTIONS OF COORDINATIVELY-UNSATURATED COMPLEXES

A study of the reaction between *trans*-$[IrCl(CO)(PMe_2Ph)_2]$ and alkyl halides has demonstrated different behaviour to that observed with $[Co(dmgH)_2py]^-$, but the details of the reaction are still unclear (Pearson and Poulos, 1979). The complications associated with the stereochemistry of the addition to the metal will be left until Section 8 and we will concentrate here on the stereochemistry observed at the carbon centre.

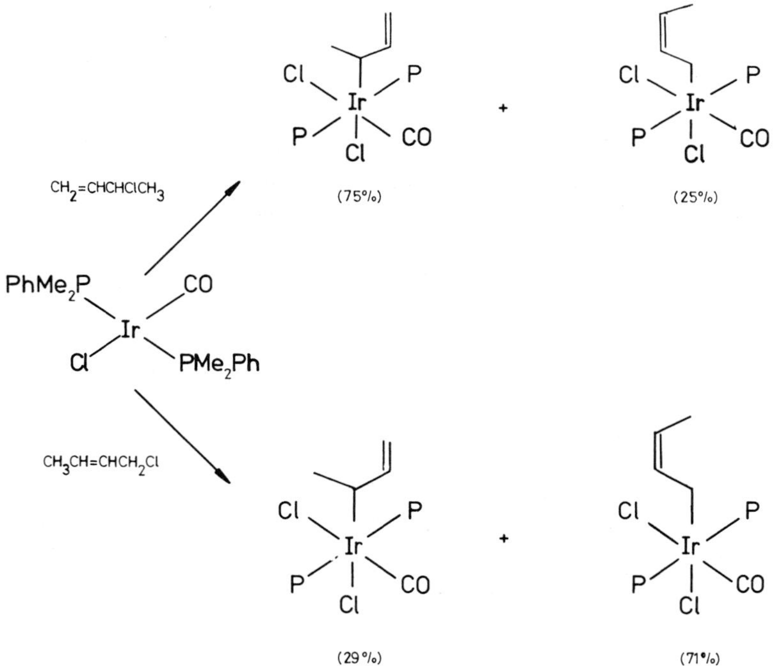

Scheme 10

The reaction of α-methylallyl or crotyl chloride with *trans*-[IrCl(CO)($PMe_2Ph)_2$] gives mixtures of products as shown in Scheme 10, the product distribution being different for the two substrates, which rules out a common intermediate. This has been interpreted in terms of a mechanism involving a two-step S_N2 process. The nucleophilic iridium atom attacks the saturated carbon centre, but some assistance is gained by π-bonding of the double bond to the metal. This gives rise predominantly to the same σ-allyl-product in which the primary carbon is bonded to iridium. It is conceivable that in methanol some π-allyl-complex formation occurs to account for the *ca* 20% isomerisation that is observed. However, it is difficult to rationalise the complete racemisation of optically active $H_2C{=}CH{-}CH(Cl)Me$ on the basis of this mechanism. It has been proposed that racemisation is induced by the iridium complex, within the encounter complex, prior to the "oxidative addition".

8 Stereochemical changes at the metal

The reactions of metal nucleophiles containing a coordinatively-saturated reaction centre are relatively trivial, and invariably occurs with little gross stereochemical change to the complex, as for instance in the reactions of $[M(dmgH)_2L]^-$ (M = Co, L = py; M = Rh, L = PPh_3) or $[M(CO)_5]^-$ (M = Mn or Re).

In the reaction of $[(\eta^5\text{-}C_5H_5)Rh(PPh_3)(\eta^2\text{-}C_2H_4)]$ with MeI, an almost colourless solid is formed initially which is believed to be the pseudo-four-coordinate cationic intermediate shown in Scheme 11. This complex could not be isolated in a pure state, possibly because of the extreme lability of the coordinated ethylene, but the analogous $[(\eta^5\text{-}C_5H_5)Rh(AsPh_3)(\eta^2\text{-}C_2H_4)Me]^+$ has been isolated and fully characterised (Oliver and Graham, 1971). Studies on $[(\eta^5\text{-}C_5H_5)M(CO)(R)(I)]$ (M = Co or Rh, R = CF_3, C_2F_5 or C_3F_7) by ^1H and ^{19}F nmr spectroscopy have demonstrated that the metal in these complexes is an asymmetric centre (McCleverty and Wilkinson, 1964).

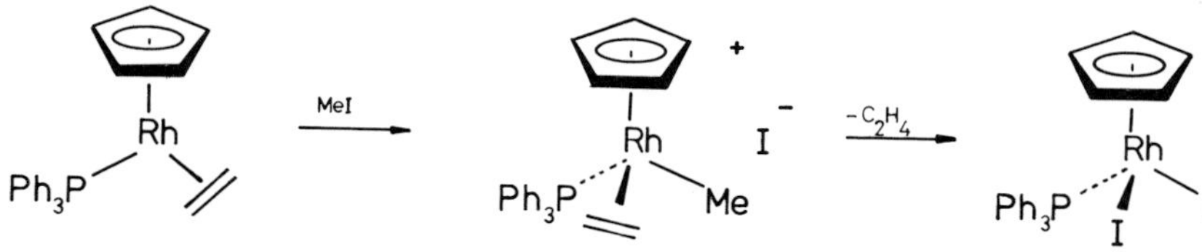

Scheme 11

In contrast to the studies with $[(\eta^5\text{-}C_5H_5)Rh(PPh_3)(\eta^2\text{-}C_2H_4)]$, where predissociation of a ligand is not essential for this species to react with alkyl halides, the reactions of $[Pd(PPh_3)_3]$ involve the 14-electron species $[Pd(PPh_3)_2]$. This behaviour is manifest in the kinetics (Stille and Lau, 1977 and references therein; Fauvarque and Pfluger, 1981). Similarly studies of $[Pt(PPh_3)_2(C_2H_4)]$ with MeI, $PhCH_2Br$ or ICH_2CH_2I shows an inverse dependence on the concentration of ethylene demonstrating that the "active species" is $[Pt(PPh_3)_2]$ (Birk *et al.*, 1968). However, other studies have shown that both $[Pt(PPh_3)_2]$ and $[Pt(PPh_3)_3]$ can react (Pearson and Rajaram, 1974).

Similarly, in the reactions of *trans*-$[IrCl(CO)(PPh_3)_2]$ with aryl iodides, the rate constant (k_{obs}) for disappearance of Ir(I) species obeys the two term rate-law shown in (26). The second term, exhibiting an inverse dependence on the concentration of PPh_3, is attributed to a highly reactive, 14-electron species $[IrCl(CO)(PPh_3)]$ (Mureinik *et al.*, 1979). In (26), K is the equilibrium constant for dissociation of PPh_3 from the complex, k_2 is the rate constant for nucleophilic attack of $[IrCl(CO)(PPh_3)]$ on ArI and k_1 refers to the solvent mediated pathway.

$$k_{obs} = k_1 + \frac{k_2 K[\mathrm{ArI}]}{(K + [\mathrm{PPh_3}])} \tag{26}$$

An interesting, if somewhat anomalous, conclusion comes from a kinetic study of the reactions of $[RhCl(PPh_3)_3]$ with H_2, O_2, C_2H_4 and MeI. Although H_2, O_2 and C_2H_4 react with the species $[RhCl(PPh_3)_2]$, MeI reacts not with this species, but rather $[RhCl(PPh_3)_3]$, and the derived dimer shown in (27). There is no apparent reason for these various reactivities (Ohtani *et al.*, 1977).

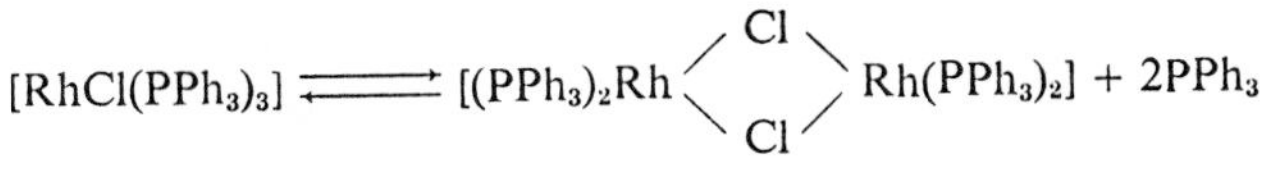

$$[\mathrm{RhCl(PPh_3)_3}] \rightleftharpoons [\mathrm{(PPh_3)_2Rh}\langle{}^{\mathrm{Cl}}_{\mathrm{Cl}}\rangle\mathrm{Rh(PPh_3)_2}] + 2\mathrm{PPh_3} \tag{27}$$

In considering the reactions of coordinatively-unsaturated complexes, the stereochemistry of the addition at the metal can be either *cis* or *trans*. Most of the stereochemical studies in this area have been conducted with complexes of the type, *trans*-$[IrX(CO)(PR_3)_2]$.

The reaction of *trans*-$[IrCl(CO)(PPh_3)_2]$ with MeI gave exclusively product [6], in which the substrate molecule has added *trans* across the plane of the complex. Furthermore, in the presence of an excess of $Bu_4N^+Cl^-$ the

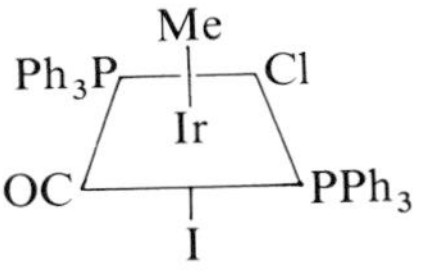

Me
Ph_3P Cl
Ir
OC PPh_3
I

[6]

exclusive product is still that shown in [6]. It was proposed that these results were a consequence of a one-step, concerted addition, consistent with theoretical considerations (Pearson and Muir, 1970). The *trans*-addition of alkyl halides to *trans*-$[IrCl(CO)(PMe_2Ph)_2]$ has been verified (Deeming and Shaw, 1969) and shown to yield the product whose stereochemistry is shown in [6] when the solvent is benzene. However, in methanol, a mixture is formed which comprises: $[IrI_2(Me)(CO)(PMe_2Ph)_2]$, ~55%; [6], ~40%;

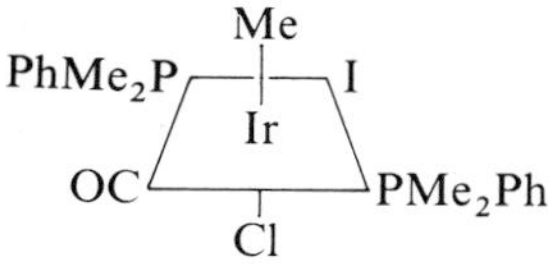

Me
$PhMe_2P$ — I
Ir
OC — PMe_2Ph
Cl

[7]

[7], ~5%. This difference in the product distribution for the two solvents has been rationalised by the pathways shown in Scheme 12.

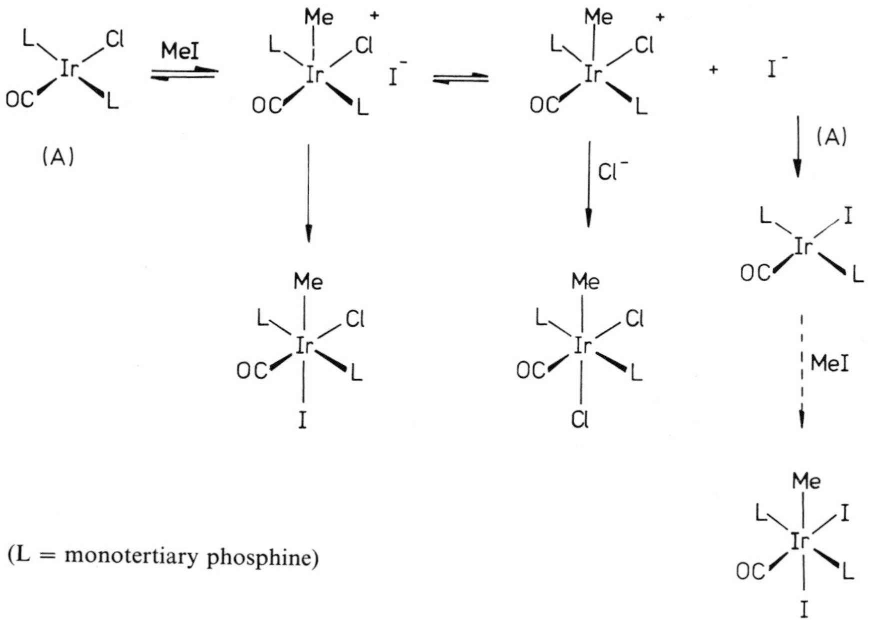

(L = monotertiary phosphine)

Scheme 12

In benzene, nucleophilic attack by the complex on MeI yields the ion-pair, and subsequent attack by iodide at the metal yields the *trans*-product.

However in the more polar methanol the ion-pair can separate, and subsequent rapid reaction of iodide with *trans*-[IrCl(CO)($PMe_2Ph)_2$] results in halide exchange and hence the corresponding mixture of products.

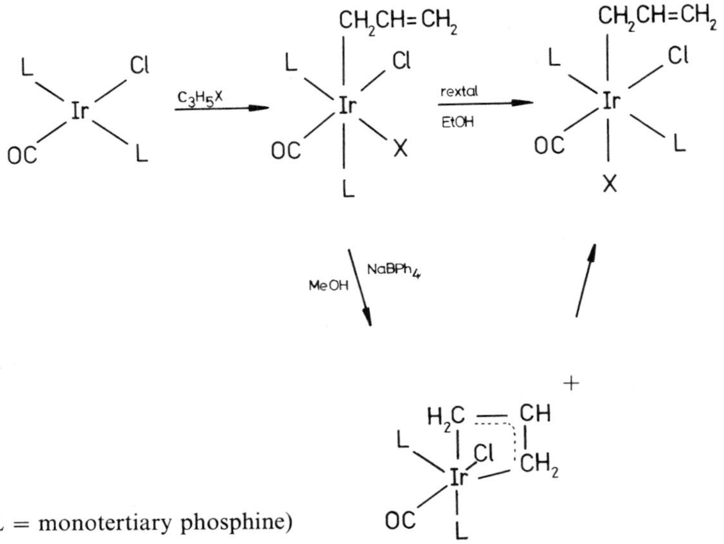

(L = monotertiary phosphine)

Scheme 13

The important role of the solvent in defining the stereochemistry of the addition is further emphasised by the reactions of *trans*-[IrCl(CO)($PMe_2Ph)_2$] with allyl halides (Deeming and Shaw, 1968a). Thus in benzene the product is formed as a consequence of *cis*-addition as shown in Scheme 13. Recrystallisation from ethanol yields the *trans*-product via the π-bonded species (which can be isolated as the BPh_4^- salt). Again this isomerisation appears to be the consequence of the lability of the halido-groups in the protic solvent. A mixture of products is formed in the reaction between *trans*-[IrBr(CO)($PMe_2Ph)_2$] and C_3H_5Cl because of the rapid exchange shown in (28), (Deeming and Shaw, 1968b). Consistent with these observations, when the reaction between *trans*-[IrBr(CO)($PMe_2Ph)_2$] and C_3H_5Cl is performed in the presence of an excess of LiBr, only one product is obtained, which contains no Ir–Cl bonds (Pearson and Poulos, 1979).

$$trans\text{-}[IrBr(CO)(PMe_2Ph)_2] + Cl^- \rightleftharpoons trans\text{-}[IrCl(CO)(PMe_2Ph)_2] + Br^- \quad (28)$$

The addition of acyl halides to *trans*-[IrCl(CO)($PMe_2Ph)_2$] also occurs with a *trans*-stereochemistry as shown in (29) (Deeming and Shaw, 1969; Kubota and Blake, 1971; Collman and Sears, 1968).

$$trans\text{-}[IrCl(CO)(PMe_2Ph)_2] + RCOBr \longrightarrow \text{PhMe}_2\text{P, Cl, OC, PMe}_2\text{Ph around Ir; C(=O)R above, Br below} \quad (29)$$

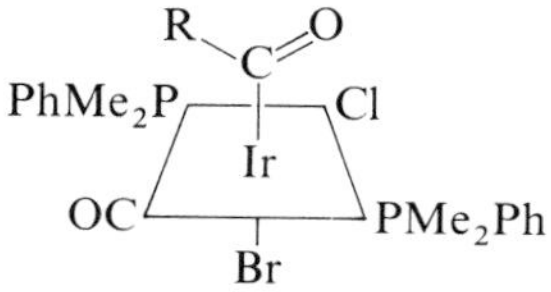

Further studies on these types of reactions, but with the more sterically-demanding PMe_2Bu^t ligand have shown that *trans*-addition of alkyl and acyl halides and *cis*-addition of allyl chloride are still the exclusive reaction modes (Shaw and Stainbank, 1972), although *trans*-[IrCl(CO)($PBu^tR_2)_2$] (R = Et, Pr^n or Bu^n) have a reduced tendency to react, and only do so with MeI.

The reactions of [MCl($PMe_2Ph)_3$] (M = Rh or Ir) with acyl chloride yield the *trans*-product. However the *trans*-chloro-group is labile and the derived PF_6^- salts yield either an equilibrium mixture of the five-coordinate complex and the alkyl complex (M = Rh) or the alkyl complex exclusively (M = Ir), as shown in Scheme 14 (Bennett *et al.*, 1981).

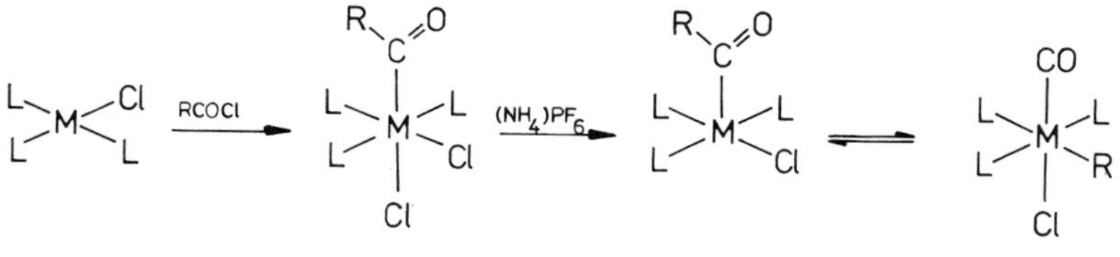

$$L_2M(Cl)L \xrightarrow{RCOCl} L_2M(COR)(L)(Cl)Cl \xrightarrow{(NH_4)PF_6} L_2M(COR)(L)(Cl) \rightleftharpoons L_2M(CO)(L)(R)Cl$$

Scheme 14

An interesting study of the reactions of [$PtMe_2$(diars)] {diars = 1,2-bis(dimethylarsino)benzene} with alkyl and acyl halides has shown that here too the acyl halide adds *trans*, but it is not possible to decide the stereochemistry of the addition of MeI. Allyl halides add *cis*, but unlike the iridium systems discussed before, the product isomer does not convert to the *trans*-form even under forcing conditions (Cheney and Shaw, 1971a).

The reactions of [$PtMe_2$(*meso*-dias)] (dias = $PhMeAsCH_2CH_2AsMePh$) with acetyl chloride, methyl iodide or allyl halides gives one isomer (Cheney and Shaw, 1971b). For acetyl chloride and methyl iodide the addition is *trans*, but it is impossible to distinguish between the isomers [8] and [9] on the basis of ir and nmr spectroscopy. With allyl halides the addition is *cis*, but again it is not possible to distinguish between the two isomers.

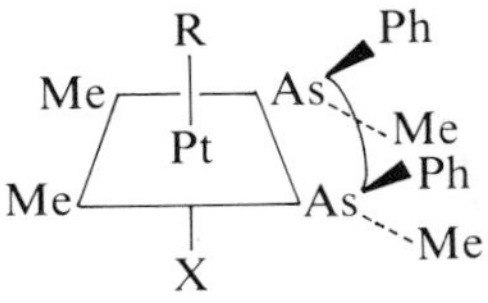

[8] 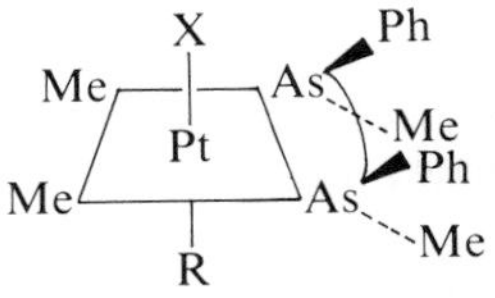 [9]

The reactions of allyl halides with $[Rh\{P(OMe)_3\}_5]^+$ gives rise to mixtures of $[RhX(\eta^3\text{-}C_3H_5)\{P(OMe)_3\}_3]^+$ and $[Rh(\eta^3\text{-}C_3H_5)\{P(OMe)_3\}_4]^{2+}$, and this has been rationalised by a mechanism involving the initial *cis*-addition of allyl halide and subsequent displacement of either a phosphite or halide ligand, respectively (Haines, 1971).

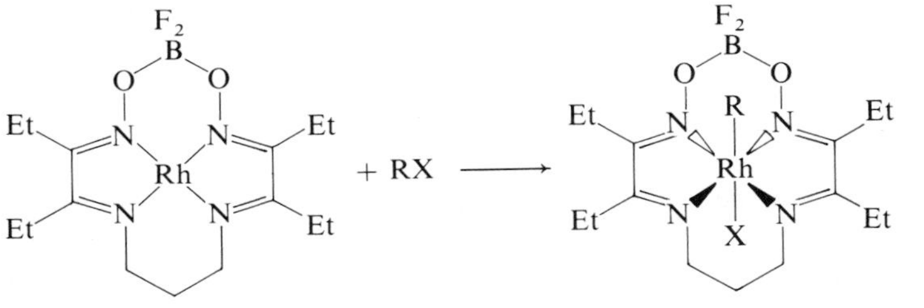

[10] + RX ⟶ (30)

In contrast to the studies with the iridium-phosphine complexes, the very reactive complex shown in [10] reacts with alkyl halides as shown in (30), but in the presence of a large excess of LiCl the reaction of Bu^nBr yields the chloro-complex under conditions where the corresponding bromo-complex does not exchange. These observations, together with the isolation of the intermediate, *trans*-$[Rh(Me)(Et_2mgBF_2)(NCMe)]^+BF_4^-$ and the reactivity order with respect to the alkyl group (Me > Et > secondary alkyl > cyclohexyl), supports an S_N2 mechanism (Collman and MacLaury, 1974).

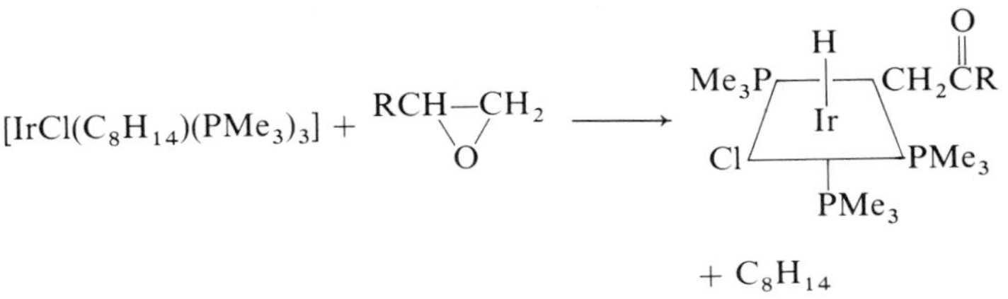

$[IrCl(C_8H_{14})(PMe_3)_3]$ + RCH—CH$_2$ (O) ⟶ (31)

+ C_8H_{14}

The reactions of epoxides with $[IrCl(C_8H_{14})(PMe_3)_3]$ or $[Ir(PMe_3)_4]^+$ yield the *cis*-hydridoalkyl-complexes shown in (31), where the stereochemistry of the products has been established (when R = H) from both ^1H nmr spectroscopy and a crystal structure determination (Milstein, 1984;

Milstein and Calabrese, 1982). However, (when R = H) there is also formed a small amount (*ca* 5%) of another isomer in which the *trans*-chloride and a *cis*-phosphine's positions are interchanged.

Finally the reaction of $Na_2[Fe(CO)_4]$ with Pr^nBr gives $[Fe(CH_2CH_2CH_3)(CO)_4]^-$, which a crystal structure determination shows has a trigonal bipyramidal structure with an apical alkyl-group (Huttner and Gartzke, 1975). The ^{13}C nmr spectrum, however, shows only a single signal, indicating a rapid scrambling process (Collman *et al.*, 1977).

9 The iodide catalysis effect

An aspect of the metal's coordination environment and nucleophilicity is the influence that iodide ion can have on the reactions of certain transition metal nucleophiles with MeI. This is an aspect of particular relevance to the homogeneously catalysed carboxylation of methanol to acetic acid which employs a rhodium iodide-promoted catalyst (Forster, 1979 and references therein).

Kinetic studies on the iodide-catalysed reaction (32) demonstrate that in methanol an equilibrium between $[Ir(cod)(phen)]^+$ and iodide (which can be studied spectrophotometrically) is established rapidly, and that both species

$$[Ir(cod)(phen)]^+ + MeI \longrightarrow [Ir(Me)(I)(cod)(phen)]^+ \qquad (32)$$

shown in Scheme 15 react with MeI to yield the alkylated product, the liberated iodide sustaining the catalysis. The iodo-complex reacts with MeI about seven times faster than the parent cation, and this is presumably because of the greater electron density at the metal centre of the former (de Waal *et al.*, 1982).

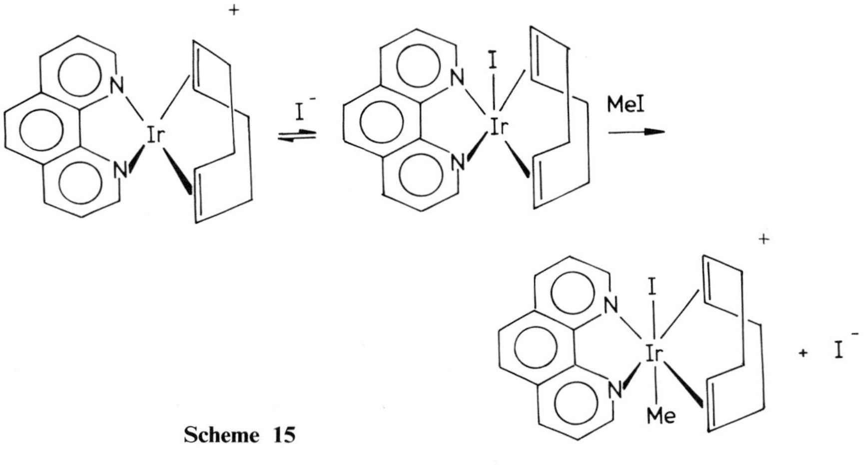

Scheme 15

Similarly the reactions of *trans*-$[RhI(CO)(MPh_3)_2]$ (M = P, As or Sb) are catalysed by iodide (Forster, 1975) and the equilibrium constants for (33)

$$trans\text{-}[RhI(CO)(MPh_3)_2] + I^- \underset{}{\overset{K}{\rightleftharpoons}} [Rh(I)_2(CO)(MPh_3)]^- + MPh_3 \qquad (33)$$

have been measured: M = P, $K < 3 \times 10^{-5}$; M = As, $K \sim 0.05$; M = Sb, $K \sim 2 \times 10^{-3}$. A lower limit for the relative rates of reactions of the anion *vs* the neutral species with MeI is 1×10^5, indicating the great increase in nucleophilicity imparted to the metal by the substitution.

Semi-quantitative studies on the reaction of *cis*-$[RhI_2(CO)_2]^-$ with MeI have shown that although increasing amounts of $Bu_4N^+I^-$ cause a small rate-increase, this is attributable to general salt effects, but $Ph_4As^+I^-$, $Ph_4As^+Cl^-$ and $Me_2im^+I^-$ (Me_2im = 1,3-dimethylimidazolium) cause a dramatic increase in the rate (Hickey and Maitlis, 1984). This has been interpreted in terms of the equilibria (34), where L = Me_2im. A low

$$[Rh_2IL(CO)_2] + I^- \rightleftharpoons [Rh(I)_2(L)(CO)_2]^- \rightleftharpoons [RhI_2(CO)_2] + L \qquad (34)$$

concentration of $[Rh(I)_2(L)(CO)_2]^-$ (which may have been detected by FT ir spectroscopy) is probably the very active nucleophile.

10 The reactions of binuclear complexes

The stereochemistry of the addition of α,γ-dihalides to $[\{(\eta^5\text{-}C_5H_5)Co\}_2(\mu\text{-}CO)_2]^-$ has already been discussed (Section 6). There are relatively few binuclear systems in which the metal centres act as a nucleophile.

The reaction of $[Pt_2(\mu\text{-dppm})_3]$ {dppm = bis(diphenylphosphino)methane} with MeI in benzene leads to the methylated product as shown in (35).

$$[Pt_2(\mu\text{-dppm})_3] + MeI \longrightarrow \left[\begin{array}{ccc} Ph_2P \frown & PPh_2 & \\ | & | & \\ Me{-}Pt{-} & {-}Pt{-}PPh_2{\frown} & \\ | & | & PPh_2 \\ Ph_2P \smile & PPh_2 & \end{array}\right]^+ + I^- \qquad (35)$$

Subsequent reaction of the methylated complex with an excess of MeI in dichloromethane yields the dimethylated product as in (36). The binuclear system appears to be more reactive than mononuclear analogues such as $[Pt(PPh_3)_3]$, and this has been ascribed to anchimeric assistance by the second platinum centre (Azam *et al.*, 1984). A similar reaction has been described for $[Pd_2(\mu\text{-dppm})_3]$ (Balch *et al.*, 1981).

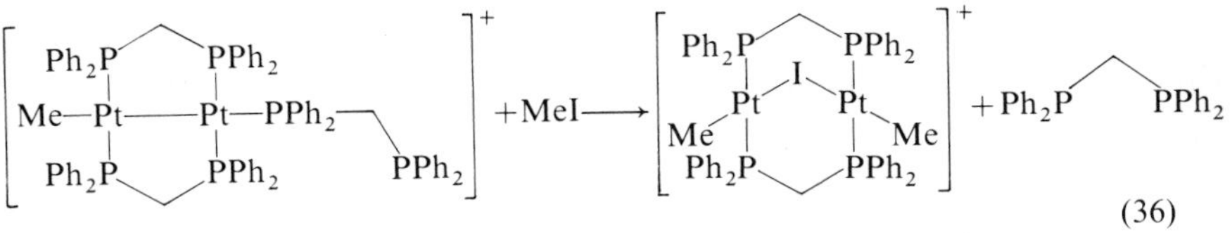

$$\left[\begin{array}{c}Ph_2P\frown PPh_2 \\ Me{-}Pt{-}Pt{-}PPh_2\frown PPh_2 \\ Ph_2P\smile PPh_2\end{array}\right]^+ + MeI \longrightarrow \left[\begin{array}{c}Ph_2P\frown PPh_2 \\ Me{-}Pt(\mu{-}I)Pt{-}Me \\ Ph_2P\smile PPh_2\end{array}\right]^+ + Ph_2P\frown PPh_2 \quad (36)$$

Direct comparison of the reactivity of a mononuclear and binuclear system is possible in the reactions of $[PtMe_2(SMe_2)_2]$ and $[Pt_2Me_4(\mu\text{-}SMe_2)_2]$ respectively with MeI (Puddephatt and Scott, 1985). The dimer reacts with MeI in acetone some twenty times faster than the monomer at $-10°C$. Furthermore, in this study, the reaction of the monomer with MeI to yield *fac*-$[PtIMe_3(SMe_2)_2]$ in C^2H_3CN was shown to proceed via the intermediate *fac*-$[PtMe_3(SMe_2)_2(C^2H_3CN)]I$ which was detected by low-temperature nmr spectroscopy.[1] The detection of this intermediate is consistent with an S_N2 mechanism.

As observed in the work on $[Pt_2(\mu\text{-dppm})_3]$,"oxidative-addition" at one metal centre deactivates the other because of electronic factors: the metal centre is now formally in a higher oxidation-state which thus attracts charge. In the absence of restraining ligands this leads to metal—metal bond formation [see (35)]. To circumvent these problems and to investigate the reactivity of a binuclear complex containing "isolated" metal centres the complexes shown in [11] and [12] (M = Rh or Ir) have been prepared (Schenk *et al.*, 1985).

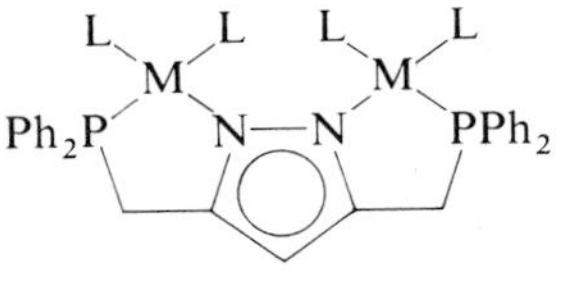

[11] 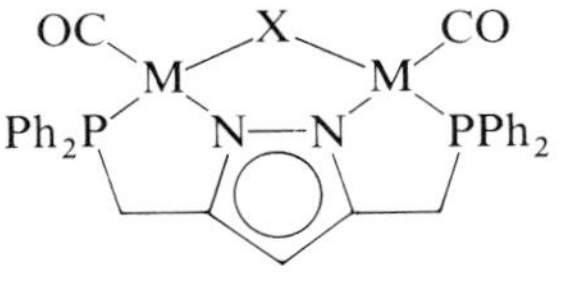 [12]

For species [12, X = PPh_2] the reactions with MeI and CH_3COCl demonstrate the greater electron-richness compared to [12, X = Cl]. The reaction with MeI proceeds according to (37); that of the rhodium complex proceeds only to equilibrium whereas that of the iridium analogue goes to completion.

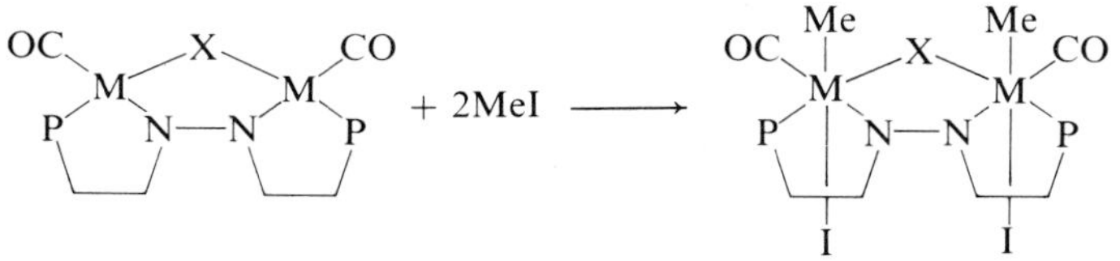

$$\text{(OC)M(}\mu\text{-X)(}\mu\text{-N–N)M(CO)} + 2MeI \longrightarrow \text{(OC)(Me)(I)M(}\mu\text{-X)(}\mu\text{-N–N)M(I)(Me)(CO)} \quad (37)$$

[1] The prefix *fac*- denotes that the three methyl groups occupy mutually *cis* positions in an octahedral structure.

The stereochemistry of the addition is not exclusively that shown in (37), although the reaction exhibits a reasonably high degree of selectivity as evidenced by nmr spectroscopy. Furthermore the monoalkylated intermediate associated with raction (37) has been isolated.

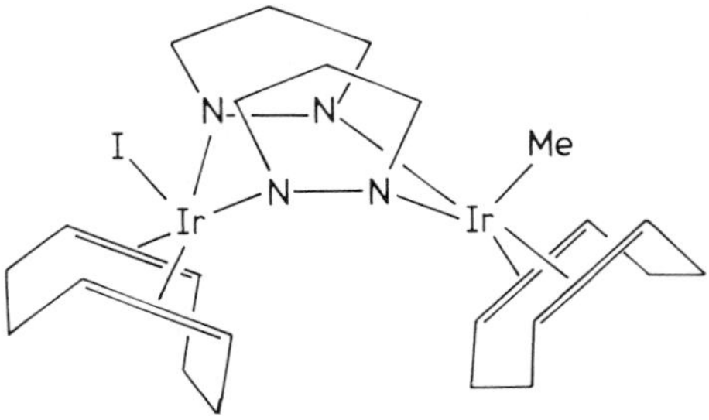

FIG. 6 Reaction product between MeI and $[Ir_2(C_3N_2H_3)_2(cod)_2]$

The only substantiated case of 1,2-addition of an alkyl halide across two adjacent metal atoms is shown in Fig. 6 (Coleman *et al.*, 1982). Finally, only one example of addition of MeI across a hetero-binuclear complex as shown in (38) has been reported (Finke *et al.*, 1983). This reaction is strongly regioselective, the addition occurring in the sense shown to the extent of more than 80%.

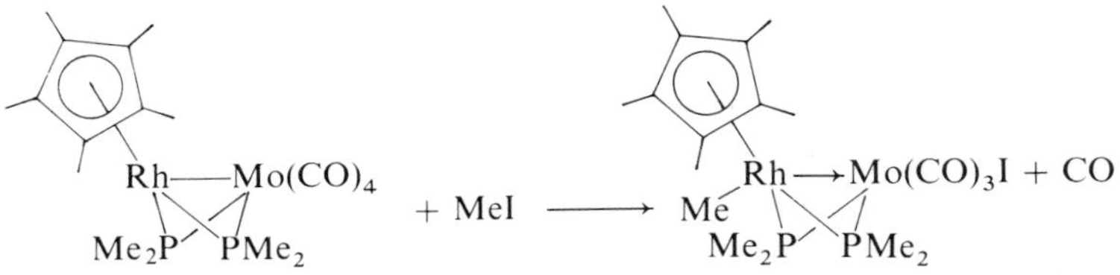

$$\text{Rh—Mo(CO)}_4\ (\mu\text{-Me}_2\text{P}\ \text{PMe}_2) + \text{MeI} \longrightarrow \text{Me—Rh}\rightarrow\text{Mo(CO)}_3\text{I}\ (\mu\text{-Me}_2\text{P}\ \text{PMe}_2) + \text{CO} \qquad (38)$$

11 The reactivity of the carbon centre

In this section are included those studies in which perturbation of the carbon centre has been used as a probe for the reaction mechanism. In the succeeding section, a further aspect of the reactivity problem will be discussed, namely, the reactions of α,ω-dihalides.

Several authors have noted, although sometimes on a purely qualitative basis, that the rates of the reactions of a particular nucleophile with an alkyl or aryl halide vary inversely with the C—X bond strength (i.e. I > Br > Cl), as would be expected for a nucleophilic displacement reaction (Ochiai *et al.*, 1969; Hart-Davis and Graham, 1970, 1971; Schrauzer and Deutsch, 1969;

Fitton and Rick, 1971; Ramasami and Espenson, 1980; Blum and Weitzberg, 1976; Collman *et al.*, 1977; Douek and Wilkinson, 1969).

The influence of the alkyl group on the reactivity has been shown to decrease with increasing chain-length. Thus in the studies of the reaction between vitamin B_{12} and alkyl halides, the relative rates in methanol at 25°C are: MeCl (180) > EtCl (1.7) > Pr^nCl (1.3) > Bu^nCl (1.0). The close similarity between these relative rates and the analogous reactions of iodide (although the latter were studied in acetone at 30°C) was considered indicative of the nucleophilic character of the former reactions (Schrauzer and Deutsch, 1968).

The surprisingly large difference between the reactivity of the methyl and ethyl halides both in the above study and in the reactions of $[(\eta^5\text{-}C_5H_5)Ir(CO)(PPh_3)]$, where MeI reacts 400–1200 times faster than EtI, has been ascribed predominantly to the steric restrictions of the ethyl group (Hart-Davis and Graham, 1970). Irrespective of the leaving group, the reaction rates of alkyl compounds with various nucleophiles are perturbed by similar amounts with increasing congestion at the carbon centre. Substitution of one or two hydrogen atoms for methyl groups at the reaction centre retards the rate of the reaction with vitamin B_{12}, whereas increasing the carbon chain length by similar amounts has little effect (Schrauzer and Deutsch, 1969). If however the hydrogen atoms are replaced by a group which can delocalise the incipient electron density on the metal, thus increasing the stability of the transition state, the reaction rate is increased. Such groups include CN, MeO, Ph, $C_{10}H_7$, $H_2NC{=}O$ and CO_2^- (although, in the last case, stabilising electronic effects are overridden by the unfavourable negative charge). Consistent with the sensitivity of these reactions towards steric effects, the replacement of one hydrogen atom, at the reaction centre, by a phenyl group leads to an increased reaction rate with vitamin B_{12}, but substitution by two phenyl groups leads to a retardation of rate. Similar steric effects have been observed in the reactions of $[Rh(dmgH)_2(PPh_3)]^-$ with $PhCH_2Cl$ ($k = 1.04 \times 10^3 dm^3 mol^{-1} s^{-1}$ in aqueous methanol at 25°C) and PhCHMeCl ($k = 96$ $dm^3 mol^{-1} s^{-1}$) (Ramasami and Espenson, 1980). In general the reactivity order observed with various alkyl halides and metal nucleophiles is: primary > secondary > neopentyl ≫ adamantyl. The fact that adamantyl bromide is unreactive towards $[Fe(CO)_4]^{2-}$ has been used as evidence that the reactions of this dianion do not proceed by an electron-transfer mechanism (Collman *et al.*, 1977).

In the reactions of aryl halides with various complexes, the effect of *para*-substituents in the benzene ring on the reaction rate has been used as a probe of the reaction pathway. The reaction of $[Pd(PPh_3)_3]$ with $p\text{-}RC_6H_4I$ as shown in Scheme 16 is influenced by the substituent; electron-withdrawing groups enhance the rate in the order: NO_2 > CN > PhCO > H. This

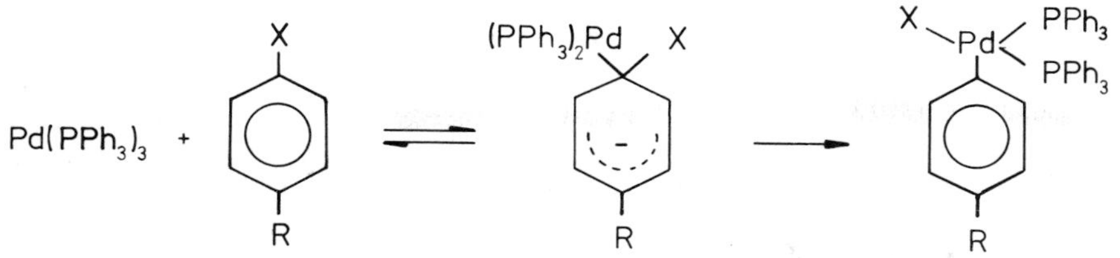

$$Pd(PPh_3)_3 + p\text{-}XC_6H_4R \rightleftharpoons (PPh_3)_2Pd(X)C_6H_4R^{-} \longrightarrow X\text{-}Pd(PPh_3)_2\text{-}C_6H_4R$$

Scheme 16

order indicates that the reaction is, in essence, an aromatic nucleophilic substitution, where breaking of the bond to the leaving group is rate-limiting (Fitton and Rick, 1971). However, the same reactivity pattern could result from an electron-transfer mechanism. A similar study (Fauvarque and Pfluger, 1981) demonstrated a good correlation of $\log_{10}k_2$ with Hammett σ ($\rho = 2.0$). This ρ-value is consistent with an increased electron density transferred to the aromatic ring in the transition state. The reaction proceeds most readily with aryl iodides, and this has been attributed to the halogen being used as a ligand in the intermediate/transition-state as shown in [13].

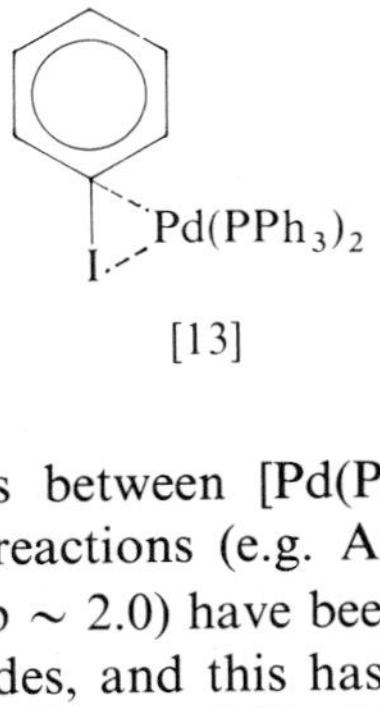

$Pd(PPh_3)_2$

[13]

Although in these reactions between $[Pd(PPh_3)_3]$ and aryl halides, no products typical of radical reactions (e.g. ArH or ArAr) were observed, similar reactivity constants ($\rho \sim 2.0$) have been observed in the reactions of $[NiBr(PPh_3)_2]$ with aryl halides, and this has been interpreted in terms of an electron-transfer reaction (Tsau and Kochi, 1979).

In a study of the reactions of $[Pt(R)_2(bipy)]$ in acetone solution, the position of the electronic transition of lowest energy in the uv/visible spectrum, which is a metal-to-ligand charge-transfer band (MLCT), was found to correlate well with the value of $\log_{10}k_2$ (where k_2 is the second-order rate constant for the reaction with MeI). The lowest energy of the MLCT band occurs when the highest electron density is at the metal. The correlation thus strongly suggests that the rate of the reaction is primarily dependent on the energy of the filled d-orbitals on platinum. The value of k_2 varies from 2.8×10^3 to $2.3\ dm^3\ mol^{-1}\ s^{-1}$, in the order: R = Me > p-$MeOC_6H_4$ > p-MeC_6H_4 > m-$MeOC_6H_4$ > C_6H_5 > p-FC_6H_4 > p-ClC_6H_4 > m-ClC_6H_4 (Jawad and Puddephatt, 1977). Interestingly, the position of the MLCT band for R = p-MeC_6H_4 is in about the same position as for o-MeC_6H_4, but this complex does not react with MeI presumably because of the unfavourable steric interactions. The Hammett correlation with values of k_2 is not very good, presumably because specific solvation effects, which are reflected in both the MLCT band and the value of k_2, are not represented in the Hammett σ-values. The reaction constant for $[Pt(Ar)_2(bipy)]$ with MeI per aryl group (ρ ~ −1.3) is very similar to that per aryl group for *trans*-$[IrCl(CO)(PAr_3)_2]$ (ρ = −1.1) (Ugo *et al.*, 1972). If such a comparison is valid, it indicates that, despite the very different distances of the aryl groups from the reaction centre in the two types of complex, very similar electronic effects are transmitted. The negative values of the reaction constants demonstrate that positive charge is formed on the metal atom in the transition state. In studies on *trans*-$[IrCl(CO)(PAr_3)_2]$ with alkyl halides (Ugo *et al.*, 1972) the value of the reaction constant for MeI (−6.4) and $PhCH_2Cl$ (−2.6) were obtained. The less negative value for the reaction constant with benzyl chloride is attributable to delocalisation of the negative charge (in the transition state) over the benzene ring.

12 The reactions of α,ω-dihalogenoalkanes

The reactions of α,ω-dihalogenoalkanes with transition-metal nucleophiles proceed by a variety of different routes. In the simplest possible pathway only one end of the alkyl-group is substituted (39); M = Mo or W, X = Br

$$[(\eta^5\text{-}C_5H_5)M(CO)_3]^- + X(CH_2)_3X \longrightarrow [(\eta^5\text{-}C_5H_5)M\{(CH_2)_3X\}(CO)_3] + X^- \qquad (39)$$

or I (King and Bisnette, 1967). However, in the presence of an excess of the metal complex, subsequent replacement of the remaining halide results in a dimetallo-species as shown in (40); M = Fe or Ru (King, 1963).

$$2[(\eta^5\text{-}C_5H_5)M(CO)_2] + Br(CH_2)_3Br \longrightarrow [\{(\eta^5\text{-}C_5H_5)M(CO)_2\}_2(CH_2)_3]^{2+} + 2Br^- \qquad (40)$$

Alternatively, after the initial alkylation step, subsequent attack by the excess of metal complex on the metal alkyl species results in migration of the

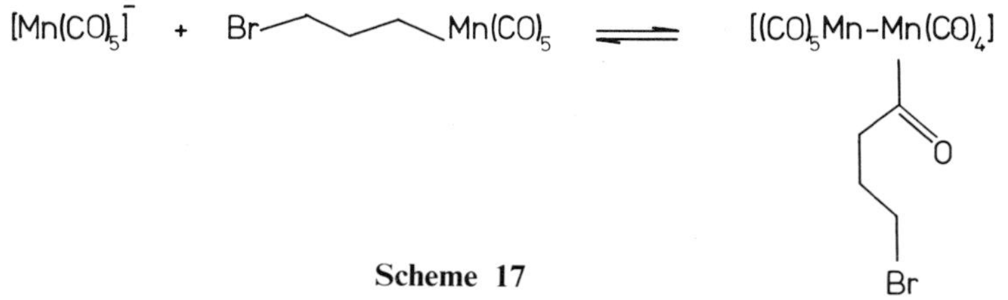

$$[Mn(CO)_5]^- + Br(CH_2)_3Mn(CO)_5 \rightleftharpoons [(CO)_5Mn\text{-}Mn(CO)_4\{C(O)(CH_2)_3Br\}]$$

Scheme 17

alkyl group to an adjacent carbonyl ligand (Scheme 17). Subsequent intramolecular attack of the carbonyl oxygen on the alkyl halide yields a carbene ligand (Casey and Anderson, 1971). The structure of the carbene complex has been established by X-ray crystallography. A second metal centre is not essential to yield the carbene complexes, however. Thus the reaction of $[(\eta^5\text{-}C_5H_5)M(CO)_3]^-$ (M = Mo or W) with 1,3-diiodopropane in THF or 1,2-dimethoxyethane at reflux yields the cyclic carbene complex, presumably

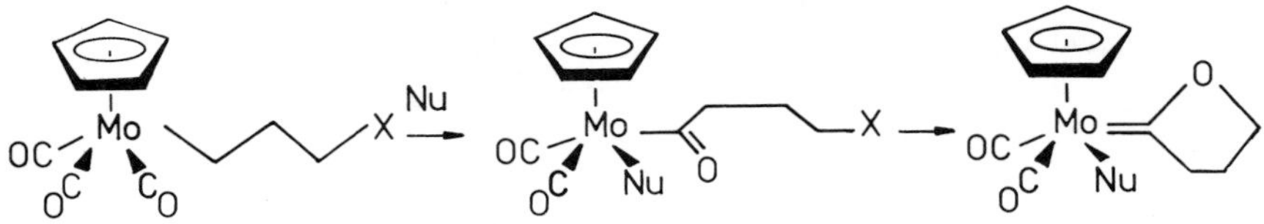

Scheme 18

by the mechanism shown in Scheme 18 (Adams *et al.*, 1984 and references therein). In the presence of a large excess of $[(\eta^5\text{-}C_5H_5)Mo(CO)_3]^-$, compound [14] is produced, presumably by a pathway analogous to that shown in Scheme 17.

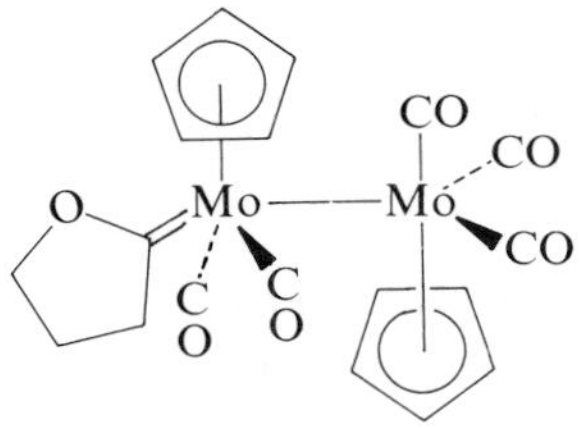

[14]

This proposed mechanism is further reinforced by the formation of the heterobimetallic species of the structure shown in [15] formed from the reaction between $[(\eta^5\text{-}C_5H_5)W(CO)_3]^-$ and $[(\eta^5\text{-}C_5H_5)Mo\{(CH_2)_3Br\}\text{-}(CO)_2]$. However, here the carbene is unexpectedly coordinated to the tungsten atom. A similar transfer of the ligand, proceeding after the alkylation but prior to the ring-closure, occurs in the reaction of $[Mn(Me)\text{-}(CO)_5]$ with $[Re(CO)_5]^-$ (Casey *et al.*, 1975).

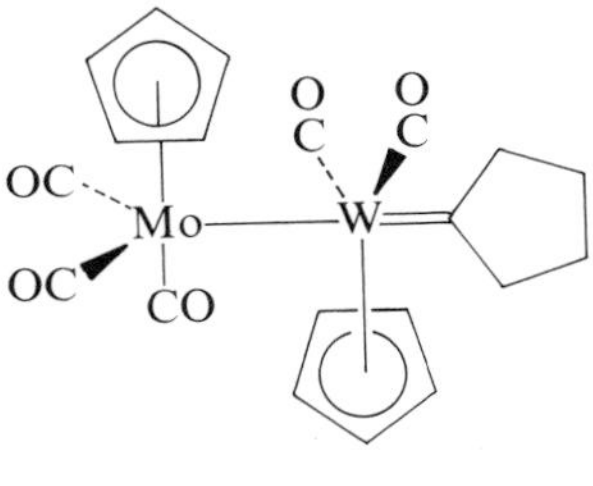

[15]

The reactions of $[(\eta^5\text{-}C_5H_5)M\{(CH_2)X\}(CO)_3]$ (M = Mo or W, X = Br or I) with transition metal nucleophiles is not a general method for making heterobimetallic carbene complexes, since the products can be homobimetallic species involving displacement of the originally alkylated metal centre as shown in Fig. 7.

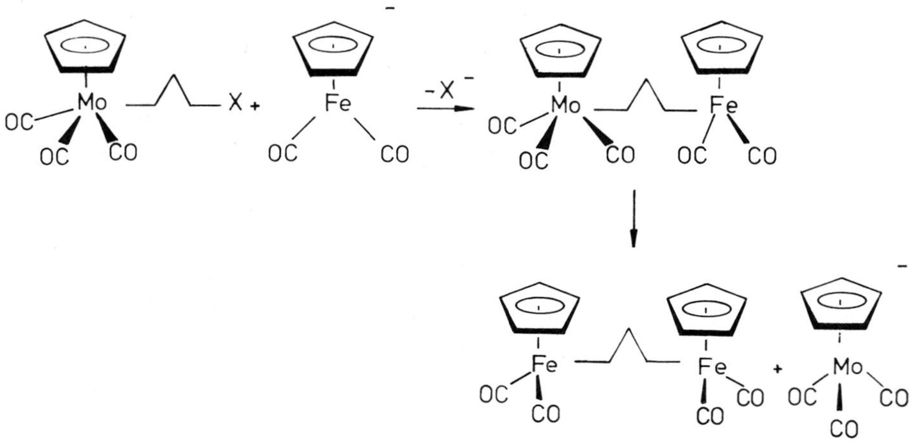

FIG. 7 Reaction between $[(\eta^5\text{-}C_5H_5)Mo\{(CH_2)_3X\}(CO)_3]$ and an excess of $[(\eta^5\text{-}C_5H_5)Fe(CO)_2]^-$

The reactions of 1,1,1-tris(halogenomethyl)ethanes, $MeC(CH_2X)_3$ (X = Br or I), with the nucleophiles $[(\eta^5\text{-}C_5H_5)Fe(CO)_2)]^-$, $[(\eta^5\text{-}C_5H_5)M(CO)_3]^-$ (M = Mo or W) $[(\eta^5\text{-}C_5H_5)Ni(CO)]$ or $[Re(CO)_5]^-$ all give the 1-methylcyclopropylmethyl-derivatives as shown in Fig. 8. No

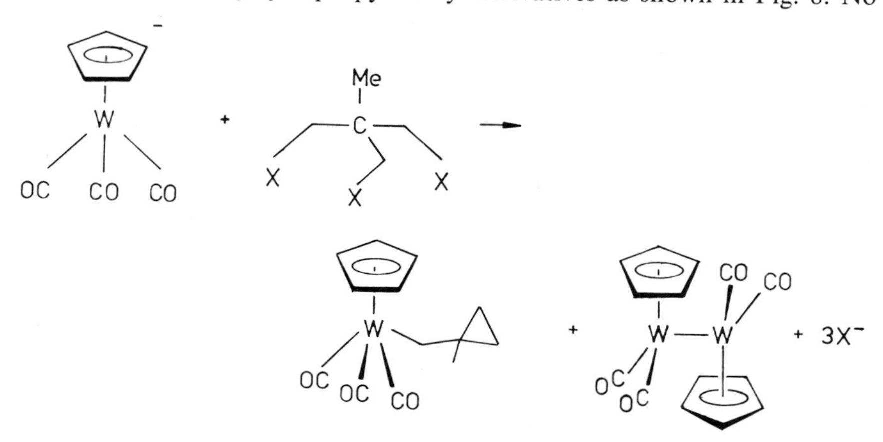

FIG. 8 Reaction between $[(\eta^5\text{-}C_5H_5)W(CO)_3]^-$ and $CH_3C(CH_2Br)_3$

reaction occurs with $[Co(CO)_4]^-$, presumably because of its poor nucleophilicity. Several mechanisms are consistent with the formation of these products including an electron-transfer process and two alternative nucleophilic pathways shown in Scheme 19 (Poli *et al.*, 1985). It is currently impossible to distinguish between these possibilities.

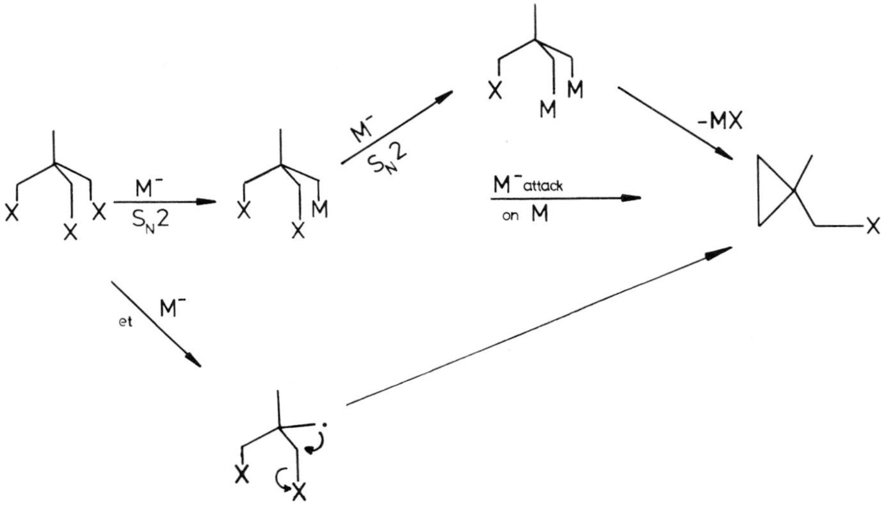

Scheme 19

The reaction of $[Re(CO)_5]^-$ with $MeC(CH_2I)_3$ also produces the 1-methylcyclopropylmethyl-derivative, but $[Re_2(CO)_9I]^-$ is also formed in appreciable amounts. In contrast, $[Mn(CO)_5]^-$ reacts with $MeC(CH_2I)_3$ to yield the "carbonyl inserted" acyl-product as shown in (41). Mechanisms have been postulated to rationalise these products but these are highly speculative.

$$4[Mn(CO)_5]^- + MeC(CH_2I)_3 \longrightarrow [Mn_3(CO)_{14}]^- + [(CO)_5Mn{-}C(=O)CH_2C_3H_4Me] + 3I^- \quad (41)$$

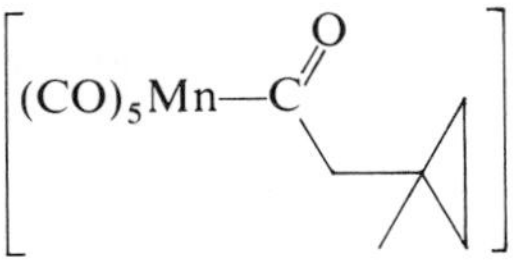

13 Activation parameters

Within this section will be discussed the activation parameters ($\Delta H^{\ddagger}$, $\Delta S^{\ddagger}$ and $\Delta V^{\ddagger}$) that have been measured for the reactions between metal nucleophiles and carbon compounds. The values of the enthalpy and entropy of activation for the reactions of many metal nucleophiles with MeI and MeOTs have been summarised (Pearson and Figdore, 1980). The parameters for the reactions of alkyl halides with metal complexes show a great deal of uniformity: $\Delta H^{\ddagger} \sim 10 \pm 5$ kcal mol^{-1} and $\Delta S^{\ddagger} \sim -30 \pm 20$ cal K^{-1} mol^{-1}. However, such values are insufficiently diagnostic to allow discrimination between the complex acting as a nucleophile or as a reductant towards alkyl halides (see Section 2) (Pearson and Figdore, 1980).

More recent data also contain activation parameters in the range indicated above. Thus the reaction of $[Fe(CO)_3(PMe_2)_2]$ with MeI has $\Delta H^{\ddagger} = 13$ kcal mol^{-1}, and $\Delta S^{\ddagger} = -33$ cal K^{-1} mol^{-1} (Bellachioma *et al.*, 1981), and the reactions of $[Ir(cod)(phen)]^+$ and [Ir(cod)(phen)I] with MeI are associated with the parameters, $\Delta H^{\ddagger} = 12.4$ kcal mol^{-1}, $\Delta S^{\ddagger} = -34.9$ cal K^{-1} mol^{-1}, and $\Delta H^{\ddagger} = 10.7$ kcal mol^{-1}, $\Delta S^{\ddagger} = -38.0$ cal K^{-1} mol^{-1} respectively (de Waal, 1982).

The solvent has a relatively large influence on the activation parameters as demonstrated in the reaction of *cis*-$[RhI_2(CO)_2]^-$ with MeI (Hickey and Maitlis, 1984). For the range of solvents, methanol, chloroform, THF and methyl acetate the values of $\Delta H^{\ddagger}$ progressively decrease from 16.5 to 11.2 kcal mol^{-1}, and $\Delta S^{\ddagger}$ decreases from -22.0 to -42.1 cal K^{-1} mol^{-1}.

The values of the activation parameters, together with the influence of the solvent on these parameters, has been interpreted in terms of a highly linear, polar transition state (Fig. 2) (Halpern, 1970 and references therein), or a transition state containing unusually stringent stereochemical restrictions. These conclusions are further substantiated by the comparable reactivities of substituted benzyl bromides towards both *trans*-$[IrCl(CO)(PPh_3)_2]$ and tertiary amines.

The large negative values of $\Delta S^{\ddagger}$ in the reactions of the iridium(I) complexes have been rationalised in terms of increased solvation of the transition state, attributable to its increased dipole. Such a dipole results not only from the interaction of the alkyl halides with the metal centre, but also from deformation of the iridium complex from planar to pseudo-octahedral geometry as would occur in a three-centre *cis*-addition (Harrod and Smith, 1970). In some cases the values of $\Delta S^{\ddagger}$ and the influence of solvent on $\Delta H^{\ddagger}$ and $\Delta S^{\ddagger}$ have been explained in terms of a polar, asymmetric, three-centre transition state, in which the interaction of the metal is predominantly with the carbon centre (Ugo *et al.*, 1972).

The characteristic values of $\Delta H^{\ddagger}$ and $\Delta S^{\ddagger}$ have been taken to indicate that a nucleophilic displacement reaction is rate-limiting in a multistep reaction. For instance reaction (42) occurs in two stages, but the intermediate which is detected is too reactive to be isolated and characterised. The

$$[RhCl(CO)_2(PPh_3)] + MeI \longrightarrow [Rh(Cl)(I)(CO)(COMe)(PPh_3)] \qquad (42)$$

first-order dependence of the kinetics of its formation on the concentration of MeI and the activation parameters, $\Delta H^{\ddagger} = 11.5\ \mathrm{kcal\ mol^{-1}}$, $\Delta S^{\ddagger} = -33\ \mathrm{cal\ K^{-1}\ mol^{-1}}$, indicate nucleophilic attack of $[RhCl(CO)_2(PPh_3)]$ on MeI to yield the intermediate $[Rh(Cl)(I)(Me)(CO)_2(PPh_3)]$. Subsequent intramolecular migration of the alkyl group on to a carbonyl carbon atom yields the acyl product (Uguagliati *et al.*, 1970).

One apparent nucleophilic reaction which has somewhat anomalous activation parameters is that of $[Pd(PPh_3)_3]$ with aryl iodides, where $\Delta H^{\ddagger} = 18.4\ \mathrm{kcal\ mol^{-1}}$ and $\Delta S^{\ddagger} = +3.1\ \mathrm{cal\ K^{-1}\ mol^{-1}}$. These values, however, may be a consequence of a slightly different transition state involving the assistance of the iodine atom acting as a ligand as shown in [13] (Fauvarque and Pfluger, 1981).

There has been only one report to date of the determination of the volume of activation ($\Delta V^{\ddagger}$) in a reaction involving a metal nucleophile and alkyl halide (Steiger and Kelm, 1973). The measured volume of activation is made up of two terms, one involving the change in volume during the formation of the transition state ($\Delta V_1^{\ddagger}$) and the other involving the accompanying change in solvation ($\Delta V_2^{\ddagger}$). The latter value can be estimated using the pressure derivative of the Kirkwood formula. Study of the reaction between *trans*-$[IrCl(CO)(PPh_3)_2]$ and MeI in a range of solvents gave $\Delta V_1^{\ddagger} = -17\ \mathrm{cm^3\ mol^{-1}}$, a value typical of a bimolecular reaction (for the Menschutkin reaction between MeI and pyridine, $\Delta V_1^{\ddagger} = -22\ \mathrm{cm^3\ mol^{-1}}$). The value is rather small for the simultaneous formation of two bonds (three-centre transition state), however, and is more consistent with a linear transition state.

14 Thermodynamics of reactions involving metal nucleophiles

There has been relatively little work on the thermodynamics of the reactions between transition-metal nucleophiles and alkyl halides. However, the enthalpies of the reactions between *trans*-$[IrCl(CO)(PMe_3)_2]$ and a variety of alkyl and acyl halides have been measured using titration calorimetry in dichloromethane (Yoneda and Blake, 1981). The $-\Delta H^{\circ}$-values calculated are: MeI, $28.1\ \mathrm{kcal\ mol^{-1}}$; EtI, 25.6; Pr^nI, 24.4; Pr^iI, 21.1; $PhCH_2I$, 22.7; MeCOI, 30.0; PhCOI, 29.0 (all values refer to the alkyl iodide in its standard state). Using these data, information about the iridium—ligand bond strengths were obtained, with the assumption that the difference between the heats of sublimation of reactants and products is approximately zero. In this way the iridium—carbon bond strengths were shown to fall in the order: $CH_3 \sim CH_3CO \sim Pr^n \sim Et \sim Pr^i > PhCH_2$, the same order as observed in the corresponding alkanes and alkyl iodides.

A further calorimetric study on the reactions of $[Pt(PPh_3)_2(C_2H_4)]$ with MeI gives $\Delta H^o = -18.9\,kcal\,mol^{-1}$. From the corresponding studies with iodine and 1,2-diiodoethane (to give $[Pt(PPh_3)_2I_2]$ in both cases), and assuming that the enthalpies of sublimation of $[Pt(PPh_3)_2(C_2H_4)]$, $[Pt(PPh_3)_2I_2]$ and $[Pt(PPh_3)_2(Me)I]$ are all approximately the same, then it can be calculated that $D(\text{Pt-Me}) - D(\text{Pt-I}) = 1.4 \pm 1.2\,kcal\,mol^{-1}$. The platinum—carbon and platinum—iodine bond strengths are thus about the same (Mortimer *et al.*, 1979).

15 Activation of carbon—hydrogen bonds

A review on the nucleophilicity of metal complexes would not be complete without some mention of the reaction of alkanes and arenes with transition-metal complexes. As recently pointed out (Halpern, 1985), a mechanism for such reactions involving the nucleophilic displacement of H^- is energetically too demanding. A much more reasonable pathway involves electrophilic displacement, one-centre concerted addition and two-centre addition. Clearly factors such as the degree of coordinative unsaturation and the thermodynamics of the reaction (attempting to trade-off the strength of the C—H bond for the M—C and M—H bonds) are of prime importance in activating carbon—hydrogen bonds. However, particularly in a one-metal centre mechanism, the nucleophilicity of the metal centre must play a significant part, although an electron-rich metal centre is not essential because of the multiple roles played by the site in these reactions (Zeimer *et al.*, 1984). Thus the complexes discussed in this section will not be the same as those encountered so far in this review. It is only in the last decade that the activation of simple hydrocarbons by a simple transition metal complex has been realised (Crabtree *et al.*, 1979; Baudry *et al.*, 1980; Green, 1978; Janowicz and Bergman, 1982). Prior to this any mechanistic information about the possible pathways had to be gleaned by investigating the microscopic reverse of the "activation" process, the more commonly encountered elimination of alkane (43). Such studies can be misleading, and led to the

$$[Pt(PPh_3)_2(R)(H)] + L \longrightarrow [Pt(PPh_3)_2L] + RH \qquad (43)$$

speculation that carbon—hydrogen activation would not be possible, on thermodynamic grounds (Sen and Halpern, 1978). With the discovery of more systems which "activate" hydrocarbons, however, mechanistic studies are beginning to appear in this area. This section is in no way supposed to represent a detailed survey of carbon—hydrogen bond activation, but concentrates only on the mechanistic studies. The literature associated with C—H and C—C bond cleavage by gas-phase transition-metal ions will not be discussed, but the interested reader is referred to the articles by Hanratty *et al.* (1985) and Houriet *et al.* (1983) and references therein.

Kinetic studies on the reductive-elimination reactions of $[(\eta^5\text{-}C_5Me_5)Rh(R)(H)(PMe_3)]$ (R = Me, Ph, *p*-MeC_6H_4, 3,4-$Me_2C_6H_3$, 2,5-$Me_2C_6H_3$ or 2,3-$Me_2C_6H_3$) have given a unique opportunity to investigate the mechanism and thermodynamics of C—H bond activation. Most importantly, this work shows that despite the 6–10 kcal mol^{-1} difference in C—H bond energies, both alkanes and arenes can be activated. The activation of alkanes and arenes clearly involves very different transition states.

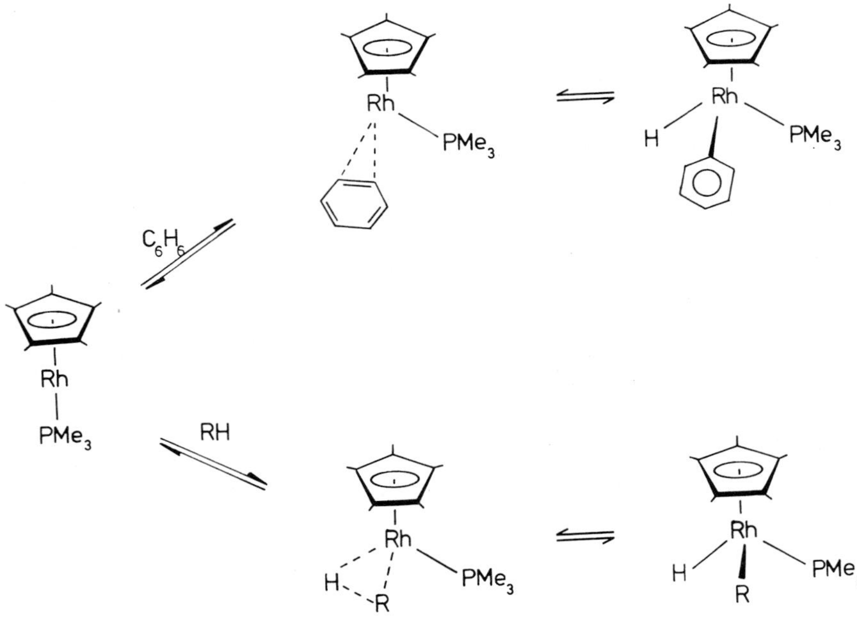

FIG. 9 Reaction pathways for $[(\eta^5\text{-}C_5Me_5)Rh(PMe_3)]$ with benzene and alkanes

The arenes initially bind in an η^2-form which is in equilibrium with the hydrido aryl complex as shown in Fig. 9, and this is in good agreement with the early postulates from *ortho*-metallation studies (Chatt and Davidson, 1964; Parshall, 1977). The activation of alkanes must be simpler, however, and involve the concerted addition of the C—H unit. Comparison of the activation barriers for the two processes shows that the transition states for alkane and arene activation are energetically very similar. The difference in energy between the Rh—Ph and Rh—Me bonds is estimated to be 10.6 kcal mol^{-1} (compared to 6 kcal mol^{-1} for C—H bonds in C_6H_6 and CH_4). This thermodynamic preference for arene over alkane activation is also reflected in the kinetics where the difference in free energies of activation for benzene over cyclopentane is $\Delta\Delta G^{\ddagger} = 0.8\ \text{kcal mol}^{-1}$ (Jones and Feher, 1984).

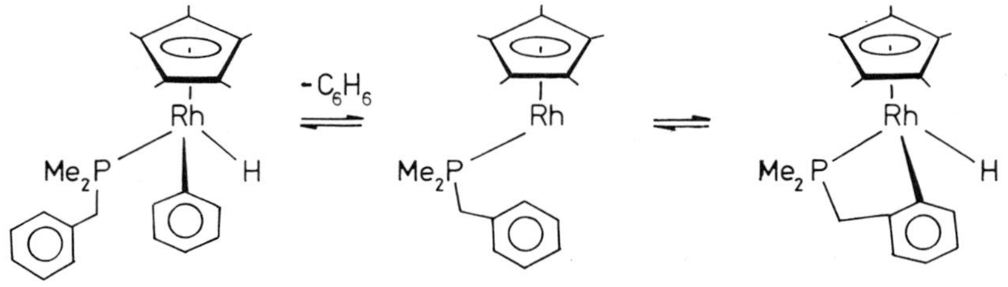

FIG. 10 Competitive equilibria used to establish the discrimination between *intra*- and *inter*molecular arene activation

It has often been suggested that entropy effects dominate the discrimination between intra- and intermolecular C—H activations (Dicosimo *et al.*, 1982 and references therein). Estimates of $-T\Delta S^{\ddagger} = 10\ \text{kcal mol}^{-1}$ have been made. However several diverse observations on the affinity of systems to intra- and intermolecular activation clearly require this point to be elaborated. This has been made possible by direct comparison on a single metal site illustrated in Fig. 10 (Jones and Feher, 1985). The intermediacy of an η^2-arene species in the intramolecular reaction can be demonstrated by the independent preparation of [16] which rearranges to [17] as shown in (44). This reaction occurs at a temperature below that required for arene dissociation and so it presumably occurs via the intermediate shown.

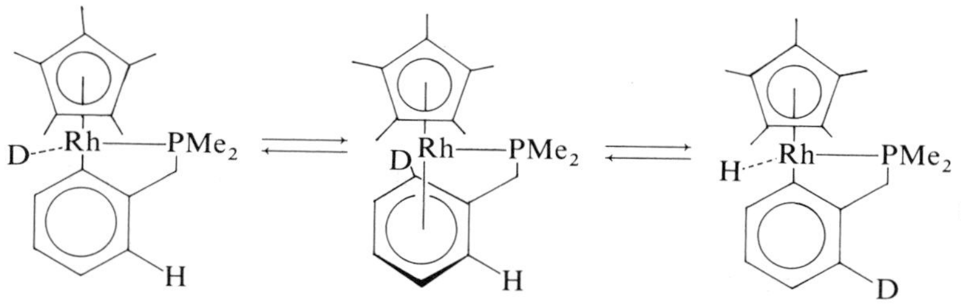

(44)

The results demonstrate that the activation parameter differences for inter- *vs* intramolecular reactions are only small ($\Delta\Delta H^{\ddagger} = 1.7\ \text{kcal mol}^{-1}$, $\Delta\Delta S^{\ddagger} = 4.5\ \text{cal K}^{-1}\ \text{mol}^{-1}$, and this corresponds to a kinetic preference for *intermolecular* activation of 1.86 : 1. The intramolecular product is the thermodynamically more favoured, however. An analogous comparison of alkane activation using $[(\eta^5\text{-}C_5Me_5)Rh(PMe_2Pr^n)]$ and propane demonstrated that, as with arenes, intermolecular activation is favoured kinetically ($\Delta\Delta H^{\ddagger} = 5.1$ kcal mol^{-1}) but not thermodynamically ($\Delta\Delta H^{\circ} = 4.5$ kcal mol^{-1}).

Direct comparison of the reactivity and selectivity of various alkanes towards $[(\eta^5\text{-}C_5Me_5)M(PMe_3)]$ (M = Rh or Ir) has shown that the rhodium complex is more discriminating than the iridium analogue. The former reacts almost exclusively with primary carbon centres, whereas the latter discriminated only weakly between primary and secondary carbon atoms (Perland and Bergman, 1984). Employing this weak discrimination and measuring the free energy change associated with equilibrium (45), it is estimated that the iridium-primary carbon bond is 5.5 kcal mol^{-1} stronger than an iridium-secondary carbon bond (Wax *et al.*, 1984).

$(\eta^5\text{-}C_5Me_5)Ir(PMe_3)(H)(C_6H_{11}) + CH_3(CH_2)_3CH_3 \rightleftharpoons (\eta^5\text{-}C_5Me_5)Ir(PMe_3)(H)(CH_2)_4CH_3 + C_6H_{12}$ (45)

Clearly a great deal of work is required before we can define the way in which the metal cleaves the carbon—hydrogen bond. Recently (Crabtree *et al.*, 1985) structural information from 18 crystal structures has been used to construct a reaction trajectory for the reaction between a metal centre and the C—H unit, in which the C—H bond approaches the metal and gradually swings and elongates the carbon—hydrogen bond to form the alkyl hydrido-product. The importance of steric effects in C—H activation has been stressed on theoretical grounds (Saillard and Hoffman, 1984).

16 Applications

The reactions between metal nucleophiles and organic compounds finds several applications in the areas of organic synthesis and catalysis. One obvious area of application is the control of stereochemistry at the carbon centre when the metal nucleophile inverts the configuration of the carbon in an S_N2 process. In this way the C-20 centre of steroids has been controlled by a double inversion sequence using $[Pd(PPh_3)_3]$ (Trost and Verhoeven, 1976). Clearly, the choice of nucleophile is important when stereochemically indiscriminate electron-transfer processes are so often energetically very comparable.

A reagent which has been referred to as a "transition-metal analogue of the Grignard reagent" is $Na_2[Fe(CO)_4]$ (Collman, 1975 and references therein). The versatility of this reagent is attributable primarily to its high stereospecificity and its specificity to particular groups, leaving many (such as carbonyls, esters, acids, etc.) completely unperturbed, thus circumventing the problems associated with the use of protecting groups. Alkyl halides can be converted to aldehydes (Cooke, 1970; Watanabe *et al.*, 1971) as illustrated in (46). Such a reaction uses to advantage the tendency of the derived

$$Cl(CH_2)_7Br \xrightarrow[AcOH]{[Fe(CO)_4]^{2-}, CO, THF} Cl(CH_2)_7CHO \tag{46}$$

alkyl complex, $[Fe(R)(CO)_4]^-$, to rearrange in the presence of carbon monoxide to yield the corresponding acyl-complex $[Fe(COR)(CO)_4]^-$. A similar strategy, this time quenching the reaction with alkyl halide, yields unsymmetrical ketones as illustrated in (47) (Collman *et al.*, 1972). This reaction is not truly general, however, since the "quenching" alkyl halide must contain a primary alkyl group.

$$Br(CH_2)_5CO_2Et \xrightarrow[EtI]{[Fe(CO)_4]^{2-}, NMP} EtC(=O)(CH_2)_5CO_2Et \tag{47}$$

Alkyl halides can be converted into carboxylic acids by exposing the reaction mixture to dioxygen (48), or the corresponding ester if the reaction

$$n\text{-}C_{12}H_{25}Cl \xrightarrow[O_2]{[Fe(CO)_4]^{2-}} n\text{-}C_{12}H_{25}CO_2H \tag{48}$$

$$n\text{-}C_3H_5Cl \xrightarrow[I_2,\ MeOH]{[Fe(CO)_4]^{2-}} n\text{-}C_3H_5CO_2Me \tag{49}$$

is quenched with an alcohol (49) (Collman *et al.*, 1973). In an analogous fashion quenching the reaction with amines gives amides.

The limitations of this reagent are associated with its basicity, which in its reactions with alkyl halides can give rise to elimination, thus prohibiting the use of tertiary alkyl halides. The other limitation of this reagent is associated with the migratory capabilities of the alkyl-group which is greatly impeded by adjacent electronegative groups on the carbon residue; thus the reactions are restricted to those of simple primary and secondary substrates.

The reactions of acyl halides with $[FeH(CO)_4]^-$ have been used to prepare aldehydes, by the (assumed) pathway shown in (50) (Cole and Pettit, 1977;

this paper contains several references to other applications of this versatile reagent).

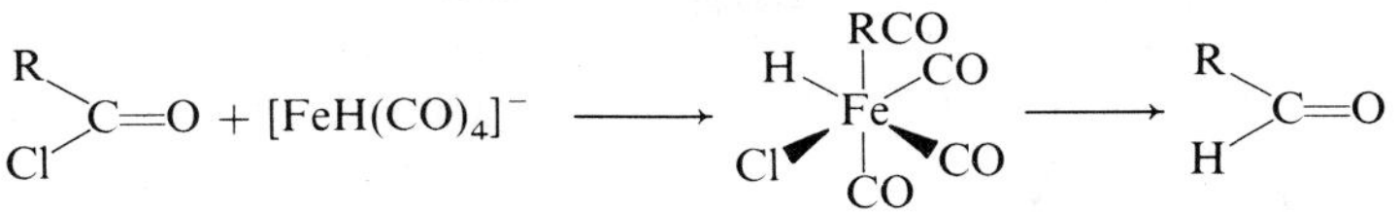

$$\mathrm{R(Cl)C{=}O} + [\mathrm{FeH(CO)_4}]^- \longrightarrow [\mathrm{FeH(RCO)Cl(CO)_3}] \longrightarrow \mathrm{R(H)C{=}O} \qquad (50)$$

The mixture of $[Fe(CO)_5]$ and NaOH in 95% methanol is a good, selective hydrogenator for α,β-unsaturated carbonyl compounds (Noyori *et al.*, 1972). It is not entirely clear what the exact nature of the "active" species is, but the wine colour of the solution may suggest the formation of $[HFe_2(CO)_8]^-$ and $[HFe_3(CO)_{11}]^-$.

The nucleophilicity of transition metal complexes towards organic molecules has found application in industrial catalysts.

The isomerisation of epoxides to ketones takes place at the complex $[RhCl(PMe_3)_3]$, as shown in Scheme 20 (Milstein, 1984 and references therein). In this catalytic cycle, the *cis*-$[RhH(CH_2COR)Cl(PMe_3)_3]$ intermediate can be isolated, and is clearly analogous to the iridium system described in Section 8.

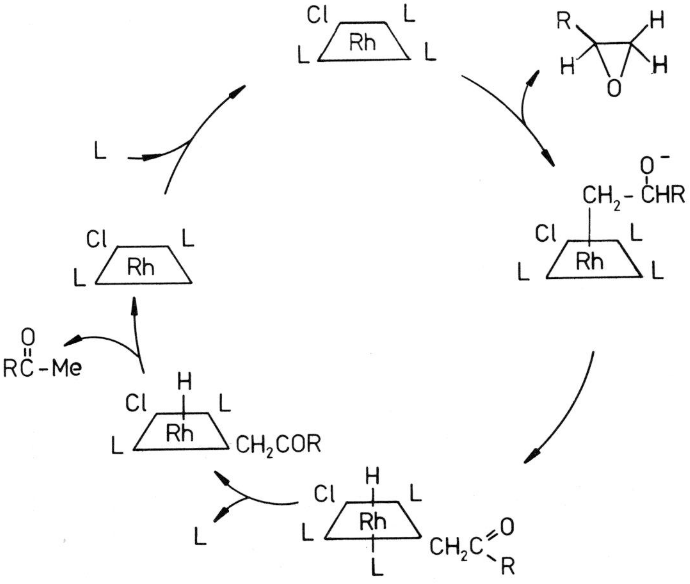

Scheme 20

A variety of reactions, such as the co-catalysed hydrogenation of aldehydes and the rhodium-catalysed decarbonylation, hydroacylation, hydrogenations and hydroformylation of aldehydes may involve the "oxidative addition" of aldehydes to the metal centre. Furthermore, although the addition of aldehydes to *trans*-$[RhCl(CO)(PPh_3)_2]$ (Landvatter and Rauchfuss, 1982), $[Ir(PMe_3)_4]^+$ (Thorne, 1980) and $[RhCl(PMe_3)_3]$ (Milstein, 1982) has been shown to yield *cis*-hydridoacyl-complexes, it is not clear that any of these reactions involve the nucleophilicity of the metal. Certainly other means of bonding of the aldehyde (via the carbonyl oxygen) can be envisaged.

Probably the most important industrial application of a transition-metal nucleophile is the Monsanto process for carbonylating methanol, using soluble rhodium-carbonyl complexes in the presence of iodide. The catalyst is in fact $[RhI_2(CO)_2]^-$ (Forster 1979, and references therein), and the catalytic cycle is shown in Scheme 21. The substrate for the rhodium catalyst is methyl iodide, which "oxidatively adds" to yield $[Rh(Me)(I)_3(CO)_2]^-$.

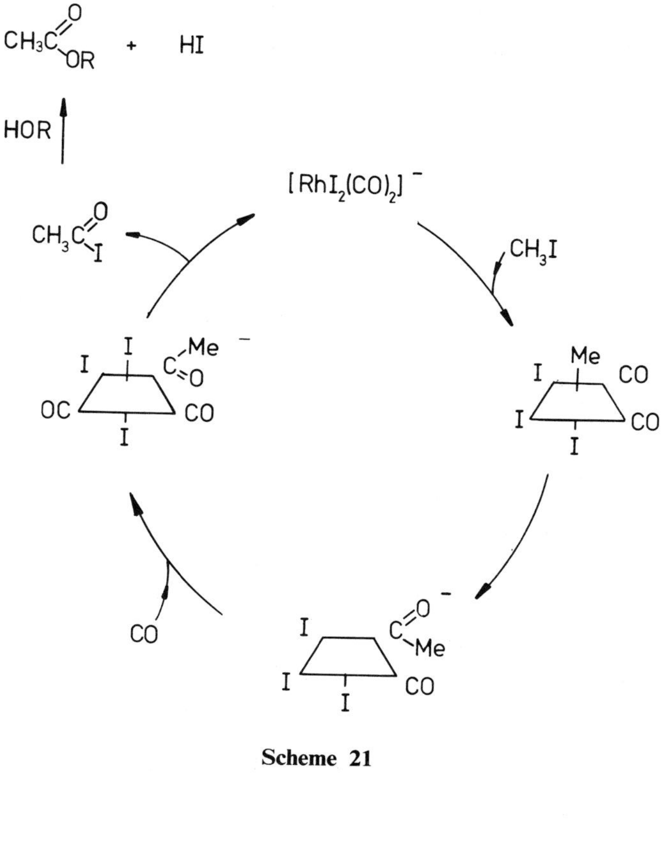

Scheme 21

Subsequent intramolecular migration of the methyl group on to the carbonyl group, and presumed reductive elimination of MeCOI, regenerates the catalyst. Solvolysis of MeCOI by either water or methanol, yields acetic acid or methyl acetate, respectively (Masters, 1981 and references therein). The role of iodide in this catalyst has already been discussed in Section 9.

The analogous iridium system is more complex, with two possible catalytic cycles, depending upon the reaction conditions, one involving $[IrI_2(CO)_2]^-$ and the other involving $[IrI(CO)_3]$ as shown in Scheme 22.

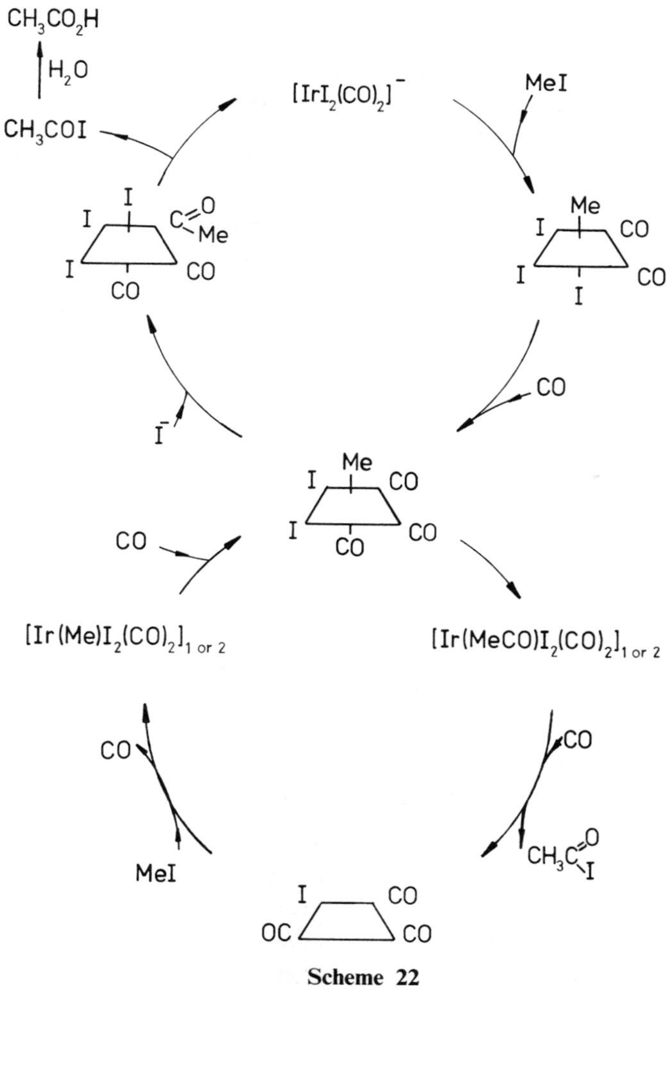

Scheme 22

The prime difference between the rhodium and iridium systems is that whereas in the former the rate-limiting step is the nucleophilic attack of $[RhI_2(CO)_2]^-$ on methyl iodide, in the iridium system the addition of methyl iodide is rapid, but the subsequent intramolecular migration is slow.

17 *Ad finem*

In this review we have indicated the sort of metal complexes that react as nucleophiles and the type of organic molecules with which they react. What then of the future? Just a casual glance through many of the sections of this review reveals that many aspects of this area of chemistry remain incomplete.

A constant dichotomy in the reactions of metal complexes with alkyl halides is whether the complex is truly reacting as a nucleophile or as a reductant, and a facile discrimination between the two roles has not been found as yet.

An area of great interest currently is the "activation" of C—H and C—C bonds by simple transition-metal complexes. The delineation of the factors involved in these stoichiometric reactions, in particular to what extent the nucleophilic character of the complex plays a role, is of fundamental importance in designing a catalytic system.

In Section 4 the problems associated with establishing a nucleophilicity scale for metal complexes was discussed; we must await, with some impatience, the establishment of such a scale.

References

Adams, H., Bailey, N. A. and Winter, M. J. (1984). *J. Chem. Soc. Dalton Trans.* 273

Alper, H. (1975). *Tet. Lett.* 2257

Anderson, S. N., Ballard, D. H., Chrzastowski, J. Z., Dodd, D. and Johnson, M. D. (1972). *J. Chem. Soc. Chem. Commun.* 685

Ashby, E. C. and DePriest, R. (1982). *J. Am. Chem. Soc.* **104**, 6144

Azam, K. A., Brown, M. P., Hill, R. H., Puddephatt, R. J. and Yavari, A. (1984). *Organometallics* **3**, 697

Balch, A. L. Hunt, C. T. Lee, C. L., Olmstead, M. M. and Farr, J. P. (1981). *J. Am. Chem. Soc.* **103**, 3764

Baudry, D., Ephritikhine, M. and Felkin, H. (1980). *J. Chem. Soc. Chem. Commun.* 1243

Becker, Y. and Stille, J. K. (1978). *J. Am. Chem. Soc.* **100**, 838

Bellachioma, G., Cardaci, G and Reichenbach, G. (1981). *J. Organomet. Chem.* **221**, 291

Bennett, M. A., Jeffery, J. C. and Robertson, G. B. (1981) *Inorg. Chem.* **20**, 323

Birk, J. P., Halpern, J. and Pickard, A. L. (1968). *J. Am. Chem. Soc.* **90**, 4491

Block, P. L. and Whitesides, G. M. (1974). *J. Am. Chem. Soc.* **96**, 2826

Block, P. L., Boschetto, D. J., Rasmussen, J. R., Deavers, J. P. and Whitesides, G. M. (1974). *J. Am. Chem. Soc.* **96**, 2814
Blum, J. and Weitzberg, H. (1976). *J. Organomet. Chem.* **122**, 261
Booth, B. L., Haszeldine, R. N. and Perkins, I. (1975). *J. Chem. Soc. Dalton Trans.* 1843
Bradley, J. S., Connor, D. E., Dolphin, D., Labinger, J. A. and Osborn, J. A. (1972). *J. Am. Chem. Soc.* **94**, 4043
Calderazzo, F. and Noack, K. (1966). *Coord. Chem. Rev.* **1**, 118
Casey, C. P. and Anderson, R. L. (1971). *J. Am. Chem. Soc.* **93**, 3554
Casey, C. P., Cyr, C. R., Anderson, R. L. and Marten, D. F. (1975). *J. Am. Chem. Soc.* **97**, 3053
Chatt, J. and Davidson, J. M. (1965). *J. Chem. Soc.* 843
Cheney, A. J. and Shaw, B. L. (1971a). *J. Chem. Soc. (A)* 3545
Cheney, A. J. and Shaw, B. L. (1971b). *J. Chem. Soc. (A)* 3549
Chock, P. B. and Halpern, J. (1966). *J. Am. Chem. Soc.* **88**, 3511
Chock, P. B. and Halpern, J. (1969). *J. Am. Chem. Soc.* **91**, 582
Cole, T. E. and Pettit, R. (1977). *Tet. Lett.* 781
Coleman, A. W., Eadie, D. T. and Stobart, S. R. (1982). *J. Am. Chem. Soc.* **104**, 922
Collman, J. P. (1975). *Acc. Chem. Res.* **8**, 342
Collman, J. P. and MacLaury, M. R. (1974). *J. Am. Chem. Soc.* **96**, 3019
Collman, J. P. and Roper, W. P. (1968). *Adv. Organomet. Chem.* **7**, 53
Collman, J. P. and Sears Jr, C. T. (1968). *Inorg. Chem.* **7**, 27
Collman, J. P., Kubota, M., Sun, J. Y., Kang, J. and Vastine, F. (1968). *J. Am. Chem. Soc.* **90**, 5430
Collman, J. P., Winter, S. R. and Clark, D. R. (1972). *J. Am. Chem. Soc.* **94**, 1788
Collman, J. P., Winter, S. R. and Komoto, R. G. (1973). *J. Am. Chem. Soc.* **95**, 249
Collman, J. P., Finke, R. G., Cawse, J. N. and Brauman, J. I. (1977). *J. Am. Chem. Soc.* **99**, 2515
Cooke, M. P. (1970). *J. Am. Chem. Soc.* **92**, 6080
Cooksey, C. J., Dodd, D., Gatford, C., Johnson, M. D., Lewis, G. J. and Titchmarsh, D. M. (1972). *J. Chem. Soc. Perkin Trans. 2* 655
Cotton, F. A. and Wilkinson, G. (1980). "Advanced Inorganic Chemistry", 4th edn, Ch. 3. Wiley Interscience, New York
Crabtree, R. H. and Hlatky, G. G. (1980). *Inorg. Chem.* **19**, 571
Crabtree, R. H., Mihelcic, J. M. and Quirk, J. M. (1979). *J. Am. Chem. Soc.* **101**, 7738
Crabtree, R. H., Holt, E. M., Lavin, M. and Morehouse, S. M. (1985). *Inorg. Chem.* **24**, 1986
Darensbourg, M. Y. and Hanckel, J. M. (1981). *J. Organomet. Chem.* **217**, C9
Darensbourg, M. Y. and Hanckel, J. M. (1982). *Organometallics* **1**, 82
Darensbourg, M. Y., Darensbourg, D. J., Burns, D. and Drew, D. A. (1976). *J. Am. Chem. Soc.* **98**, 3127
Darensbourg, M. Y., Jimenez, P. and Sackett, J. R. (1980). *J. Organomet. Chem.* **202**, C68
Darensbourg, M. Y., Jimenez, P., Sackett, J. R., Hanckel, J. M. and Kump, R. L. (1982). *J. Am. Chem. Soc.* **104**, 1521
Deeming, A. J. (1972). *M.T.P. International Review of Science*, Ser. 1. (ed. M. L. Tobe) **9**, 117
Deeming, A. J. (1974). *M.T.P. International Review of Science*, Ser. 2. (ed. M. L. Tobe) **9**, 271

Deeming, A. J. and Shaw, B. L. (1968a). *J. Chem. Soc. (D)* 751
Deeming, A. J. and Shaw, B. L. (1968b). *J. Chem. Soc. (A)* 1562
Deeming, A. J. and Shaw, B. L. (1969). *J. Chem. Soc. (A)* 1128
Deeming, A. J. and Shaw, B. L. (1971). *J. Chem. Soc. (A)* 1802
Dessy, R. E., Pohl, R. L. and King, R. B. (1966). *J. Am. Chem. Soc.* **88**, 5121
de Waal, D. J. A., Gerber, T. I. A. and Wynand, J. L. (1982). *Inorg. Chem.* **21**, 1529
DiCosimo, R., Moore, S. S., Sowinski, A. F. and Whitesides, G. M. (1982). *J. Am. Chem. Soc.* **104**, 124
Douek, I. C. and Wilkinson, G. (1969). *J. Chem. Soc. (A)* 2604
Edgell, W. F. and Chanjamsri, S. (1980). *J. Am. Chem. Soc.* **102**, 147
Edgell, W. F., Hedge, S. and Barbetta, A. (1978). *J. Am. Chem. Soc.* **100**, 1406
Fauvarque, J. F. and Pfluger, F. (1981). *J. Organomet. Chem.* **208**, 419
Finke, R. G., Gaughan, G., Pierpont, C. and Noordik, J. H. (1983). *Organometallics* **2**, 1481
Fitton, P. and Rick, E. A. (1971). *J. Organomet. Chem.* **28**, 287
Forster, D. (1975). *J. Am. Chem. Soc.* **97**, 951
Forster, D. (1979). *Adv. Organomet. Chem.* **17**, 255
Franks, S. and Hartley, F. R. (1981). *Inorg. Chim. Acta* **49**, 227
Franks, S., Hartley, F. R. and Chipperfield, J. R. (1981). *Inorg. Chem.* **20**, 3238
Gaylor, J. R. and Senoff, C. V. (1972). *Proc. XIV. Int. Coord. Conf. Chem.*
Green, M. L. H. (1978). *Pure Appl. Chem.* **50**, 27
Haines, L. M. (1971). *Inorg. Chem.* **10**, 1693
Halpern, J. (1970). *Acc. Chem. Res.* **3**, 386
Halpern, J. (1985). *Inorg. Chim. Acta* **100**, 41
Hanratty, M. A., Beauchamp, J. L., Illies, A. J. and Bowers, M. T. (1985). *J. Am. Chem. Soc.* **107**, 1788
Harrod, J. F. and Smith, C. A. (1970). *J. Am. Chem. Soc.* **92**, 2699
Hart-Davis, A. J. and Graham, W. A. (1970). *Inorg. Chem.* **9**, 2658
Hart-Davis, A. J. and Graham, W. A. (1971). *Inorg. Chem.* **10**, 1653
Hayashi, T., Hagihara, T., Konishi, M. and Kumada, M. (1983). *J. Am. Chem. Soc.* **105**, 7767
Hickey, C. M. and Maitlis, P. M. (1984). *J. Chem. Soc. Chem. Commun.* 1609
Hill, R. H. and Puddephatt, R. J. (1985). *J. Am. Chem. Soc.* **107**, 1218
Houriet, R., Halle, L. F. and Beauchamp, J. L. (1983). *Organometallics* **2**, 1818
Huttner, G. and Gartzke, W. (1975). *Chem. Ber.* **108**, 1373
Jawad, J. K. and Puddephatt, R. J. (1976). *J. Organomet. Chem.* **117**, 297
Jawad, J. K. and Puddephatt, R. J. (1977). *J. Chem. Soc. Dalton Trans.* 1466
Janowicz, A. H. and Bergman, R. G. (1982). *J. Am. Chem. Soc.* **104**, 352
Jensen, F. R. and Davis, D. D. (1971). *J. Am. Chem. Soc.* **93**, 4048
Jensen, F. R. and Knickel, B. (1971). *J. Am. Chem. Soc.* **93**, 6339
Jensen, F. R., Madan, V. and Buchanan, D. H. (1970). *J. Am. Chem. Soc.* **92**, 1414
Jensen, F. R., Madan, V. and Buchanan, D. H. (1971). *J. Am. Chem. Soc.* **93**, 5283
Johnson, M. D. (1970). *J. Chem. Soc. (D)* 1027
Johnson, R. W. and Pearson, R. G. (1970). *J. Chem. Soc. (D)* 986
Jones, W. D. and Feher, F. J. (1984). *J. Am. Chem. Soc.* **106**, 1650
Jones, W. D. and Feher, F. J. (1985). *J. Am. Chem. Soc.* **107**, 620
Jordan, R. F. and Norton, J. R. (1982). *J. Am. Chem. Soc.* **104**, 1255
King, R. B. (1963). *Inorg. Chem.* **2**, 531
King, R. B. (1975). *J. Organomet. Chem.* **100**, 111
King, R. B. and Bisnette, M. B. (1967). *J. Organomet. Chem.* **7**, 311

Kitching, W., Olszowy, H., Waugh, J. and Dodrell, D. (1978). *J. Org. Chem.* **43**, 898
Kochi, J. K. (1978). "Organometallic Mechanisms and Catalysis". Academic Press, New York, San Francisco, London
Koermer, G. S., Hall, M. L. and Traylor, T. G. (1972). *J. Am. Chem. Soc.* **94**, 7205
Kubota, M. and Blake, D. M. (1971). *J. Am. Chem. Soc.* **93**, 1368
Kubota, M., Kiefer, G. W., Ishikawa, R. M. and Benacia, K. E. (1973). *Inorg. Chim. Acta* **7**, 195
Kuivila, H. G. and Alnajjar, M. S. (1982). *J. Am. Chem. Soc.* **104**, 6146
Kuivila, H. G., Considine, J. L. and Kennedy, J. D. (1972). *J. Am. Chem. Soc.* **94**, 7206
Labinger, J. A. and Osborn, J. A. (1980). *Inorg. Chem.* **19**, 3230
Labinger, J. A., Braus, R. J., Dolphin, D. and Osborn, J. A. (1970). *J. Chem. Soc. (D)* 612
Labinger, J. A., Kramer, A. V. and Osborn, J. A. (1973). *J. Am. Chem. Soc.* **95**, 7908
Labinger, J. A., Osborn, J. A. and Colville, N. J. (1980). *Inorg. Chem.* **19**, 3236
Landvatter, E. F. and Rauchfuss, T. B. (1982). *Organometallics* **1**, 506
Lappert, M. F. and Lednor, P. W. (1976). *Adv. Organomet. Chem.* **14**, 345
Lau, K. S. Y., Fries, R. W. and Stille, J. K. (1974). *J. Am. Chem. Soc.* **96**, 4983
Lau, K. S. Y., Wong, P. K. and Stille, J. K. (1976). *J. Am. Chem. Soc.* **98**, 5832
Louw, W. J., de Waal, D. J. A., Gerber, T. I. A., Demanet, C. M. and Copperthwaite, R. G. (1982). *Inorg. Chem.* **21**, 1667
Masters, A. F. (1981). "Homogeneous Catalysis, A Gentle Art", p. 97. Chapman and Hall, London
McCleverty, J. and Wilkinson, G. (1964). *J. Chem. Soc.* 4200
Milstein, D. (1982). *Organometallics* **1**, 1549
Milstein, D. (1984). *Acc. Chem. Res.* **17**, 221
Milstein, D. and Calabrese, J. C. (1982). *J. Am. Chem. Soc.* **104**, 3773
Moro, A., Foa, M. and Cassar, L. (1980). *J. Organomet. Chem.* **185**, 79
Mortimer, C. T., Wilkinson, M. P. and Puddephatt, R. J. (1979). *J. Organomet. Chem.* **165**, 265
Mureinik, R. J. Weitzberg, M. and Blum, J. (1979). *Inorg. Chem.* **18**, 915
Nitay, M. and Rosenblum, M. (1977). *J. Organomet. Chem.* **136**, C23
Noyori, R., Umeda, I. and Ishigami, T. (1972). *J. Org. Chem.* **37**, 1542
Ochiai, E. I., Long, K. M., Sperati, C. R. and Busch, D. H. (1969). *J. Am. Chem. Soc.* **91**, 3201
Ohtani, Y., Fujimoto, M. and Yamagishi, A. (1977). *Bull. Chem. Soc. Jap.* **50**, 1453
Oliver, A. J. and Graham, W. A. (1971). *Inorg. Chem.* **10**, 1165
Pannell, K. H. and Jackson, D. (1976). *J. Am. Chem. Soc.* **98**, 4443
Parshall, G. (1977). *Catalysis* **1**, 334
Pearson, R. G. (1985). *Chem. Rev.* **85**, 41
Pearson, R. G. and Figdore, P. E. (1980). *J. Am. Chem. Soc.* **102**, 1541
Pearson, R. G. and Gregory, C. D. (1976). *J. Am. Chem. Soc.* **98**, 4098
Pearson, R. G. and Muir, W. R. (1970). *J. Am. Chem. Soc.* **92**, 3519
Pearson, R. G. and Poulos, A. T. (1979). *Inorg. Chim. Acta* **34**, 67
Pearson, R. G. and Rajaram, J. (1974). *Inorg. Chem.* **13**, 246
Perland, R. A. and Bergman, R. G. (1984). *Organometallics* **3**, 508
Poli, R., Wilkinson, G., Motevalli, M. and Hursthouse, M. B. (1985). *J. Chem. Soc. Dalton Trans.* 931
Pribula, C. D. and Brown, T. L. (1974). *J. Organomet. Chem.* **71**, 415
Puddephatt, R. J. and Scott, J. D. (1985). *Organometallics* **4**, 1221

Ramasami, T. and Espenson, J. H. (1980). *Inorg. Chem.* **19**, 1846
Saillard, J. Y. and Hoffmann, R. (1984). *J. Am. Chem. Soc.* **106**, 2006
San Filippo Jr, J. and Silbermann, J. (1981). *J. Am. Chem. Soc.* **103**, 5588
San Filippo Jr, J., Silbermann, J. and Fagan, P. J. (1978). *J. Am. Chem. Soc.* **100**, 4834
Schenck, T. G., Milne, C. R. C., Sawyer, J. F. and Bosnich, B. (1985). *Inorg. Chem.* **24**, 2338
Schrauzer, G. N. and Deutsch, E. (1968). *J. Am. Chem. Soc.* **90**, 2441
Schrauzer, G. N. and Deutsch, E. (1969). *J. Am. Chem. Soc.* **91**, 3341
Schrauzer, G. N., Deutsch, E. and Windgassen, R. J. (1968). *J. Am. Chem. Soc.* **90**, 2441
Sen, L. A. and Halpern, J. (1978). *J. Am. Chem. Soc.* **100**, 2915
Shaw, B. L. and Stainbank, R. E. (1972). *J. Chem. Soc. Dalton Trans.* 223
Shriver, D. F. (1970). *Acc. Chem. Res.* **3**, 231
Sisido, K., Kozima, S. and Takizawa, K. (1967). *Tet. Lett.* 33
Slack, D. A. and Baird, M. C. (1974). *J. Chem. Soc. Chem. Commun.* 701
Slack, D. A. and Baird, M. C. (1976). *J. Am. Chem. Soc.* **98**, 5539
Sokolov, V. I. (1976). *Inorg. Chim. Acta* **18**, L9
Steiger, H. and Kelm, H. (1973). *J. Phys. Chem.* **77**, 290
Stille, J. K. and Lau, K. S. Y. (1976). *J. Am. Chem. Soc.* **98**, 5841
Stille, J. K. and Lau, K. S. Y. (1977). *Acc. Chem. Res.* **10**, 434
Thorne, D. L. (1980). *J. Am. Chem. Soc.* **102**, 7109
Toscano, P. J. and Marzilli, L. G. (1984). *Prog. Inorg. Chem.* **31**, 105
Trost, B. M. and Strege, P. E. (1977). *J. Am. Chem. Soc.* **99**, 1649
Trost, B. M. and Verhoeven, T. R. (1976). *J. Am. Chem. Soc.* **98**, 630
Trost, B. M. and Verhoeven, T. R. (1978). *J. Am. Chem. Soc.* **100**, 3435
Trost, B. M. and Verhoeven, T. R. (1980). *J. Am. Chem. Soc.* **102**, 4730
Tsou, T. T. and Kochi, J. K. (1979). *J. Am. Chem. Soc.* **101**, 6319
Uguagliati, P., Palazzi, A., Deganello, G. and Belluco, U. (1970). *Inorg. Chem.* **9**, 724
Ugo, R., Pasini, A., Fusi, A. and Cenini, S. (1972). *J. Am. Chem. Soc.* **94**, 7362
Vaska, L. (1968). *Acc. Chem. Res.* **1**, 335
Van Duong, K. N. and Gandemer, A. (1970). *J. Organomet. Chem.* **22**, 473
Watanabe, Y., Mitsudo, T., Tanaka, M., Yamamoto, K., Okajima, T. and Takegami, Y. (1971). *Bull. Chem. Soc. Jap.* **44**, 2569
Watson, P. L. and Bergman, R. G. (1979). *J. Am. Chem. Soc.* **101**, 2055
Wax, J., Stryker, J. M., Buchanan, M., Kovac, C. A. and Bergman, R. G. (1984). *J. Am. Chem. Soc.* **106**, 1121
Weber, J. H. and Schrauzer, G. N. (1970). *J. Am. Chem. Soc.* **92**, 726
Whitesides, G. M. and Boschetto, D. J. (1969). *J. Am. Chem. Soc.* **91**, 4313
Whitesides, G. M. and Boschetto, D. J. (1971). *J. Am. Chem. Soc.* **93**, 1529
Wong, P. K., Lau, K. S. Y. and Stille, J. K. (1974). *J. Am. Chem. Soc.* **96**, 5956
Yang, G. K. and Bergman, R. G. (1983). *J. Am. Chem. Soc.* **105**, 6045
Yoneda, G. and Blake, D. M. (1981). *Inorg. Chem.* **20**, 67
Zeimer, E. H. K., DeWit, D. G. and Caulton, K. G. (1984). *J. Am. Chem. Soc.* **106**, 7006
Ziegler, T. (1985). *Organometallics* **4**, 675

Isotope Effects on nmr Spectra of Equilibrating Systems*

HANS-ULLRICH SIEHL

Institute of Organic Chemistry, University of Tübingen, D-7400 Tübingen, Germany

1 Introduction

The investigation of isotope effects on chemical reaction rates and equilibria is a well-established tool in physical organic chemistry, and nmr spectroscopy has become a standard technique for the investigation of structure and dynamics of molecules and persistent intermediates. It is the purpose of this chapter to describe a method and its applications to chemical problems

* Dedicated to the memory of the late Victor Gold. He was among the first to use nuclear magnetic resonance for the determination of isotope effects on equilibrating systems.

ADVANCES IN PHYSICAL ORGANIC CHEMISTRY
VOLUME 23 ISBN 0 12 033523 9

which is based upon a combination of equilibrium isotope effects and nmr spectroscopy. The vast majority of applications deal with deuterium isotope effects; only a few ^{13}C-equilibrium isotope effects are discussed to demonstrate the scope of the method.

The ^{1}H-nmr spectroscopic determination of a deuterium isotope effect on the fast conformational equilibria of cycloheptatriene (Jensen and Smith, 1964) and the measurement of the deuterium fractionation factor between t-butyl alcohol and water under conditions of slow exchange in the ^{1}H-nmr spectrum (Gold, 1968) are early examples of the application of the technique to equilibrium isotope effects. The effect of isotope perturbation of fast degenerate equilibria on nmr spectra was first applied by Saunders and coworkers (Saunders *et al.*, 1971; Saunders and Vogel, 1971a,b) to study carbocation rearrangements and subsequently was developed into a tool to distinguish rapidly equilibrating molecules from symmetric molecules using ^{13}C spectroscopy (Saunders and Kates, 1977; Saunders *et al.*, 1977a,b). Since then it has been realised that this method has broad general applicability and the area has progressed to such an extent that an overview of the method and some unification of the various applications is desirable. Introductory reviews with selected applications are available (Saunders, 1979; Kalinowski, 1984). The method is presented in reviews concerned with carbocation rearrangements (Saunders *et al.*, 1980; Ahlberg *et al.*, 1983a). Some applications are mentioned in a series on general isotope effects (Forsyth, 1984) and more references can be found in a review of isotope effects on nuclear shielding (Hansen, 1983). The method and some examples have found their way already into an introductory textbook on reactive intermediates (Vogel, 1979) and into recent books on ^{13}C nmr spectroscopy (Kalinowski *et al.*, 1984) and carbocation chemistry (Vogel, 1985).

The first part of this chapter gives an introduction to the nmr technique for measuring isotope effects on degenerate equilibria that is most often used, namely, that which is based on chemical shift differences. Sufficient background information is given to follow the discussion of the applications in the second part. The applications are not organised by classes of compounds throughout but for the line of argument and according to general dynamic and structural features. The selection was made primarily to illustrate the different aspects of the subject. It has been considered better to discuss some applications more thoroughly than others, and the chapter is not intended to be a complete survey of the field.

Some areas which are not covered are isotope effects on proton and deuterium exchange with solvent, for example, the water-hydronium ion system (Saunders *et al.*, 1984), deuterium isotope effects on acid and base strength (Halevi *et al.*, 1979), on amino acids (Petersen and Led, 1979) and on hydration of cobalt (II) (Saunders and Evilia, 1985). Solvent-dependent isotope effects on equilibria involving hydrogen bonds in carbohydrates and other compounds containing exchangeable hydrogen atoms (Reuben 1984, 1985a,b,c, 1986a,b) are not considered. Isotope effects on complexation equilibria of shift reagents with methyl ethers (DePuy *et al.*, 1976, 1977) and allenic methyl esters (Hansen and Lang, 1980) are also not further discussed.

For convenience and internal comparison and consistency, most of the equilibrium reactions involving isotopic molecules are written as proceeding downhill from the initial state (reactant) on the left-hand side to the final state (product) on the right-hand side of the equilibrium. This gives equilibrium constants, K, which are larger than unity. ΔG^{o} and ΔH^{o} are accordingly given with a negative sign in the favoured direction of the equilibrium reaction. For most reactions considered in this chapter $K_H = 1$ for the unlabelled molecules. The equilibrium isotope effect is defined here as K_D/K_H and, with $K_H = 1$, is then equal to K_D. This definition is the inverse of common usage; kinetic isotope effects, for example, are usually defined as the rate ratio k_H/k_D.

In the discussion of specific applications, generally only the evidence from isotope effect investigations is presented; this does not always imply that at least some of the conclusions could not have been reached using other evidence. An attempt has been made to give consistent interpretations within the framework of existing isotope effects without doing violence to the Bigeleisen formalism and the Born–Oppenheimer approximation.

EQUILIBRIUM ISOTOPE EFFECTS

An equilibrium isotope effect is observed when the equilibrium constant of a chemical reaction is different for isotopomeric compounds. When the bond to an isotopic atom is broken or formed during the course of the reaction the isotope effect is termed primary and when the isotopic bond is neither being broken nor formed the effect is called a secondary one. Secondary isotope effects may be further classified as α-, β-, etc. effects depending on the distance to the reaction centre.

The theory of isotope effects is well established and has been presented in detail in the books by Collins and Bowman (1970), Melander (1960), Melander and Saunders (1980) and Willi (1983). Only some general principles of isotope effects on chemical equilibria are presented here mainly to introduce the formulations and parlance of isotope effect theory. A frame of reference is given which is intended to allow the interpretations of the isotope effect studies to be followed with regard to the specific problems described.

The Born–Oppenheimer approximation is the cornerstone of theories dealing with the effect of isotopic substitution on molecular properties. This approximation states that electronic and nuclear motion of a molecule can be separated. The electronic energy of a molecule depends on the nuclear charges of the constituent atoms and on the number of electrons in the system but is independent of the masses of the nuclei. The electronic energy can be evaluated as a function of the fixed position of the nuclei, and the resulting electronic energy surface is the potential surface for the motion of the nuclei. Isotope effects are then nuclear mass effects resulting from motion of different mass nuclei on the same (isotope-dependent) potential energy surface (Wolfsberg, 1969).

Model calculations for simple isotopic exchange equilibria taking into account corrections to the Born–Oppenheimer approximation indicate that, although there is no *a priori* reason to neglect electronic isotope effects (Wolfsberg and Kleinman, 1973, 1974a,b), the predicted magnitude of electronic isotope effects is very small. For equilibrium isotope effects of organic molecules no experimental data are available which have the required accuracy to confirm or deny failures of the Born–Oppenheimer approximation.

For some chemical reactant R_H which is in equilibrium (1) with some

$$R_H \rightleftharpoons P_H \tag{1}$$

product P_H the equilibrium constant K_H, neglecting activity coefficients, can be expressed as (2). If R_D and P_D are the corresponding species containing a

$$K_H = [P_H]/[R_H] \tag{2}$$

specific isotopic substituent, i.e. deuterium, the corresponding equilibrium constant is K_D given by (3). The equilibrium isotope effect is then usually

$$K_D = [P_D]/[R_D] \tag{3}$$

defined by (4), where K_{HD} is really the equilibrium constant for the isotope

$$K_{HD} = K_H/K_D = [P_H][R_D]/[P_D][R_H] \tag{4}$$

exchange reaction (5).

$$R_H + P_D \rightleftharpoons R_D + P_H \tag{5}$$

The equilibrium constant may be expressed in terms of a ratio of complete partition functions Q of the species involved in the equilibrium as in (6). The

$$K_{HD} = \frac{Q(P_H)Q(R_D)}{Q(R_H)Q(P_D)} \tag{6}$$

reduced partition functions ratios for nuclear motion can then be used to calculate the isotope effect (Halevi, 1963; Wolfsberg, 1972).

According to the formalism developed by Bigeleisen and Mayer (1947) and also by Melander (1960) calculation of isotope effects requires only the knowledge of the frequencies of all the normal modes of vibration of both isotopic forms in the initial and in the final state. The Bigeleisen equation can be written in general form as (7). The MMI (mass, moment of inertia)

$$K_{HD} = \Delta\,\mathrm{MMI} \times \Delta\,\mathrm{EXC} \times \Delta\,\mathrm{ZPE} \tag{7}$$

term is the contribution due to the ratio of ratios of translational and rotational partition functions between the two isotopic species in the initial and in the final state. The EXC or excitation term factor is the effect caused by population of vibrational energy levels above the zero level. The ZPE or zero point energy term accounts for the difference in vibrational zero point energy between the reactants and the products of the isotopic molecules. For molecules also containing heavier atoms such as carbon, the mass and the moment of inertia (MMI term) are scarcely affected by substitution with isotopic hydrogen. Room temperature and below are close enough to absolute zero temperature that most of the C—H or C—D bonds are in their lowest (zero point) vibrational energy level; thus the EXC term makes no significant contribution. For organic molecules and for hydrogen isotopes, where vibrational frequencies and therefore zero point energy are large, it has been found that the ZPE term is the only major contributor to the isotope effect. This is especially valid for secondary isotope effects of hydrogen which are predominantly in this review. Zero point vibrational energy and the corresponding vibrational frequencies and force constants are the focus of all qualitative discussions explaining secondary deuterium isotope effects.

Zero point energy effects are generally represented in terms of potential energy diagrams. The potential energy curve for a one dimensional harmonic oscillator with zero point energy levels for stretching vibrations only (Fig. 1) is a considerable oversimplification of the real multidimensional energy surface. Other vibrations which are sensitive to isotopic substitution have to be taken into account for a quantitative evaluation of the isotope effect.

The potential energy curve for a C—H and a C—D bond for a given bonding situation are identical according to the Born–Oppenheimer approximation. The shape of the bottom of the potential energy curve governs the force constant for the vibrations. A greater force constant (the force constant being the curvature of the potential energy surface) results from a steeper, more curved potential energy well (Fig. 1*a*), i.e. from a more closely confined vibrational motion. The restraining forces may originate from the bonds which hold the hydrogen or deuterium to the rest of the molecule or from non-bonding steric repulsive interactions.

For a given bond force constant vibrational frequencies and zero point energy $ZPE = \frac{1}{2}h\nu$ are lower for C—D bonds than for C—H bonds

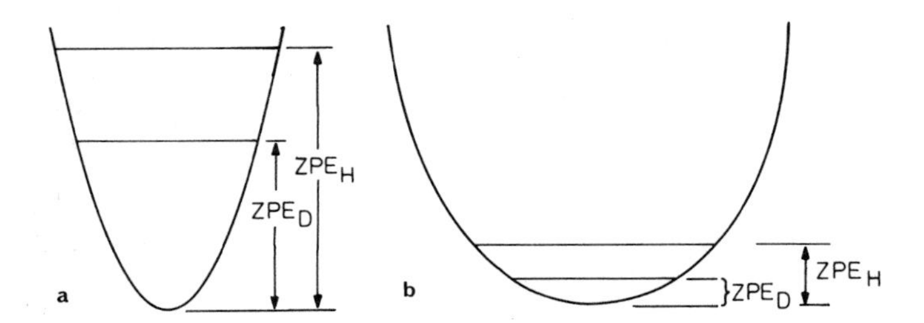

FIG. 1 One dimensional representation of the multidimensional potential energy well showing zero point energy levels for C—H and C—D vibrations for two different bonding situations, (*a*) high vibrational frequencies and large zero point energy difference, (*b*) lower vibrational frequencies and smaller zero point energy difference

because of the reciprocal square root mass dependence of vibrational frequencies. In the harmonic oscillator approximation the force constant F for a stretching vibration frequency is related to the vibrational frequency by the Hooke's law expression (8), where μ is the reduced mass for hydrogen

$$\nu \sim (F/\mu)^{\frac{1}{2}} \tag{8}$$

and deuterium respectively. The difference in zero point energy (ΔZPE) between C—H and C—D increases with increasing vibrational force constant for a given C—H bond.

In a chemical reaction where protons and deuterons are allowed to equilibrate among two or more sites having different bonding situations and thus different zero point energy determined by different vibrational force constants, the heavier isotope deuterium tends to accumulate in those positions where it is most closely confined by potential barriers, that is, where force constants, vibrational frequencies and zero point energy are larger. The converse applies for hydrogen which tends to concentrate at those sites in the equilibrium where force constants are smaller (Fig. 2). This can be formulated as a rule of thumb, a first law of isotopic chemistry, as follows. Having the choice, the heavier isotope prefers to make the stiffer bonds. The term "stiffer" binding (Shiner and Hartshorn, 1972) includes restraints to all vibrational modes and is preferred to "tight" which connotes dissociation energies, or to "strong" which might be associated with a short bond and stretching motions only. Recent calculations, for example, have shown that α-secondary isotopic rate effects on reactions leading to carbocations are largely the result of changes in angle-bending force constants, whereas β-secondary isotope effects on such processes arise from changes in bond stretching and angle-bending motions (Hehre *et al.*, 1983).

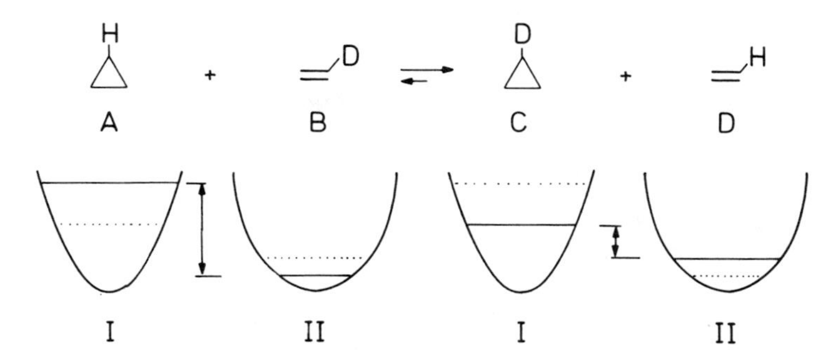

FIG. 2 Potential energy curves for an isotopic exchange equilibrium

No isotope effect on a chemical equilibrium is observed when there is no force constant change between initial and final state for a coordinate involving the isotopic atom. For example, the equilibrium constant for the fast 2,3-hydride shift in 2,3-di-trideuteriomethyl-2-butyl cation [1] is unity due to symmetry (Saunders and Vogel, 1971b).

$$(D_3C)(H_3C)\overset{+}{C}-CH(CD_3)(CH_3) \rightleftharpoons (D_3C)(H_3C)CH-\overset{+}{C}(CD_3)(CH_3) \qquad (9)$$

[1]

For secondary deuterium isotope effect at not too high temperatures the zero point energy difference between initial and final state for an isotope exchange equilibrium is directly related to the standard free energy change and has an exponential effect on the equilibrium constant. The principal possibilities of the temperature dependence of equilibrium isotope effects are discussed in the literature (Melander and Saunders, 1980; Collins and Bowman, 1970). For most practical purposes to evaluate the temperature dependence of secondary isotope effects it has been found that approximate solutions are sufficient. The low temperature approximation (i.e. when the populations of upper vibrational levels are negligible) of the temperature dependence of the equilibrium isotope effect has the form (10) (Stern *et al.*,

$$\ln K = A - B/T \qquad (10)$$

1968). With $A = \Delta S^\circ/\boldsymbol{R}$ and $B = \Delta H^\circ/\boldsymbol{R}$ this gives (11). The normal behaviour of the isotope effect is a smooth monotonic decrease with increasing

$$\ln \mathrm{K} = -\Delta H^\circ/\boldsymbol{R}T + \Delta S^\circ/\boldsymbol{R} \qquad (11)$$

temperature. At indefinitely high temperatures the equilibrium isotope effect vanishes and approaches the classical limit of unity.

A hypothetical exchange equilibrium between cyclopropane and ethylene (Fig. 2) may serve as an example to demonstrate the qualitative zero point energy analysis of isotope effects. The total reaction $A + B \rightleftharpoons C + D$ is degenerate in the absence of an isotope because A and C and also B and D are identical pairs. The two bonding situation for C—H bonds which are interchanged are represented by two different energy wells I and II symbolising a larger and a smaller vibrational force constant for the bond to the cyclopropyl carbon and to the olefinic carbon respectively. The force constant change in the forward reactions A to D (I to II) and B to C (II to I) is opposite to that in the back reactions D to A (II to I) and C to B (I to II) but the total change is the same in the forward and backward direction. If one proton, i.e. the olefinic proton in B, is substituted by deuterium, the potential energy surface with the site-specific force constants is unchanged, but the degeneracy of the equilibrium is lifted because now different total zero point energies apply for both sides of the equilibrium. The total zero point energy of the left-hand side (ZPE of H in well I + ZPE of D in well II) is larger than on the right-hand side (ZPE of D in well II + ZPE of H in well I). The zero point energy difference between H and D vibrations on the right side is less than on the left side of the equilibrium as the hydrogen of A (in well I) loses more energy on conversion to D (in well II) than deuterium is raised in energy on going from well II in B to well I in C. A shift of the equilibrium towards the right-hand side leads to a net decrease in vibrational energy. As expected deuterium accumulates in the stiffer bonding position I which has the higher zero point energy whereas the proton prefers position II which has the lower force constant and zero point energy.

The fractionation factor, which is the equilibrium constant $K_F = [C][D]/[A][B]$, was calculated using the Bigeleisen equation as $K_F = 1.11$ at 25°C (Shiner and Hartshorn, 1972). This reaction might serve as a model for a maximum expected α-deuterium isotope effect in which a $CH_2{=}CHD$ group is changed to a cyclopropyl-D group. An experimental example is the equilibrium isotope effect $K = 1.109$ at 29°C in 1,5-dimethyl-2-deuteriosemibullvalene [2] (Askani *et al.*, 1982)

H D ⇌ H D (11)

[2]

The investigation of equilibrium isotope effects on chemical reactions is a study of how different masses of nuclei affect motion on the same potential surface. Thus there is a clear and logical difference between the ordinary kind of substituent effect, where the energy surface changes for each substituent, and an isotopic substitution where it does not. In referring to inductive or steric contribution to an isotope effect the origin of chemical isotope effects in vibrational energy difference should be kept in mind. It should also be remembered that, within the validity of the Born–Oppenheimer approximation, isotopic molecules execute their motions on the same potential energy surface to a very high degree of accuracy.

Equilibrium isotope effects constitute a subtle tool to study the structure and reaction dynamics of molecules because the isotope hardly changes the object under examination. An observed isotope effect is rationalised by first relating the observed direction and magnitude to certain vibrational force constant changes and these changes can then be related to geometry and molecular structure.

ISOTOPE EFFECTS ON NMR CHEMICAL SHIFTS OF STATIC MOLECULES

Isotope effects on vibrational amplitudes are important to rationalise isotope effects on such properties as dipole moment, nmr chemical shift and fine structure in epr spectra (Wolfsberg, 1969).

In contrast to thermodynamic isotope effects on chemical equilibria which result from changes in vibrational force constants and zero point energies between the initial and the final state, nmr isotope shifts of static molecules are single-state physical properties, intrinsic to that molecule and its particular set of vibrational force constants. Intrinsic isotope shifts normally do not change much with temperature whereas equilibrium isotope shifts show large temperature dependences.

The only conceivable source of intrinsic isotope shifts within the Born–Oppenheimer approximation lies in the dependence of zero point vibrational motion on isotopic mass (Forsén *et al.*, 1978). All nmr shifts are averages over vibrational motion on the potential surface rather than values for a fixed nuclear configuration (Batiz-Hernandez and Bernheim, 1967). In an isotopically substituted molecule, the average frequencies can in principle be different due to the different amplitudes of zero point motion. The motion then covers a different portion of the same vibrational energy surface and therefore the average shift is changed (Saunders *et al.*, 1984).

The majority of deuterium induced isotope shifts of carbon signals are to higher field and decrease rapidly with the number of bonds separating the observed nuclei from the deuterium. Recently an increasing number of upfield and downfield long range deuterium isotope shifts have been reported (Hansen, 1983; Aydin and Günther, 1981). For one-bond intrinsic isotope shifts, theoretical explanations based on the vibrational origin of isotope effects are available (Jameson, 1977; Jameson and Osten, 1984a,b). The understanding of long range isotope shifts is still scant and the field is under active investigation. In accord with the vibrational origin of isotope effects it has been suggested that for cyclohexanes and many other compounds efficient coupling between certain vibrations of the intervening bond system plays the major role in determining the magnitude of long range isotope shifts (Günther *et al.*, 1984).

Recently the first isotope effect of tritium on ^{13}C chemical shifts has been determined (Kresge *et al.*, 1986). For monolabelled acetone the ratio of tritium to deuterium isotope effects for both the upfield α-effect on the methyl group resonance and the downfield β-effect on the carbonyl group resonance are the same, although the isotope effects on the two resonances are believed to be of different origin (i.e. the so-called "inductive" effect for the methyl group and isotopic perturbation of hyperconjugation for the carbonyl group). The measured ratios (1.424 and 1.41) are close to 1.44. This is the value for the exponent relating tritium and deuterium kinetic isotope effects which was derived regarding the isotope effect as a pure zero point vibrational energy effect (Swain *et al.*, 1958).

Long range intrinsic isotope shifts often seem to indicate a difference in electron distribution produced by the isotopic substitution as if an electronic substituent effect were responsible. Isotope effects are sometimes discussed not in the framework of general isotope effect theory but in the more familiar "electronic" parlance of physical organic chemistry. Without denying the ultimately vibrational origin of isotope effects, Halevi (1963) has discussed deuterium isotope effects qualitatively in much the same way as chemical substituent effects. Wolfsberg (1969) has pointed out that such an approach has meaning in terms of amplitude isotope effects and has shown (Wolfsberg, 1972) how this can be rationalised within the framework of isotope effect theory. Recently attempts have been made to model the magnitude of intrinsic isotope shifts using molecular orbital calculations (Servis and Domenick, 1986; Forsyth and Yang, 1986; Forsyth *et al.*, 1986). A perturbation induced by artificial shortening of C—H bonds at the site of isotopic substitution is used to simulate the vibrational differences of isotopic isomers.

Intrinsic isotope shifts have been discussed in terms of isotopic perturbation of resonance (Saunders and Kates, 1977), through space perturbation of resonance (Forsyth and MacConnell, 1983), perturbation of hyperconjugation (Servis and Shue, 1980; Servis and Domenick, 1985; Günther and Wesener, 1982; Ernst *et al.*, 1983; Forsyth *et al.*, 1984), angular dependence and steric effects (Yashiro *et al.*, 1986; Majerski *et al.*, 1985). Deuterium induced intrinsic isotope shifts have been correlated empirically with hybridisation (Günther and Wesener, 1985; Günther *et al.*, 1985) and in terms of polar electronic influences in the same way as the substituent-induced chemical shift of nonisotopic substituents (Berger and Künzer, 1985; Berger and Diehl, 1986).

EFFECT OF CHEMICAL EQUILIBRIA ON NMR SPECTRA

Dynamic nmr spectroscopy is well suited for the investigation of chemical exchange processes. The method and its application have been extensively treated elsewhere (Jackman and Cotton, 1975; Sandström, 1982; Oki, 1985); thus only a brief summary is given in this subsection.

In chemical equilibria several molecular states or several molecular species frequently co-exist. This is the case, for example for the individual conformations taken by a given molecule, for molecules being in a tautomeric equilibrium, for carbocations undergoing hydride shifts, etc. When magnetic nuclei experience different magnetic fields because of different chemical environments, a chemical process which allows the nucleus successively to occupy sites of different environment can produce changes in the nmr spectrum even when the system is at equilibrium and no net chemical reaction occurs. This is especially useful for the investigation of degenerate systems, in which the exchange leads to molecules indistinguishable from the original ones.

The following changes are generally observed in the nmr spectrum when a reaction interchanges nuclei among different sites. When the process occurs slowly, all lines expected for that particular structure are visible. As the reaction becomes more rapid, as a result of increased temperature, the lines broaden and eventually overlap and coalesce into one or more broad lines. As the rate becomes still faster, these broad lines sharpen and finally one observes again sharp lines but fewer than in the absence of the rate process. The positions of these lines correspond to the weighted averages of the line positions in the static species.

The site exchange processes exhibited by the nuclear spins in systems at equilibrium can be investigated by means of band shape analysis. If the frequency difference (Δ) between the two exchanging peaks is given in Hz the fast limit beyond which increasing rate does not sharpen the averaged lines appreciably is about (Δ^2). Depending on this frequency separation and the temperature range accessible, barriers as low as 3 kcal mol^{-1} have been measured in some cases, as for example in 2,3-dimethyl-2-butyl cation [3] where a large shift difference is present (Saunders and Kates, 1978).

$(H_3C)_2C^+{-}CH(CH_3)_2 \rightleftharpoons [\,(H_3C)_2C{\cdots}H{\cdots}C(CH_3)_2\,]^{+\ddagger} \rightleftharpoons (H_3C)_2CH{-}C^+(CH_3)_2$ (12)

[3a] [4] [3b]

If the barrier to rearrangement goes to zero, there is no longer a set of equilibrating structures like [3a] and [3b] but a single intermediate structure which might have higher symmetry, for example [4], would then be the minimum energy structure. The dynamics of molecular rearrangements is thus connected to the question of molecular structure. The fundamental question is whether an internal rearrangement is occurring over a barrier on a multiple minimum energy surface or whether a hybrid structure with a single energy minimum is the stable species. Beyond the limit of fast exchange, dynamic nmr spectroscopy cannot generally distinguish between time-averaged symmetry caused by a dynamic process with a very low barrier and time-independent symmetry of a static structure with a single energy minimum.

INFLUENCE OF ISOTOPIC PERTURBATION ON NMR SPECTRA OF EQUILIBRATING SYSTEMS

Two types of chemical equilibria may be distinguished depending on whether the equilibrium is degenerate or nondegenerate in the absence of the isotope. A number of isotope effects on nondegenerate equilibria have been investigated by nmr spectroscopy (Robinson and Baldry, 1977b; Lloyd, 1978; Booth and Everett, 1980b; Reuben, 1984; Forsyth and Pan, 1985). The major part of applications, however, apply to degenerate equilibria where $K_H = 1$ and the measurement of the equilibrium isotope effect reduces to a measurement of K_D.

Strictly speaking a degenerate rearrangement, for example the cation equilibrium (12), can be nondegenerate with respect to symmetry from the viewpoint of isotopes in natural abundance. Depending on the reporter nuclei, which might be a ^{13}C isotope used to monitor by nmr spectoscopy the equilibrium isotope effect of another isotope which could be also ^{13}C, the rearrangements of certain isotopomeric molecules containing two ^{13}C isotopes can be nondegenerate and isotope effects can be measured even without isotopic enrichment. For example, a composite primary-secondary ^{13}C equilibrium isotope effect has been determined in 2,3-dimethyl-2-butyl-cation [5] from the shift difference observed in single- and double-quantum ^{13}C nmr spectra at natural abundance (Olah *et al.*, 1985a). The sensitivity of this experiment is of course severely limited by the low natural abundance of

$$(H_3C)_2\overset{+}{C}-\overset{*}{C}H(CH_3)\overset{*}{C}H_3 \rightleftharpoons (H_3C)_2CH-\overset{*}{\overset{+}{C}}(CH_3)\overset{*}{C}H_3 \qquad (13)$$

[5]

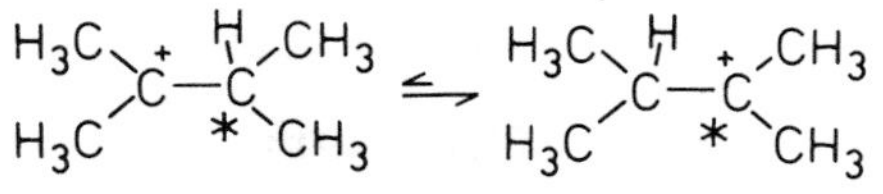

$$(CH_3)_2\overset{+}{C}-\overset{*}{C}H(CH_3)_2 \rightleftharpoons (CH_3)_2CH-\overset{+*}{C}(CH_3)_2 \qquad (14)$$

[6]

the vicinal ^{13}C-^{13}C isotopomers [5]. The primary equilibrium isotope effect on (14), essential to extract the pure secondary effect in (13) is not accessible by this approach but requires specific labelling (Saunders *et al.*, 1981).

If the rate of a dynamic process is slow relative to the nmr time scale, separate signals for the isotopic isomers can be observed and the equilibrium constant, i.e. the thermodynamic deuterium isotope effect, can be obtained directly from the value of the ratio of integrals at equilibrium. As integration is normally less precise than frequency measurements, the equilibrium constant can be obtained only with lower precision from area methods than from chemical shift methods (Saunders *et al.*, 1980c; Booth and Everett, 1980b). Recently, however, it has been demonstrated that the area method is also capable of measuring relatively small equilibrium isotope effects from ^1H spectra with high accuracy using the newest generation of very high field spectrometers under very carefully controlled conditions (Rappoport *et al.*, 1985).

The chemical shift method for the determination of equilibrium isotope effects is based on the chemical shift difference (the equilibrium isotope shift) between nmr signals for nuclei which are time-averaged to equivalence in the absence of the isotopic perturbation (Saunders, 1979).

A rapidly equilibrating system, for example, a two-fold degenerate rearrangement in carbocation [7] shows an nmr spectrum with averaged resonances for those atoms which are interchanged by the dynamic process (15).

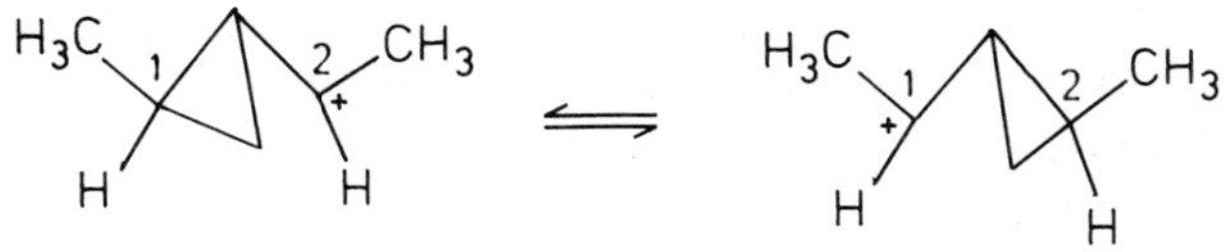

(15)

[7a] [7b]

The signal for carbon-1 in [7a] and [7b], which has an average shift of the two positions in [7a] and [7b], and the corresponding averaged signal for carbon-2 in [7a] and [7b] appear at the same position (f_{aver}) because the concentrations of [7a] and [7b] are equal ($K = 1$) in the nondeuteriated cation. Unsymmetrical introduction of deuterium breaks the symmetry of the twofold degenerate rearrangement (16). The equilibrium is perturbed if there are zero point energy differences resulting from differences in vibrational force constants for the isotopic isomers [8a] and [8b].

(16)

A [8a] ⇌ B [8b]

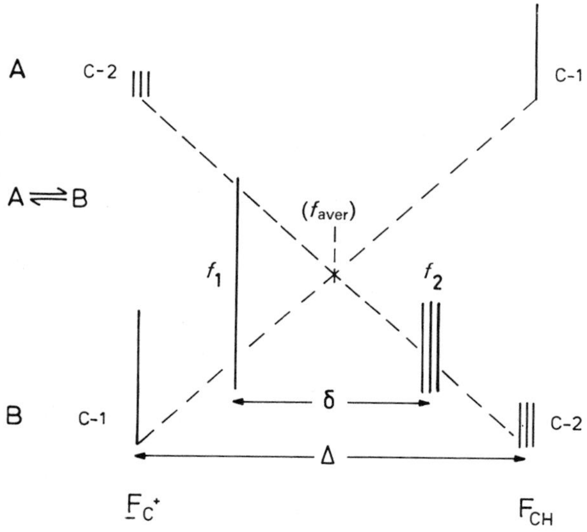

The change in free energy which is related to the zero point energy difference will be different from zero and with $\Delta G^\circ = -\boldsymbol{RT}\ln K$ and $K = [B]/[A]$ where [A] and [B] are the concentrations of [8a] and [8b] the equilibrium constant K is not equal to unity. The resulting concentration difference between [8a] and [8b] leads to a population difference of the exchanging sites for C-1 and C-2 in (16) and this lifts the chemical shift degeneracy of the averaged C-1 and C-2-signals which was present in (15).

The chemical shift for C-1 is the concentration-weighted average f_1 of the two shifts $F_1 = F_{C+}$ and $F_2 = F_{CH}$ of the two sites in [8a] and [8b] and is given by (17). If the equilibrium is shifted to the right preferring [8b], site A is

$$f_1 = (F_2[A] + F_1[B])/([A] + [B]) \tag{17}$$

less populated than site B. The averaged signal for C-1 is observed offset from the unperturbed averaged position $f_{aver} = \frac{1}{2}(F_{C+} + F_{CH})$; the shift is downfield towards the chemical shift $F_1 = F_{C+}$. Analogously, the averaged signal f_2 of C-2 in [8a] and [8b], given in (18), is offset from the unperturbed

$$f_2 = (F_1[A] + F_2[B])/([A] + [B]) \tag{18}$$

position in the opposite direction, shifted upfield towards the chemical shift $F_2 = F_{CH}$ of C-2 in [8b]. The difference $\delta = f_1 - f_2$ is the equilibrium isotope shift difference. The ratio of the isotope shifts for the peak in the downfield position ($f_1 - f_{aver}$) and in the upfield position ($f_{aver} - f_2$) of the peaks of [8] at positions f_1 and f_2 relative to the peak of [7] at f_{aver} is 1 : 1, because a two site fast exchange is taking place between a singly populated highfield and a singly populated low field site.

Depending on the distance of the deuterium substitution from the other exchange site appropriate corrections for the deuterium-induced intrinsic chemical shift (δ_{intr}) must be taken into account to evaluate the splitting δ ($\delta = \delta_{obs} - \delta_{intr}$) caused by the chemical isotope effect.

The size of the splitting δ between the averaged lines is related to the chemical shift difference ($\Delta = F_1 - F_2 = F_{C+} - F_{CH}$) in the spectrum of the cation at the slow exchange limit and to the isotopic equilibrium constant $K = [B]/[A]$. The equation relating the isotope splitting δ to the equilibrium constant K may be derived as follows. From (17) and (18), δ is given by (19),

$$\delta = \frac{(F_2[A] + F_1[B]) - (F_1[A] + F_2[B])}{[A] + [B]} \tag{19}$$

and substituting $K = [B]/[A]$ one obtains (20). This equation can be simpli-

$$\delta = \frac{F_1K - F_2K - F_1 + F_2}{1 + K} \tag{20}$$

fied using $\Delta = F_1 - F_2$ yielding (21) and (22). Equation (22) relates the

$$\delta = \frac{K\Delta - \Delta}{1 + K} \tag{21}$$

$$K = \frac{\Delta + \delta}{\Delta - \delta} \tag{22}$$

observed equilibrium isotope effect shift difference δ to the isotopic equilibrium constant K in a two site fast exchange equilibrium where two nuclei are interchanged between two singly populated sites.

An analogous calculation for the methyl groups or the averaged methine carbons in the triply degenerate rearrangement (23) of 1-deuterio-1-(β,β-dimethylcyclopropyl)ethyl cation [9] which is a two site fast exchange

[9a]

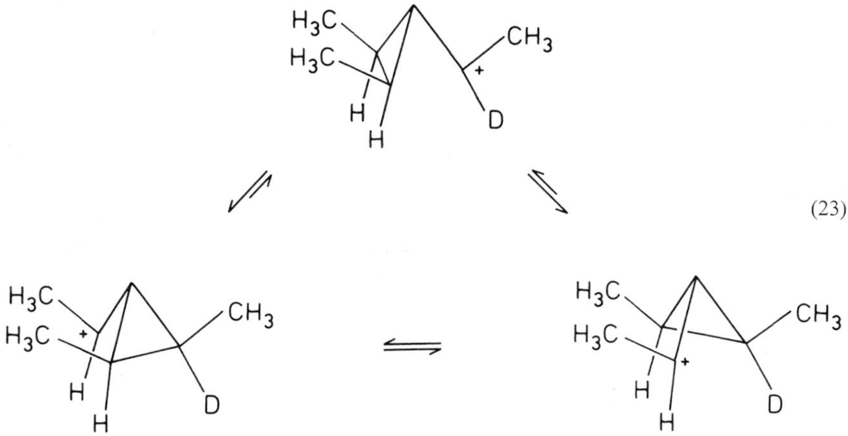

H_3C H_3C CH_3 H H D

(23)

H_3C H_3C CH_3 H H D

[9b]

H_3C H_3C CH_3 H H D

[9c]

interchanging three nuclei between a singly populated (low field) site and a doubly populated (high field) site yields (24).

$$K = \frac{\Delta + \delta}{\Delta - 2\delta} \tag{24}$$

If the fast dynamic process can be frozen out on the nmr time scale by lowering the temperature, Δ can be determined directly. Otherwise it must be estimated using suitable reference data from compounds which might serve as a model.

There are several possible sources of error in the determination of K. One source of error is Δ which might not be accurate either because the lines under slow exchange conditions are still broad or the slow exchange area is not accessible and the estimate for Δ from model compounds may not be accurate enough. The error in K then depends on the relative ratio of Δ and δ. Large values of Δ and small values of δ give the smallest error in K. It is often advantageous to use a reporter nucleus which has a large shift range such as ^{13}C, ^{19}F, etc. Also the precision may be increased by using high field spectrometers. As the isotope splitting δ normally increases with the number of deuterium atoms at equivalent positions in the molecule, the error in K increases for multi-deuteriated compounds. The accurate determination of δ can be a source of error when the δ-values which are dependent on the size of Δ and on the equilibrium constant K are small. High field measurements can increase the accuracy but when the lines are kinetically broadened the accessible temperature range for accurate determination of δ may be smaller.

The corrections necessary to consider intrinsic isotope shifts may be another source of error, especially in those cases when δ is small.

Sometimes several different signals can be observed which are averaged by the same dynamic process. The same equilibrium constant K applies then for all isotope splittings observed, but each set of averaged signals has a different Δ and thus different δ. The size of δ depends on the size of Δ and the equilibrium isotope effect K which measures the difference between local force fields at the exchanging sites and can be very large. The 1-trideuterio-2-trideuteriomethyl-3-methyl-2-butyl cation [10], for example, shows an isotope splitting of $\delta = 167.7$ ppm for the averaged $C^+/C—H$ carbons in the ^{13}C spectrum at $-138°C$ (Kates, 1978).

$$(CD_3)_2\overset{+}{C}—CH(CH_3)_2 \rightleftharpoons (CD_3)_2CH—\overset{+}{C}(CH_3)_2 \qquad (25)$$

[10]

The isotope splitting δ varies strongly with temperature and this is an important diagnostic characteristic that an equilibrium isotope effect is being observed. From the temperature dependence of δ the temperature dependence of K and $\Delta G° = -\boldsymbol{R}T \ln K$ can be determined. If $\ln K$ is found to vary linearly with $1/T$ in the temperature region studied $\Delta H°$ and $\Delta S°$ can be determined from a van't Hoff plot of $\ln K = -\Delta H°/\boldsymbol{R}T + (\Delta S°/\boldsymbol{R})$ by regression analysis. For systematic errors in the determination of thermodynamic parameters for equilibrium isotope effects using these chemical shift methods, criteria apply similar to those discussed in the literature on the determination of activation parameters from dynamic nmr measurements. Precision and accuracy are dependent on the temperature range studied and on the accuracy of the temperature measurement. $\Delta S°$ exhibits a strong tendency to become zero with improvement of the accuracy of the experiment.

In special cases an inverse temperature dependence of an equilibrium isotope effect increasing at higher temperatures has been observed (Kates, 1978; Walter, 1985). When an equilibrating system, that exhibits a small isotope effect, is separated by only a small energy barrier from a structurally different system and this interchange causes a larger isotope effect, the observed effect will increase at higher temperatures.

The kind of information obtainable through the use of the isotopic perturbation method is manifold and the problems which can be investigated cover a broad range. Fundamental questions connected to the interplay between molecular dynamics and structure like the distinction between time-averaged symmetry of molecules with a multiple minimum energy surface and molecules having a single energy minimum can be made using this technique.

The occurrence of sizeable temperature-dependent isotope splittings resulting from a thermodynamic isotope effect is characteristic of a dynamic equilibrium. In molecules which have a single minimum energy surface, for example, a resonance system like 1-deuteriocyclohexenyl cation [11], there is no equilibrium and hence no thermodynamic isotope effect is observed (Saunders and Kates, 1977). The intrinsic isotope shifts in single state structures are much smaller and not very sensitive to changes in temperature.

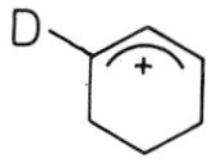

D

[11]

As long as the activation energy for equilibration in a multiple energy minimum system is not below the zero point vibrational energy difference of the isotopomeric molecules, the height of the energy barrier has no direct bearing on the strength of an equilibrium isotope effect. The nmr test for isotope perturbation of degenerate equilibria is still decisive even when no kinetic line-broadening is observable in systems rearranging over a very low barrier.

Tautomerism, valence-isomerism, bridging and hypercoordination[1] have structural implications where the question of whether an observed atom equivalence is time-averaged or not is relevant, and representative examples have been submitted to the perturbation of degeneracy test.

The method is not limited to the purpose of demonstrating the existence or the absence of an isotope effect. Provided that the chemical shift difference Δ is accessible, it is a very accurate method for the determination of different kinds of equilibrium isotope effects. The isotope effect is calculated from nmr frequency differences which can be measured with high precision. The method does not depend on knowing the exact amount of isotope incorporation in starting materials or products, as in other methods for determining equilibrium isotope effects. Its usefulness is limited however to those molecules undergoing degenerate rearrangement rapidly enough to give averaged spectra.

The species investigated are stable molecules or persistent intermediates which are structurally well defined by nmr spectroscopy. Data obtained by

[1] A carbon or hydrogen atom is hypercoordinated if the number of attached atoms exceeds four or one respectively.

this method may thus be useful as an experimental test of isotope effect theory and calculation procedures and can be used to determine isotopic fractionation factors, for example for molecule-solvent hydrogen exchange (Saunders and Jarret, 1985, 1986).

Equilibrium isotope effects serve as an empirical calibration of kinetic isotope effects. β-Secondary deuterium isotope effects in equilibrating carbocations have been used especially to prove the interpretation of the hyperconjugative origin and the dihedral angle dependence of β-kinetic deuterium isotope effects in nucleophilic substitution.

Besides the answer to fundamental questions and the test of general principles this method can give access to molecular structural details which are not easily accessible otherwise. The applicability of the isotope method to study conformational equilibria rests on the fact that the increase in force constant and frequency for all isotopes of hydrogen at sterically more constrained positions increases the separation between zero point energy levels of isotopic oscillators.

The isotopic perturbation method is very useful for studying fractional bonding and hypercoordination in coordinatively unsaturated and electron deficient compounds such as transition metal complexes or carbocations. The 2-norbornyl cation and the bicyclobutonium cation are the most prominent examples of carbocations whose structures have led to much controversial discussion (the so-called classical-nonclassical ion controversy). The application of the isotopic perturbation methed is likely to be the most decisive piece of nmr evidence for the hypercoordinated structure of these two cations in solution.

In this context it is very useful that this technique provides some general evidence on exchange characteristics which can be evaluated from the spectra of the deuteriated isomers independent of whether the slow exchange region is accessible or not. Distinction between equilibria which are degenerate or not in the absence of the isotope can often be made from the temperature dependence of chemical shifts of the unlabelled molecule but also from the fact that no isotopic perturbation is observed for nuclei positioned on a symmetry element common to all averaged deuteriated isomers. The isotopic splitting patterns allow the determination of the number of exchanging sites and their relative population.

In most cases the direction of an isotope effect can be extracted from ^{1}H or ^{13}C nmr spectra even though quantitative evaluation of the isotope effect is not always possible. The direction of the isotope effect gives important information on the relative stiffness of vibrational bond force constants which in turn can be related to characteristic features of the alternative structures under consideration.

2 Applications

PROTON TAUTOMERIC SYSTEMS

2,4-Pentanedione

Different possibilities for the structure and dynamics of the *cis*-enol of 2,4-pentanedione have been suggested: the symmetrical structure [13] (Shapet'ko *et al.*, 1975; Karle *et al.*, 1971), an equilibrium between two structures [12a] and [12b] (Robinson *et al.*, 1977; Forsén *et al.*, 1978) or an equilibrium between all three structures (Chan *et al.*, 1970; Leipert, 1977). In the case of a potential energy surface with shallow minima, deuteriation of —OH could lead to large temperature-dependent intrinsic isotope shifts caused by an amplitude isotope effect on the vibrational frequencies. Therefore the test for isotopic perturbation of symmetry is a more conclusive distinction between a single- and a double-well potential energy surface.

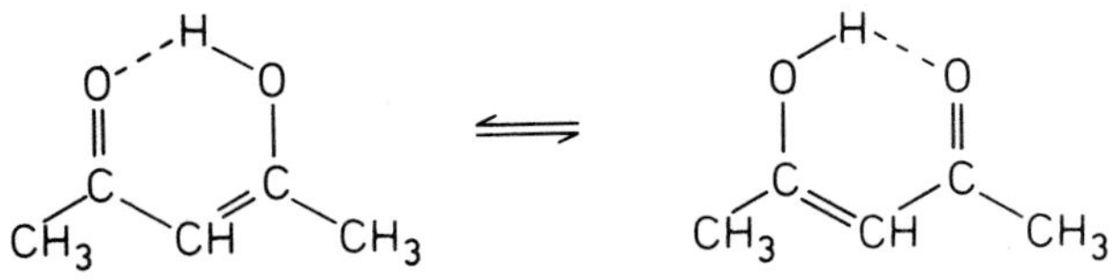

[12a] [12b]

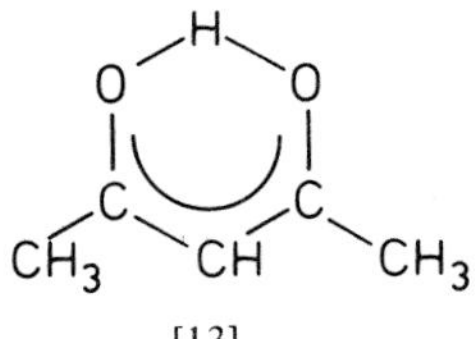

[13]

The ^{13}C nmr spectrum of the enol tautomer of 2,4-pentanedione in $CDCl_3$ shows a single time averaged peak for the carbonyl carbons. Unsymmetrical introduction of deuterium was used by Saunders and Handler (1985) to break the symmetry. In a mixture of 1-mono-, di- and trideuterio-2,4-pentanedione [14] ($x = 1, 2, 3$) a pair of peaks was observed for the carbonyl carbons in each isotopic isomer. The splitting at 23°C was 0.167 ppm per deuterium and was found to be temperature-dependent, decreasing with increasing temperature. This is characteristic of an equilibrium isotope effect and indicates a double minimum energy surface and an equilibrium (26) between [14a] and [14b]. Analysis of the temperature dependence yielded $\Delta H^o = -14.4$ cal mol^{-1} per D.

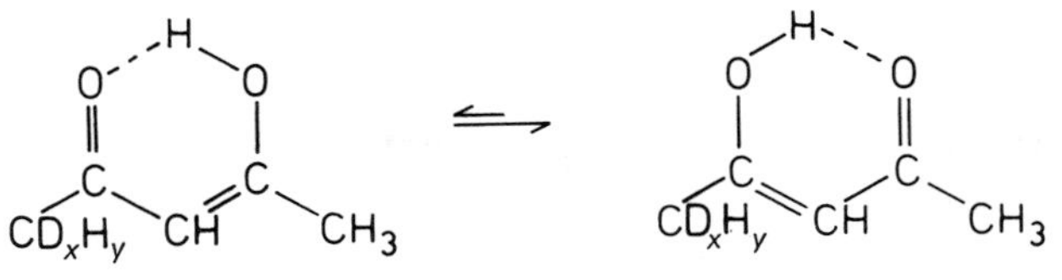

(26)

[14a] [14b]

Conclusive proof comes from an investigation of the unsymmetrically deuteriated bis-2,4-pentanedione complex of zinc. The carbonyl resonance of this complex is a single peak and no splittings were observed. This shows that the anion of the zinc complex is a symmetrical single energy minimum and confirms the conclusion that the enol is a rapidly equilibrating mixture of [12a] and [12b].

The direction of the equilibrium isotope effect in the enol was determined from the observation, that for [14] (x = 1, 2, 3) the downfield signal of each isotopic peak pair was a singlet and the upfield signal showed deuterium couplings. The singlet signal results from the averaged carbons which are further from the deuterium. The signals found shifted downfield towards the ketone position indicate that the isomer [14b] with the deuteriated methyl group at the enol carbon is preferred in the equilibrium.

Thiol-hydroxy and amino-hydroxy tautomerism

Hansen (1983) has reviewed a number of primary deuterium isotope effects in other tautomeric systems and only some recent examples are given here.

Deuterium isotope shifts over at least six bonds were observed in rapidly interconverting β-thioxoketone tautomers [15] using ^{13}C-nmr spectroscopy (Hansen *et al.*, 1982). The observed effects are caused by a shift of the fast enol-enethiol equilibrium (27) when the chelating proton is substituted by deuterium. Deuterium prefers attachment to oxygen as compared to sulphur

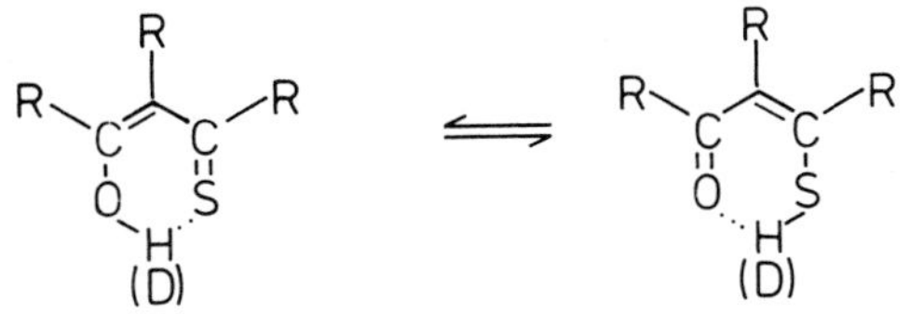

(27)

[15]

because oxygen—hydrogen(deuterium) bonds are shorter and thus stiffer than thiol S—H(D) bonds. The magnitude (up to 8 ppm) and the sign of the isotope shift depend on the chemical shift differences of the exchanging sites and on K_H for the nondegenerate equilibria. β-Thioxo-esters and thiolesters show only small deuterium isotope effects (Hansen and Duus, 1984); in β-thioacids the tautomerism does not take place and only intrinsic isotope effects are observed.

Deuterium isotope effects in nondegenerate tautomeric equilibria where the chelating H(D) is bonded either to an oxygen or a nitrogen atom have been investigated in enaminoketones [16] (Shapiro *et al.*, 1981) and Schiff's bases of salicylaldehydes [17] where the nonaromatic tautomer is strongly disfavoured in equilibrium (29) (Hansen *et al.*, 1982). An intrinsic origin for the small effects (± 0.4 ppm) in enaminoketones [16] was envisaged by Shapiro, whereas Hansen suggested an equilibrium perturbation effect or a combination of both influences.

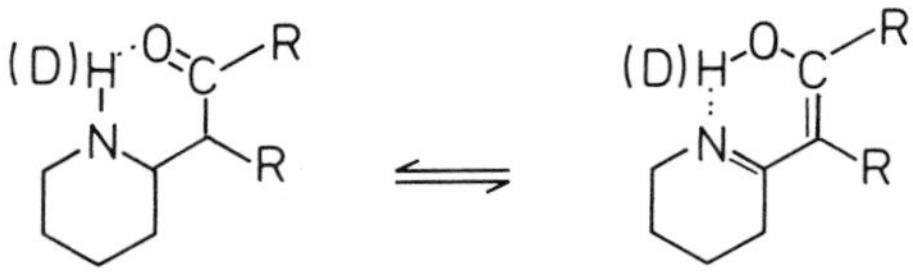

[16] (28)

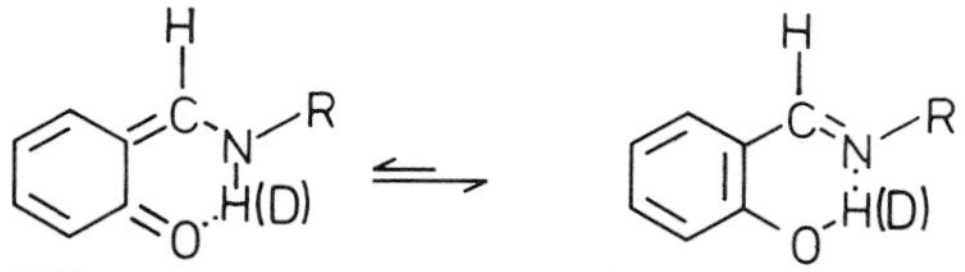

[17] (29)

Deuterium isotope shifts have been observed in the α-ketohydrazone/azoenolic tautomeric compound [18] by ^{13}C and ^{15}N nmr spectroscopy (Hansen and Lyčka, 1984). Besides large intrinsic isotope shifts, long range equilibrium isotope effects were observed which are caused by perturbation of the tautomeric equilibrium (30). The shifts increased on lowering the

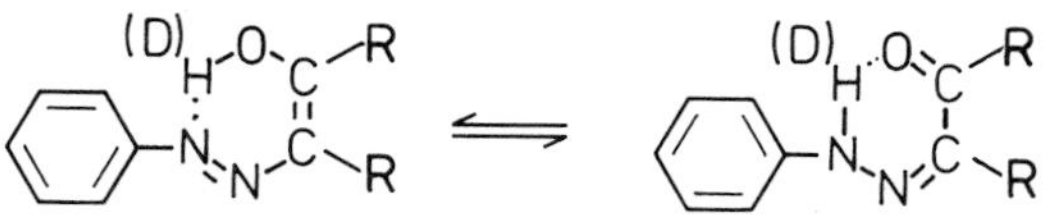

[18] (30)

temperature as expected for equilibrium isotope effects. Deuterium substitution shifted the equilibrium in all cases further in the direction of the tautomer which is predominant in the protio compound.

METALOMERIC AND VALENCE ISOMERIC COMPLEXES AND CARBANIONS

Cyclopentadienyl complexes

Rapid migration of the metal around the ring in many η^1-cyclopentadienyl complexes results in averaged carbon and proton signals for the ring atoms. If the barrier of this process is low, no line-broadening can be observed even at very low temperatures. In this case nmr spectroscopy cannot distinguish between pentahapto coordination and a fast equilibrium of monohapto-coordinated species.

Two key systems [19] and [20] for which the structures are known were investigated by Saunders *et al.* (1980b) to show that a distinction can be made between a fluxional η^1-cyclopentadienyl ring and a η^5-cyclopentadienyl ring using the isotopic perturbation technique. The proton coupled ^{13}C spectrum of a mixture of 1,1′-bisdeuterioferrocene and nonlabelled compound showed a triplet for the deuteriated carbons shifted 0.178 ppm upfield from the centre of the doublet ($J_{CH} = 177$ Hz) which was assigned to the methine carbons in the deuteriated and unlabelled ferrocene. The shift is found to be temperature-independent as expected for a normal intrinsic isotope shift in a static system and is thus consistent with symmetrically bound pentahapto coordination and a single minimum structure of ferrocene [19].

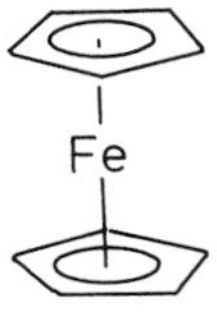

Fe

[19]

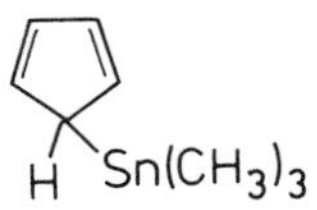

H $Sn(CH_3)_3$

[20]

The spectrum of a mixture of monodeuterio and perprotiocyclopentadienyltrimethyltin on the contrary showed for the triplet of the deuteriated carbons a downfield shift of 0.324 ppm at 25°C relative to the centre of the methine carbon doublet from the nondeuteriated ring. More accurate isotope shifts and better resolved signals for the β- and γ-carbons were obtained from proton-decoupled ^{13}C nmr spectra. Significant increases in the isotope shifts were observed when the temperature was lowered, indicating a definite equilibrium isotope effect in cyclopentadienyltrimethyltin. The tin complex is an equilibrium mixture of three different isotopic isomers [21a-c]. The fast exchange takes place between isomer [21a] with deuterium at the metal-substituted α-carbon and isomers [21b] and [21c] with deuterium at the β- or γ-olefinic carbons (31). A peak for the single α-carbon showed deuterium coupling and was found shifted downfield, whereas the

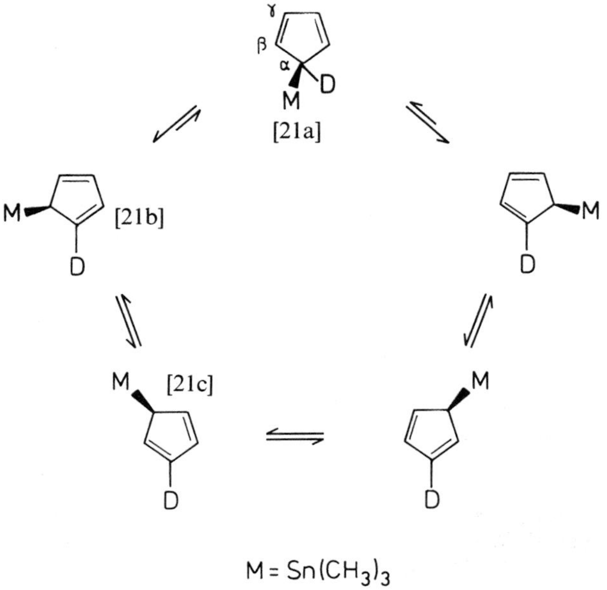

[21a] [21b] [21c] (31)

$M = Sn(CH_3)_3$

peaks for the two β- and two γ-carbons are shifted upfield relative to the shift of the unlabelled complex [20]. The chemical shifts for the metal-substituted α-carbon and the β- and γ-olefinic carbons in a static cyclopentadienyl tin complex can be estimated by considering the chemical shifts in limiting slow exchange spectra of suitable model compounds. The metal-substituted carbon in [22], [23] and [24] have chemical shifts of 30–50 ppm and the olefinic carbons appear at 120–150 ppm. Comparison with the observed isotope shifts in the tin complex [21] shows that the deuterium prefers the olefinic carbon relative to the metal-substituted carbon. The isomers [21b] and [21c] with deuterium at the olefinic position are therefore preferred in the equilibrium.

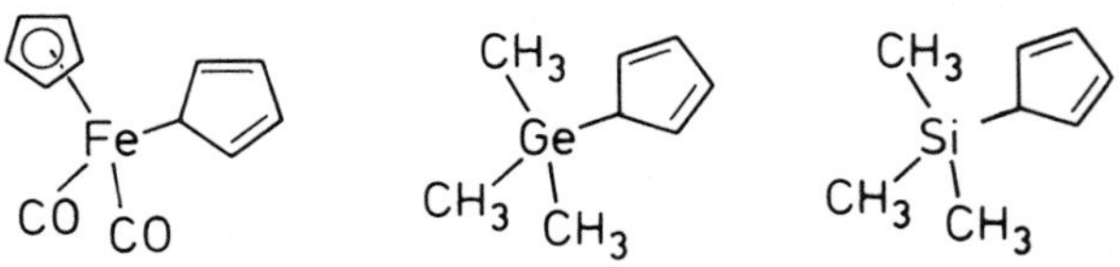

[22] [23] [24]

The direction of this equilibrium isotope effect is the same as in the allylmagnesium bromide [26] (see below) and is the reverse of what is observed in isotopic exchange equilibria of unsubstituted hydrocarbons with similar structural differences. In the Cope rearrangement (32) of the

D_2 D_2 ⇌ D_2 D_2 (32)

[25a] [25b]

tetradeuteriated-1,5-hexadiene [25], deuteriums bonded to terminal sp^2-hybridised carbons are interchanged with deuteriums bonded to sp^3-hybridised carbons and preferred at the sp^3-hybridised carbon site in [25b] (Sunko *et al.*, 1970).

Since deuterium isotope effects are determined in large measure not only by the number but also by the kind of atoms directly attached to the carbon with the exchanging hydrogen or deuterium, it might be anticipated that the valence bending motions of C—H and C—D bonds which are predominantly responsible for α-secondary isotope effects are very different at the carbon bonded to a metal in [20] as compared to an sp^3-hybridised aliphatic carbon in [25].

If it is assumed that there is no force constant difference and thus no equilibrium isotope effect between the β- and γ-olefinic positions on [21], the different upfield shift observed for these carbons is due to different chemical shifts for the two positions in the "frozen" static spectrum. The isotopic exchange would then occur between a single populated downfield site, the α-position, having different vibrational force constants from the upfield side which has four times the population (the β,β′ and γ,γ′-positions). The spectra indicate that the sum of the upfield shift of the two olefinic positions is half the size of the downfield shift of the deuteriated α-carbon when corrections for intrinsic shifts are taken into account.

The observed shift difference between the deuteriated carbon and the β-carbon was corrected for intrinsic isotope shifts which were taken from model compounds. This gave an equilibrium isotope splitting of $\delta = 0.589$ ppm at 25°C. The approximate formula for calculating the equilibrium constant is $K = (\Delta + \delta)/(\Delta - 4\delta)$ assuming equivalent olefinic positions. Using $\Delta = 90$ ppm as an approximate shift difference between the metal-substituted and olefinic carbons, one obtains $K = 1.034$ at 25°C.

Allyl derivatives of the alkali metals and alkaline earth metals

Dynamic nmr spectroscopy even at very low temperatures shows apparent symmetry for the allylmetal compounds [26] indicating either a very low barrier for the metalomeric equilibrium (33) between degenerate monohapto species [26a] and [26b] or the presence of a π-bridged species [26c] with time-independent symmetry. Deuterium isotope effects on terminal mono- and geminal-dideuteriated allyl compounds of the alkali and alkaline earth metals were investigated to allow distinction between these two possibilities

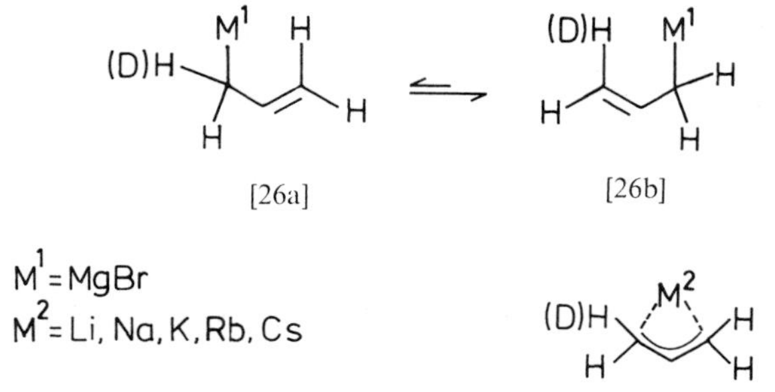

(33)

[26a] [26b]

M^1 = MgBr
M^2 = Li, Na, K, Rb, Cs

[26c]

(Schlosser and Strähle, 1980, 1981; Schleyer and Neugebauer, 1980; Bywater *et al.*, 1980).

In monodeuteriated allylmagnesium bromide the averaged terminal carbons give two separate signals, a triplet for the deuteriated carbon shifted downfield and a singlet for the nondeuteriated carbon shifted upfield with respect to the peak of the unlabelled compound. The splitting is about 1.9 ppm at room temperature when an intrinsic upfield shift of 0.4 ppm for the deuteriated carbon is taken into account. At lower temperatures the splitting increases as expected for an equilibrium isotope effect.

The alkali metal allyl compounds show splittings that are one order of magnitude smaller. No temperature-dependence was reported or is evident from comparison of different investigations. It was concluded that the allyl magnesium bromide has a σ-covalent unsymmetrical structure with a monohapto ligand-metal interaction. A fast, reversible equilibrium (33) interconverts structure [26a] and its tautomeric structure [26b]. In contrast the alkali metal allyl compounds can be described as more or less symmetric π-complexes [26c] with trihapto interaction between metal and ligand.

The equilibrium constant in terminal monodeuteriated allylmagnesium bromide was estimated to be $K = 1.04$–1.08 at 24°C using a value of 70 ppm for the shift difference between the olefinic (estimated 92 ppm) and the metal-bound (estimated 22 ppm) carbons. The direction of the isotope effect is evident from the downfield shift of the deuteriated carbon towards the olefinic position. The equilibrium is shifted so that the isotopic isomer with deuterium at the terminal olefinic position is preferred in the equilibrium.

The values for the small isotope shifts and splittings in the deuteriated alkali metal allyl compounds vary slightly in different investigations, most probably due to different experimental conditions i.e. concentration and temperature. In a 90 MHz ^{13}C nmr reinvestigation of alkali metal allyl compounds, Schlosser and Strähle (1981) report small upfield shifts for both the deuteriated and the nondeuteriated terminal carbon signals with respect to the peak in the corresponding unlabelled compounds. In the d_1-lithium compound, the deuteriated carbon triplet is found downfield, whereas in the potassium and caesium compounds this peak is upfield relative to the nondeuteriated carbons. For the lithium compound it is suggested that bonds of unequal strength bridging the two terminal carbons of the allyl unit would give an unsymmetrical distorted π-complex structure. Very small distortions in the potassium and caesium π-complexes which might even be absent in the unlabelled species were considered to explain these observations.

Although the effects observed are small, it is reasonable to envisage a gradual change going from monohapto coordination for magnesium to unsymmetrical bridging for lithium and symmetrical bridging for the higher alkali metals. The situation is similar to that with carbocations where it has been shown that there is no dichotomous contrast between fast rearranging and bridged species, i.e. between a multiple minimum energy surface and a single minimum. Because of the shallow potential wells of these species, small changes in energy can change the structure, and it is best to think of the energy surface as showing a continuous variation.

Bicyclo[3.2.1]octa-3,6-dien-2-yl anion

The bicyclo[3.2.1]octa-3,6-dien-2-yl anion [27] was proposed as the prototype of bishomoaromatic anions. The isotopic perturbation method has been applied to this system to obtain experimental evidence on its nature (Christl *et al.*, 1983).

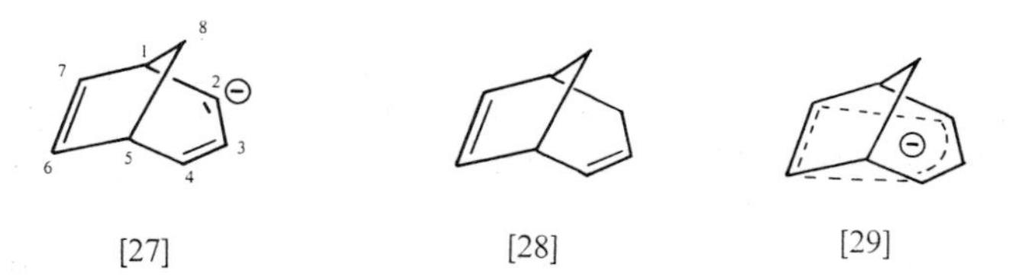

Two interpretations have been advanced to explain the upfield shift observed for the C-6,7 carbons (91 ppm) in [27] compared to the corresponding shifts of 130.36 and 139.82 ppm (C-6;C-7) in the parent hydrocarbon [28]. If the anion is a bishomoaromatic static molecule [29], delocalisation of charge from the allyl anion moiety to C-6,7 could account for the observed upfield shift. A multiple minimum energy surface with a rapidly equilibrating mixture of allyl anion [30] and the tricyclic and tetracyclic isomers [31] and [32] would give C-6,7 in part carbanion and part cyclopropyl-carbon character and could also explain the observed upfield shift.

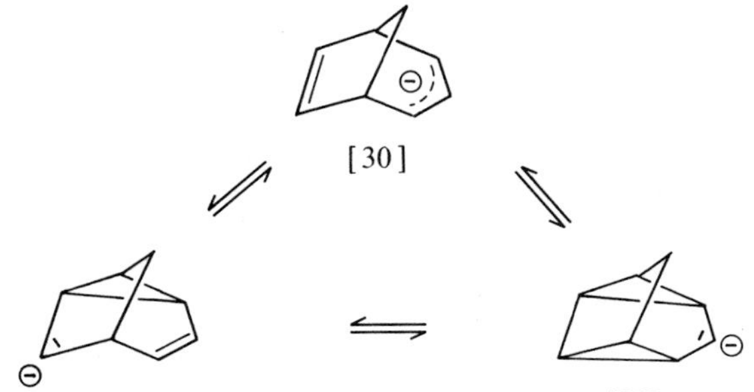

[30]

[31] [32]

(34)

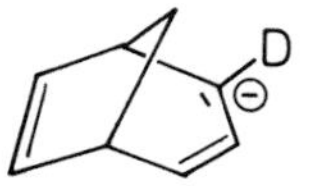

[33] 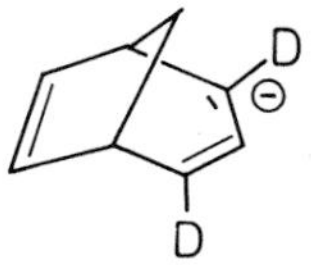 [34]

If the anion were equilibrating the deuteriated anions [33] and [34] would be expected to show equilibrium isotope effects since the C—H(D) vibrational bond force constants are different in the valence isomers [30]–[32]. The structures [31] and [32] containing cyclopropyl rings should be favoured in the equilibrium because C—H(D) vibrations are more confined in a cyclopropane position than at an sp^2-hybridised carbon. Only small upfield shifts were observed in the deuteriated anions [33] and [34]. The maximum shift was 0.49 ppm for the deuteriated carbons C-2,4 in the dideuteriated anion [34]. The small magnitude of the shifts and the typical decrease of the long range effects characterise these as being intrinsic rather than resulting from an equilibrium perturbation. The single energy minimum bishomoaromatic structure [29] is in accord with this result.

The intrinsic upfield shift at C-6,7 in [33] and [34] is opposite to the downfield shift observed in the 2,4,4-trideuteriated hydrocarbon [28]. This has been taken as evidence for a direct bonding interaction between the allyl anion terminus and the ethylenic unit.

Titanacyclobutanes

The important species in the fundamental step of olefin metathesis and related catalytic reactions have been described as either rapidly equilibrating metal-alkylidene complexes or metallacyclobutanes. X-ray crystallographic studies on several substituted dicyclopentadienyltitanium complexes have shown that these complexes have a near planar titanacyclobutane structure in the solid state. Deuterium isotope effects have been investigated by Grubbs *et al.* (1981) in order to establish the structure in solution.

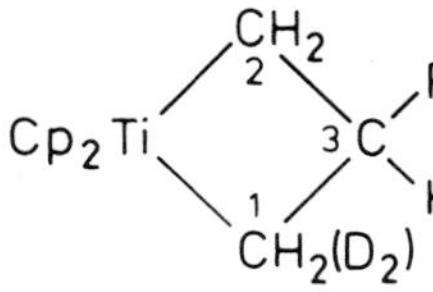

[35]

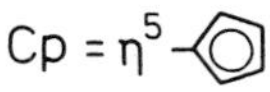

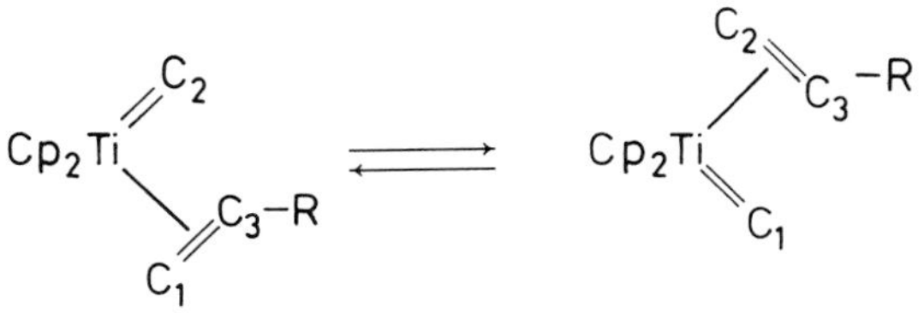

$Cp = \eta^5$-

(35)

[36]

The ^{13}C nmr spectra of C-1-dideuteriated [35] showed two peaks for the C-1 and C-2-methylene groups. The quintet of the deuteriated methylene C-1 was shifted upfield by 0.791–0.758 ppm compared to the corresponding protio compound, whereas the nondeuteriated methylene group C-2 was shifted downfield by 0.121, 0.060 and 0.045 ppm depending on the C-3 substituent. All other signals showed intrinsic isotope shifts upfield.

The temperature-independence of the shifts and the small ratio of the observed isotope splitting to the estimated shift difference (100 ppm) between the averaged C-1/C-2 carbons in a titanium-alkylidene-olefin structure [36] were interpreted as being consistent with a symmetrical but easily distorted titanacyclobutane structure [35] resting at the minimum of a broad shallow potential energy surface which allows easy distortion towards a transition state for the metathesis reaction.

VALENCE ISOMERISM

Barbaralone

At room temperature barbaralone undergoes a rapid Cope rearrangement (36) between two identical structures [37a and b] which leads to averaged signals for H-1/H-5 and for H-2,H-8/H-4,H-6.

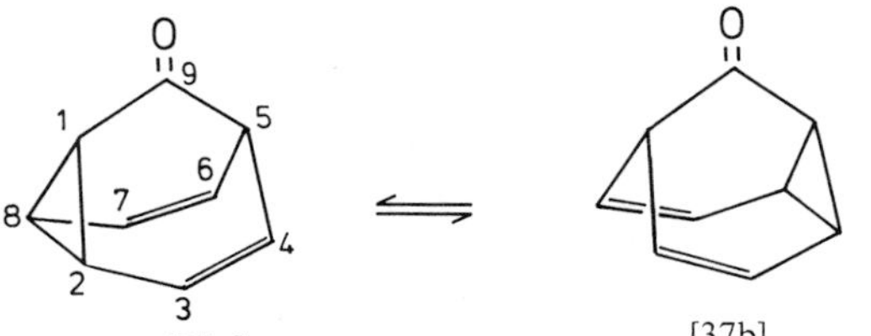

[37a] ⇌ [37b] (36)

In 1-deuteriobarbaralone [38], deuterium lifts the degeneracy of the equilibrium (37) and two peaks for the averaged H-2,H-8/H-4,H-6 protons were observed separated by 0.168 ppm at room temperature (Schleyer *et al.*, 1971). From the limiting, slow exchange spectrum of [37] an intrinsic shift difference of 3.04 ppm between the H-2,H-8 and H-4,H-6 protons was determined. Integration of the signals for the cyclopropyl proton at 2.46 ppm provided the direction of the equilibrium isotope effect and showed that deuteriums are preferentially attached at the bridgehead position in [38b]. The equilibrium constant can be calculated using the formula $K = (\Delta + \delta)/(\Delta - \delta)$; with $\Delta = 3.04$ ppm and $\delta = 0.168$ ppm, $K = 1.117$ at room temperature, corresponding to a ratio of 53 : 47 for [38a] : [38b] and $\Delta G^o = -65\ \text{cal mol}^{-1}$ at 25°C.

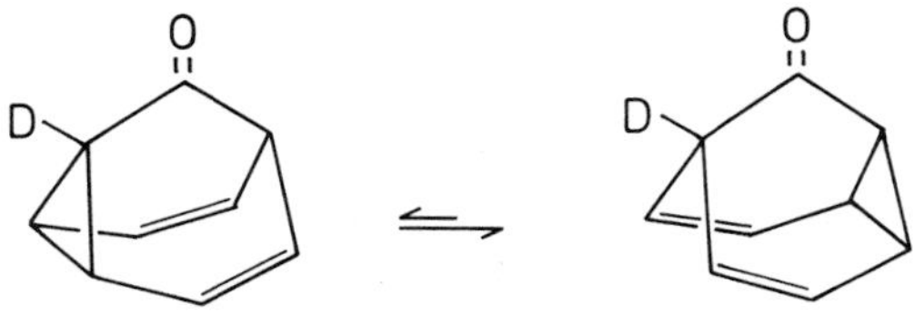

[38a] ⇌ [38b] (37)

The equilibrium constant recalculated from the shift values reported is in reasonable agreement with the fractionation factor for a simple model reaction of isotopic exchange equilibrium (38). The fractionation factor, $K_f = 1.07$ (25°C), for d_1-cyclopropane relative to 2-d_1-propane [39] (Shiner and Hartshorn, 1972) might serve as a model value for the maximum expected α-deuterium isotope effect.

$$c\text{-}C_3H_5D + CH_3CH_2CH_3 \rightleftharpoons CH_3CHDCH_3 + c\text{-}C_3H_6 \qquad (38)$$

[39]

The equilibrium isotope effect in [38] is a differential α- *vs* γ-effect. In 1-d_1-semibullvalene [45] a comparable exchange situation exists for deuterium bonded either to a cyclopropyl carbon or to a bridgehead aliphatic carbon and similar isotope effect data have been reported by Kalinowski (1984). The degenerate Cope rearrangement of bicyclo[5.1.0]octa-2,5-diene (3,4-homotropylidene) which is a non-bridged analogue of [37] is perturbed by deuterium. Analysis of 100 MHz ^{1}H nmr spectra of an unsymmetrically octadeuteriated isomer revealed that CD_2 is favoured in the diallyl position and CH_2 in the cyclopropyl position in accord with the results obtained for [38] and [45] (Günther *et al.*, 1975).

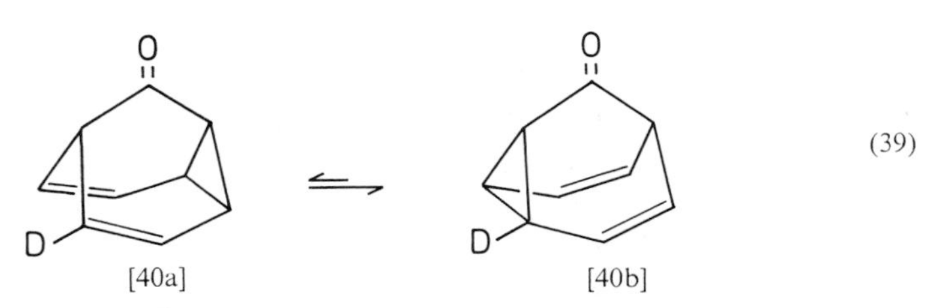

Although quantitative evaluation of the equilibrium isotope effect in 2-deuteriobarbaralone [40] was not possible, the direction of the isotope effect favouring deuterium bonded to the cyclopropyl carbon in [40b] over the olefinic position in [40a] could be established. The sign of the equilibrium isotope effect is in agreement with what would be predicted from force constant calculations of model exchange reactions and also with the equilibrium isotope effect observed in 1,5-dimethyl-2-deuteriosemibullvalene [46] (Askani *et al.*, 1982).

9-Barbaralyl cation

The 9-barbaralyl cation ($C_9H_9{}^+$) can be regarded as cationic analogue of barbaralone. This cation has been reviewed by Ahlberg *et al.* (1983a) and only a brief summary of the decisive experiments is given here. The ^{13}C nmr spectrum of $C_9H_9{}^+$ at $-135°C$ shows only one broad peak indicating a dynamic process which averages all nine carbons. At $-151°C$ this peak is split into two peaks at 101 ppm and 152 ppm with an intensity ratio of 6 : 3. In the octadeuteriated cation $C_9D_8H^+$ [42] both peaks show large isotope shifts as compared to the unlabelled ion, indicating that the spectrum at $-151°C$ does not result from a static symmetrical cation but that both peaks are still averaged by a fast equilibration process which is perturbed by deuteriums (Ahlberg *et al.*, 1981). The lowest energy cation [41] has thus a structure which undergoes a partial (six-fold) degenerate rearrangement, and the averaging process with the lowest barrier is not a Cope-type rearrangement but a divinylcyclopropylmethyl-divinylcyclopropylmethyl cation rearrangement (40).

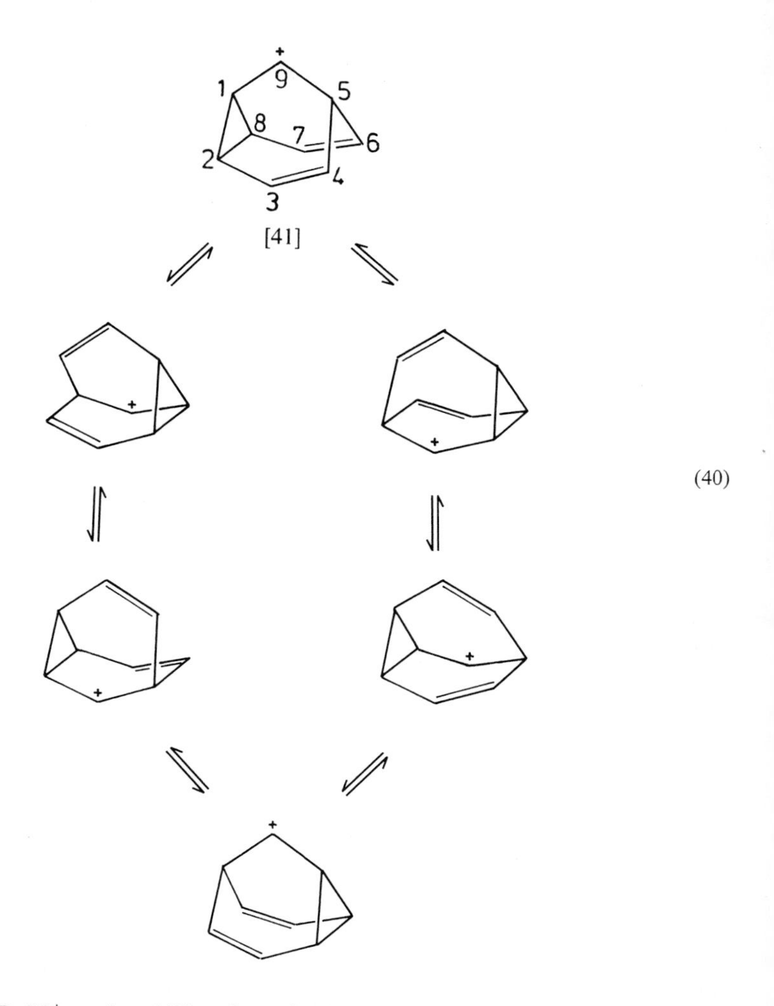

In the $C_9D_8H^+$-cation [42] a downfield shift of 6 ppm is observed for the six averaged carbons which appear at 101 ppm in $C_9H_9^+$. The three averaged carbons at 152 ppm in $C_9H_9^+$ are observed 1 ppm upfield in $C_9D_8H^+$. These equilibrium isotope shifts can be rationalised by taking into account the differences of the C—H bond vibrational force constants of the exchanging carbons (Ahlberg *et al.*, 1983b). The upfield signal results from the averaging of six carbons (C-1, C-5, C-2, C-4, C-8, C-6), two of them olefinic and the other four saturated C—H bonds. Since vibrational force constants are lower in olefinic C—H bonds than saturated C—H bonds, C—H will be preferred in the former and C—D in the latter positions. By comparison with the 9-CH_3-substituted cation, the olefinic carbons are

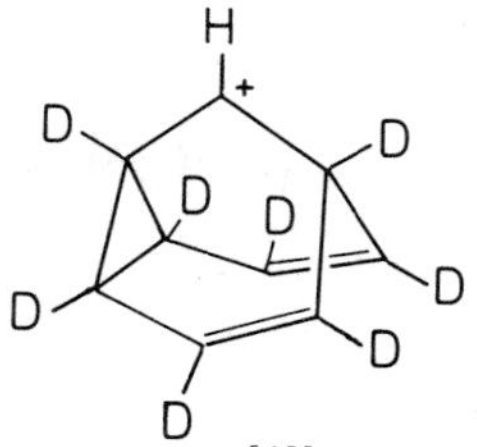

[42]

expected to absorb at lower field than the aliphatic carbons in a static 9-barbaralyl cation. Therefore a downfield shift for the C—H signals of the six averaged carbons in the $C_9D_8{}^+$-ion is observed. The smaller upfield shift (1 ppm) for the three averaged carbons, C-9, C-3 and C-7 (one charged carbon and two olefinic carbons) at 152 ppm has been suggested to result from a slight preference for the C—H groups to occupy the olefinic positions.

At -135°C the totally degenerate rearrangement having a higher barrier occurs *via* [43] as an intermediate or transition state (Ahlberg *et al.*, 1983c) and this averages the two equilibrium isotope effects, resulting in a downfield isotope shift of 4.5 ppm.

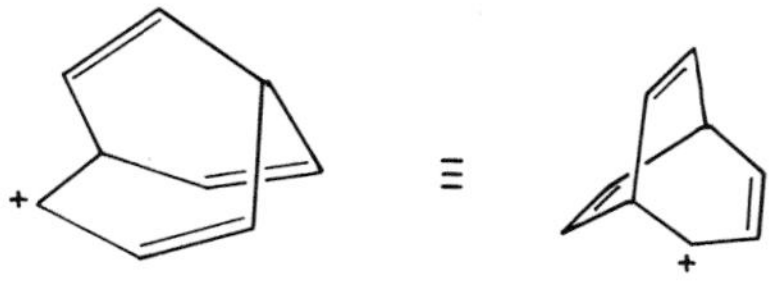

[43]

Semibullvalene

Semibullvalene [44] has a very low barrier to Cope rearrangement (41) which averages C-1/C-5 and C-2,C-8/C-4,C-6 and leaves C-3 and C-7 unaffected. Below -160°C the exchange is slow on the nmr time scale and the averaged carbon and corresponding protons signals are split, giving five resonances as expected. The carbon shift differences are 5.8 ppm and 89.6 ppm for C-1/C-5 and C-2,C-8/C-4,C-6 respectively. The cyclopropyl carbons C-1 and C-2,C-8 are shifted upfield (Anet *et al.*, 1974).

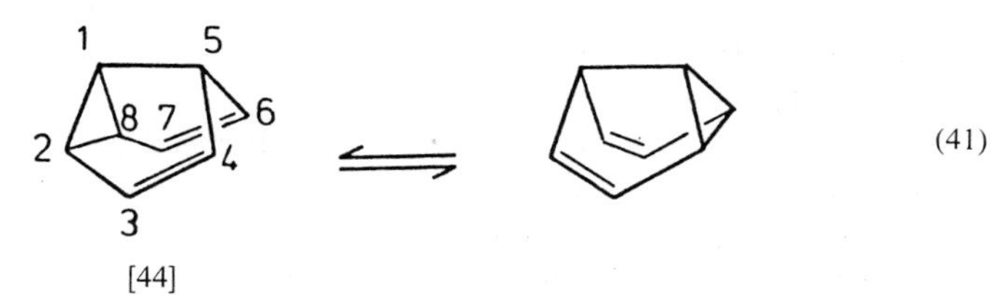

[44]

(41)

Equilibrium isotope effects in deuteriated derivatives of [44] were investigated by ^{13}C nmr spectroscopy by Askani *et al.* (1982, 1984) and have been reviewed in detail (Kalinowski, 1984), As expected, the largest isotope splittings are observed for the averaged C-2,C-8/C-4,C-6 carbons which have the largest shift difference at slow exchange. The direction of the shift of the equilibrium was determined from one bond or long range carbon-deuterium coupling constants.

In 1-deuteriosemibullvalene [45] the deuterium shows the same preference for bonding to the aliphatic C-5 bridgehead position over the cyclopropyl position as was found in 1-d_1-barbaralone [38]. The equilibrium constant is $K = 1.094 - 1.144$ between $+29$ and $-44°C$ giving $\Delta H° = -84.3$ cal mol^{-1} and $\Delta S° = -0.10$ cal mol^{-1} K^{-1}.

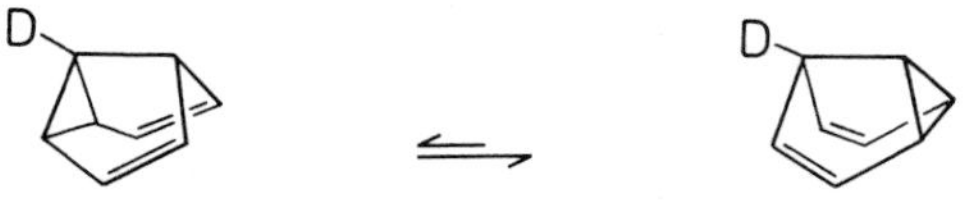

D D (42)

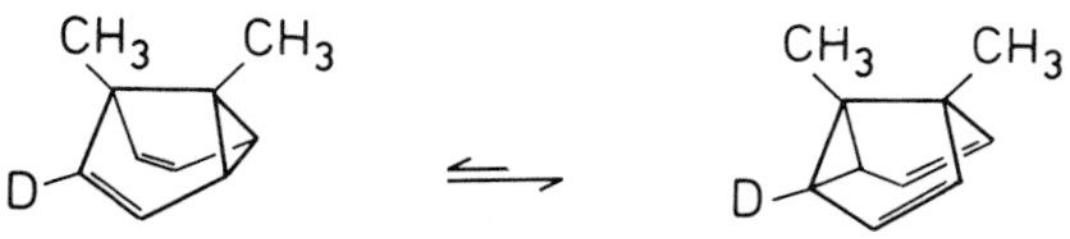

[45]

CH_3 CH_3 CH_3 CH_3 (43)

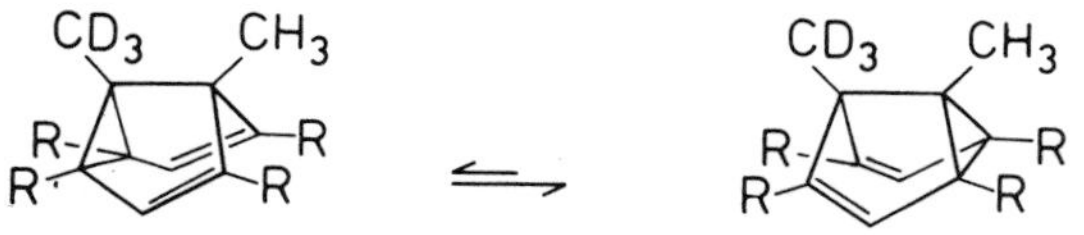

D D

[46]

CD_3 CH_3 CD_3 CH_3

R R R R (44)

R R R R

[47]

In 1,5-dimethyl-2-deuterio-semibullvalene [46] the isotopic equilibrium (43) was found to be shifted in favour of the isotopomer with deuterium bonded to a cyclopropyl carbon and thus confirms the direction of the equilibrium shift in 2-D-barbaralone [40]. The isotope effect data are $K = 1.109–1.191$ between $+29$ and $-53°C$, $\Delta H° = -115.9$ cal mol^{-1} and $\Delta S° = -0.17$ cal mol^{-1} K^{-1}.

An equilibrium isotope effect was also observed, when the deuterium is not directly attached to the averaged carbons but is two bonds removed. 1-Trideuteriomethyl-5-methylsemibullvalene [47] (R = H) ($K = 1.011–1.021$ between $+29°C$ and $-86°C$, $\Delta H° = -9.6$ cal mol^{-1} per D and

$\Delta S^{\circ} = -0.01$ cal mol^{-1} K^{-1} per D) and the corresponding 2,4,6,8-tetramethylcarboxylate [47] (R = $COOCH_3$) ($K = 1.015$–1.035 between $+29$°C and -92°C, $\Delta H^{\circ} = -5.9$ cal mol^{-1} per D and $\Delta S^{\circ} = -0.01$ cal mol^{-1} K^{-1} per D) show the same direction of perturbation of the Cope equilibrium (44) as in 1-deuterio-semibullvalene [45] although the effect is much smaller. The data for the tetramethylcarboxylate [47] (R = $COOCH_3$) may be less accurate because the static shift difference for the frozen out species may be different from that in [44]. The mono- and di-deuteriated 1,5-dimethylsemibullvalenes [47] have correspondingly smaller isotope shifts and thermodynamic parameters indicating that the isotope effect is approximately additive.

Cyclobutadiene

A direct spectroscopic proof of the valence isomeric rectangular structure of cyclobutadiene [48] was obtained from the investigation of 1,2,3-tri-t-butylcyclobutadiene [49] in which one outer t-butyl group was perdeuteriated (Maier *et al.*, 1982). The peak for the doubly populated olefinic site (C-1, C-3) in [49] shows a temperature dependent splitting of 0.453–0.297 ppm between -96°C and -62°C, indicating isotopic perturbation of a fast equilibrium (46) of valence isomers. Carbon C-1 substituted with the deuterated t-butyl group is shifted upfield.

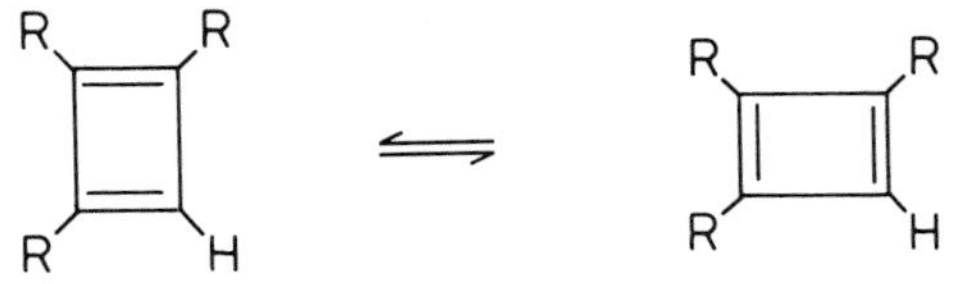

[48] (45)

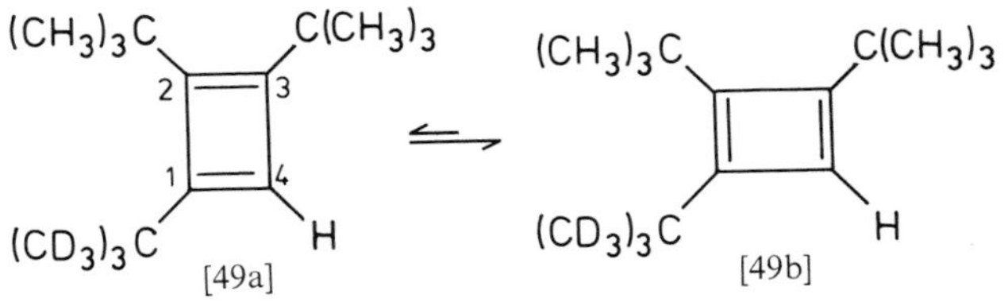

[49a] [49b] (46)

The model compounds 2,3,5-tri-t-butylcyclopentadienone [50] and 2,3,4-tri-t-butylcyclopentadienone [51] show that in [50] an upfield shift of 8.5 ppm is observed for C-3, which is connected to the other t-butyl substituted olefinic carbon (C-2) *via* the double bond, compared to [51], where C-4 is connected to the t-butyl substituted olefinic carbon C-3 *via* a single bond.

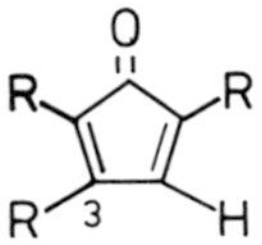

[50] 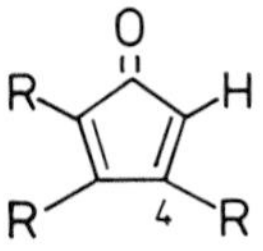 [51]

The upfield shift of C-1 in [49] shows that the isomer [49b] in which C-1 is connected to C-2 via the double bond is favoured in the equilibrium. The equilibrium constant $K = 1.113$–1.073 between -96°C and -62°C; values of $\Delta H° = -84.6\,\text{cal mol}^{-1}$ and $\Delta S° = -0.26\,\text{cal mol}^{-1}\,\text{K}^{-1}$ were calculated using 8.5 ppm as shift difference for the frozen-out static structures. In accord with the vibrational origin of equilibrium isotope effects, the observed direction of the equilibrium shift can be rationalised as follows. Steric hindrance of two adjacent t-butyl groups connected *via* the shorter bond, i.e. the double bond, is more pronounced than steric hindrance of those t-butyl groups which are connected by the longer single bond. The potential energy for C—H/C—D vibrations in the t-butyl groups is perturbed by this hindrance. Vibrational frequencies and the zero point energy for C—H/C—D vibrations in the t-butyl-groups connected *via* the shorter bond are higher. The deuteriated t-butyl group is raised in energy in the sterically more constrained bonding situation less than a protiated t-butyl group; hence the equilibrium is shifted towards [49b].

CONFORMATIONAL EQUILIBRIA

1,1,3,3-Tetramethylcyclohexane

At temperatures where the ring inversion (47) is fast, 1,1,3,3-tetramethylcyclohexane shows only one averaged methyl signal in its ^{13}C nmr spectrum. The degeneracy of the fast conformational equilibrium is lifted in [52] which has one deuteriated methyl group. This leads to a splitting of the averaged

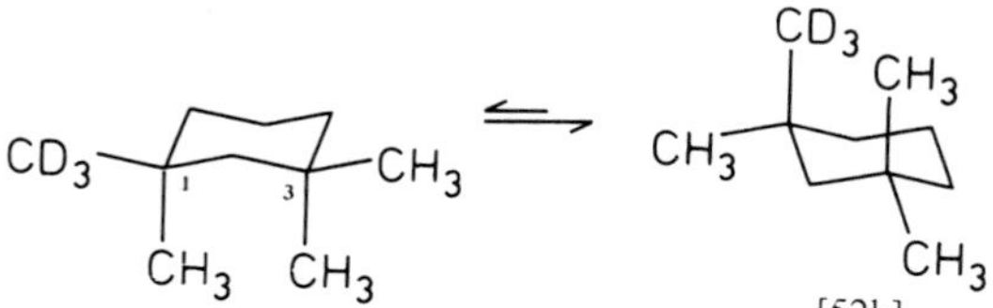

[52a] [52b] (47)

resonances and in the methyl region three lines are observed in the high field ^{13}C nmr spectrum of [52] (Saunders *et al.*, 1980a). Two peaks separated by 0.184 ppm were assigned to the two methyl resonances at C-3. Since axial methyl groups absorb at higher field than equatorial methyl groups, the high field line was assigned to the C-3 methyl group which spends more time in the axial position and conversely the low field line represents the C-3 methyl group which spends more time in the equatorial position.

A broader peak in between the two outer resonances arises from the CH_3-group at C-1. The line-broadening is due to unresolved long range deuterium couplings. This peak is shifted 0.08 ppm upfield from the lowfield C-3 methyl peak as a result of the intrinsic upfield shift of the geminal CD_3-group. The fact that this line is upfield of only that one of the two C-3 methyl peaks which was assigned to the averaged equatorial methyl groups shows that CH_3 at C-1 is preferred in the equatorial position and the geminal CD_3-group in the axial position. The equilibrium (47) is shifted in favour of conformation [52b].

The CD_3-group at C-1 should show a large one bond upfield intrinsic isotope shift and should appear about 0.5 ppm upfield of the CH_3-signals but it was not observed under the experimental conditions.

The equilibrium constant $K = 1.024$ at 17°C was calculated using $\delta = 0.184$ ppm for the equilibrium isotope splitting and $\Delta = 9.03$ ppm for the shift difference of the axial and equatorial C-3 methyl groups at slow exchange (−100°C). The free energy difference is $\Delta G^o = -24\ \text{cal mol}^{-1}$ in favour of the conformer with the axial CD_3-group.

The reason for the conformational preference of the CD_3-group in the axial position is that 1,3-interactions between two axial methyl groups are sterically more hindered than in the 1,3-diequatorial arrangement. The hydrogen vibrations in the axial methyl groups are thus more closely confined resulting in a steeper and more curved potential energy well compared to the equatorial situation. This leads to greater force constants, increased C—H vibrational frequencies and higher zero point energy for the axial methyl groups. The CD_3-group has a lower zero point energy compared to CH_3-groups and thus is raised less in energy in positions which have higher zero point energy. The axial preference for the CD_3-group gives the lowest total zero point energy and lowest total energy.

Substituted piperidines, cyclohexanes and dioxanes

Similar deuterium isotope effects on the conformational equilibrium (48) in N-trideuteriomethyl-N,3,3-trimethylpiperidinium ion [53] were studied by

CD3 N+ CH3 CH3 CH3 CH3 ⇌ CD3 CH3 N+ CH3 CH3 (48)

[53]

Robinson and Baldry, (1977a). The deuteriated methyl group is preferred in the axial position. The equilibrium constant at 27°C is $K = 1.028$ giving a free energy difference $\Delta G^\circ = -16.7\ \mathrm{cal\ mol^{-1}}$.

It was also shown that the *syn* 1,3- and 1,5-diaxial interaction of one methyl group at C-1 with two hydrogens at C-3 and C-5 is sufficient to observe a conformational equilibrium isotope effect in N-trideuterio-methyl-N-methylpiperidinium ion [54] and in the deuteriated *trans*-1,3-dimethylcyclohexane [55] and *trans*-2,6- and *trans*-3,5-dimethylcyclo-hexanones [56] and [57] (Robinson and Baldry, 1977b).

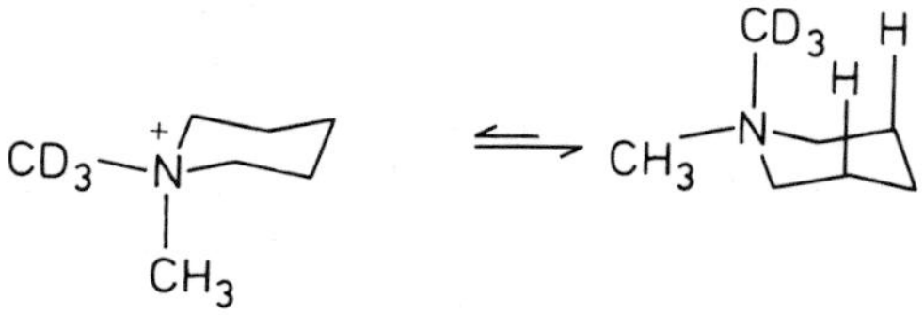

[54] (49)

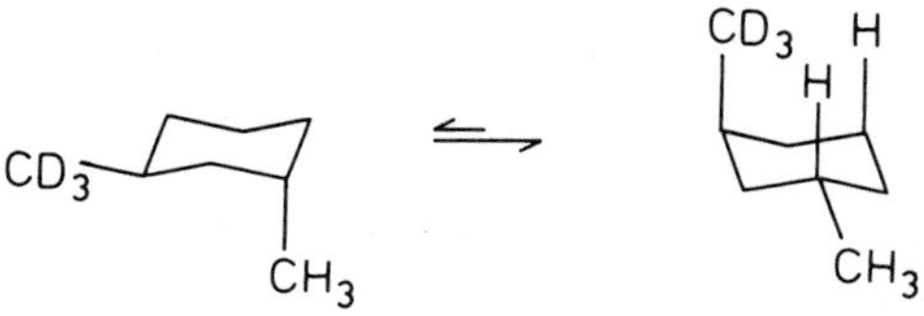

[55] (50)

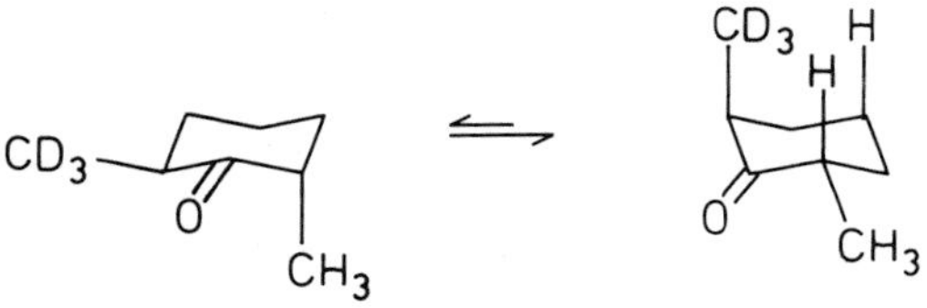

[56] (51)

[57] (52)

In these cases the equilibrium constants are not far from unity ($K = 1.01$–1.03 and $\Delta G^\circ = -19$ to $-11\ \mathrm{cal\ mol^{-1}}$. Uncertainties in the ^{13}C-spectral measurements, which were done at 23 MHz, and possible errors in the estimation of static shift differences and intrinsic isotope shifts were discussed; they preclude in part comparison of these small effects. In *cis*-1-ethyl-4-trideuteriomethylcyclohexane [58], where the conformational equilibrium (53) is non-degenerate ($K_H = 1.08$ at 25°C) in the unlabelled

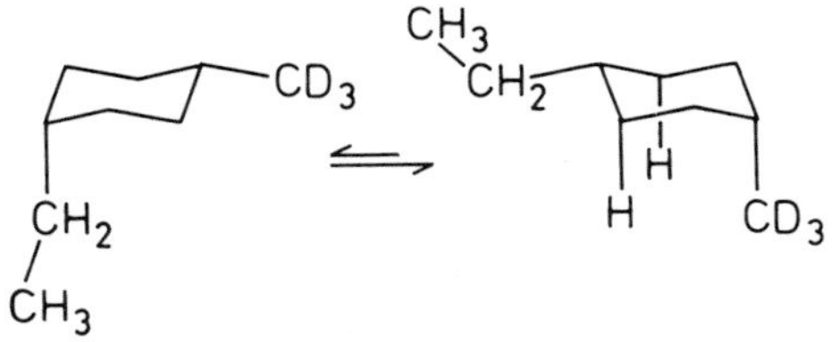

(53)

[58]

compound, the same order of magnitude for the deuterium isotope effect was observed using high field ^{13}C nmr spectroscopy (Booth and Everett, 1980b). The ethyl-CH_2 was used to trace the equilibrium isotope effect because this carbon is six bonds removed from deuterium and shows no intrinsic isotope shifts and also has a reasonably large shift difference at slow exchange which makes this position particularly sensitive to any change in the position of the equilibrium. The isotope effect, $K = K_D/K_H = 1.02$ and $\Delta G^o = -11.9 \text{ cal mol}^{-1}$ at 25°C, indicates a slightly greater preference for the conformer with axial CD_3-orientation compared to axial CH_3 orientation in the unlabelled compound.

In [^{13}C-1-methyl]-*cis*-1,4-dimethylcyclohexane [59] no isotopic perturbation was observed within experimental error. It was concluded that a ^{13}C-isotope effect on the degenerate equilibrium (54) in *cis*-1,4-dimethylcyclohexane is either too small to be detected or non-existent (Booth and Everett, 1980a).

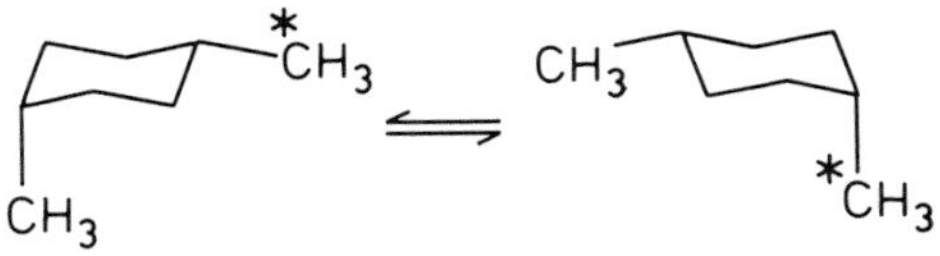

(54)

[59]

The conformational equilibrium of *trans*-3-methyl- and 3-hydroxymethyl-1-trifluoromethylcyclohexane is shifted at room temperature in favour of the conformer with an equatorial trifluoromethyl group. Isotopic perturbation of these nondegenerate equilibria (55) was observed in the deuteriated compounds [60] (R = D or OH) using CF_3 as a reporter group to monitor the isotope effect by ^{19}F-nmr spectroscopy (Robinson and Baldry, 1977b). In [60] the ^{19}F-resonance is shifted further upfield compared to the corresponding unlabelled compounds showing that the equilibria are shifted further in favour of conformations [60b] with the CF_3-group in the

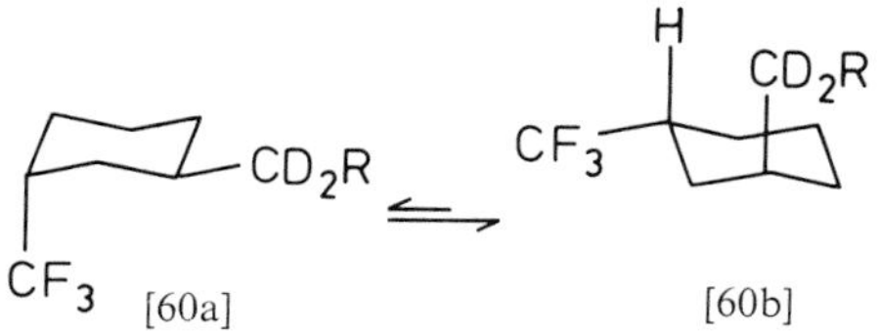

[60a] ⇌ [60b] (55)

equatorial position. The deuteriated methyl- and hydroxymethylene-groups are found more frequently than the nondeuteriated groups in the axial position which has a higher zero point energy than the equatorial position due to more confined C—H(D) vibrations caused by the *syn*-axial interactions with the hydrogens at C-1 and C-5. The quantitative interpretation of the small effects is somewhat limited by experimental errors involved in the measurements at low field and the uncertainty of the estimation of static shift differences. The calculated isotope effects per deuterium were found to be constant for successive deuteriation within experimental error. This is consistent with the interpretation of conformational equilibria in cyclohexanes that the strain associated with an axial group -CHXY is largely attributable to repulsion between the hydrogen in -CHXY and the syn 1,3-diaxial ring H-atoms.

2,6,6-Trideuterio-2-methylcyclohexanone shows isotope effects on the ^{13}C chemical shifts which could not be explained assuming intrinsic effects only (Wehrli and Wirthlin, 1976). Saunders *et al* (1980a) have suggested that the origin of these effects is an equilibrium isotope effect. Conformational isotope effects have been observed in deuteriated 1,4-dioxanes (Jensen and Neese, 1971) and in substituted 1,3-dioxanes (Robinson, 1971).

A long range intrinsic deuterium isotope shift in 2,2-bis(trideuteriomethyl-5,5-dimethyl-1,3-dioxane [61] was observed by Anet and Dekmezian (1979). This molecule does not show an equilibrium isotope effect because the two rapidly interconverting chair isomers have exactly the same energy. 2-D-5,5-Dimethyl-1,3-dioxane, however, exhibits an isotope effect on the conformational equilibrium (Anet and Kopelevich, 1986a). Deuterium is preferred in the equatorial position ($\Delta G^o = -49\,\text{cal mol}^{-1}$ at 25°C).

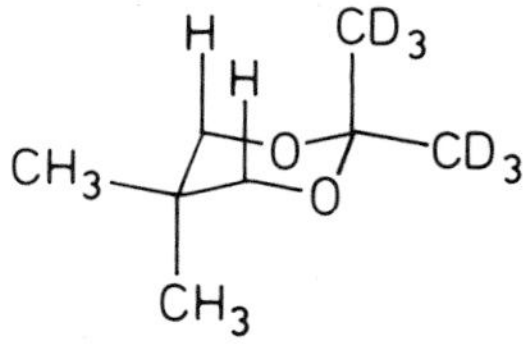

[61]

cis-*2-Decalone*

The conformational equilibrium (56) between the "steroid-like" [62a] and "non steroid-like" [62b] conformation of *cis*-2-decalone is shifted in favour of [62b] because this conformation has one less hydrogen-mediated gauche butane type interaction (3-alkylketone effect).

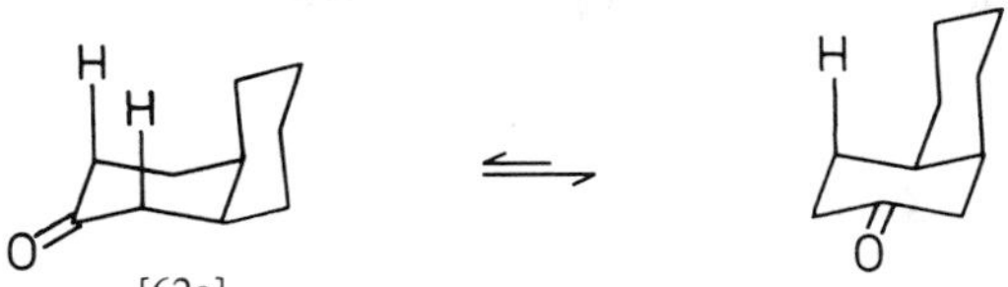

(56)

[62a] [62b]

(57)

[63a] [63b]

The ^{13}C nmr spectrum of β,β′-tetradeuteriated *cis*-2-decalone [63] shows intrinsic isotope shifts over up to four bonds as well as equilibrium isotope effects over five bonds (Lippmaa *et al.*, 1982). Under fast exchange conditions at room temperature, no net isotope shifts are observed at carbons C-7 and C-9, which are separated from the deuterium by four bonds, because the intrinsic isotope shift and the equilibrium isotope effect on the chemical shift happen to have the same magnitude but different sign, thus compensating each other. At −70°C the equilibrium is frozen out and consequently equilibrium isotope effects disappear. The remaining four bond intrinsic isotope shifts for C-7 and C-9 are 0.02 ppm downfield and are only observed in conformation [63a] indicating the presence of deuterium-mediated gauche interactions to C-4 and C-2 which are not present in [63b]. The equilibrium isotope shifts in the fast exchange spectra must be 0.02 ppm upfield for C-7 and C-9 to give the apparent zero effect at room temperature. This direction of the equilibrium isotope shift is towards the chemical shift of C-7 and C-9 in [62a] which are upfield compared to [62b].

The equilibrium isotope effect is observed unmasked by an intrinsic isotope shift at C-8 which is separated by five bonds from deuterium. At room temperature, C-8 is shifted 0.02 ppm to lower field and has an identical shift to the unlabelled compound in the frozen-out spectrum at −70°C. This confirms the equilibrational origin of the isotope shift of C-8 observed under conditions of fast exchange.

At −70°C the C-8 carbon appears at 26.6 ppm and at 20.5 ppm in conformation [62a] and [62b] respectively. The fact that the averaged chemical shift of C-8 is more downfield in the deuteriated compound [63] compared to the unlabelled ketone [62] shows that the equilibrium in [63] is shifted towards conformation [63a] where the C—D vibrations are more confined due to one additional gauche butane type interaction as compared to conformation [63b].

Cyclodecanone

Deuterium isotope shifts over up to six bonds have been observed in the ^{13}C nmr spectrum of deuteriated cyclodecanones (Wehrli *et al.*, 1978). The equilibrational origin of the observed long range effects was briefly discussed by Anet and Dekmezian (1979) and was explained in detail by Whipple *et al.* (1981). In the low temperature spectra of deuteriated cyclododecane isotopomers Anet and Rawdah (1978) have also observed deuterium isotope effects which are likely to have a conformational origin and to arise from the lack of precise D_4-symmetry in the preferred conformation of that hydrocarbon.

Low temperature nmr measurements have shown that cyclodecanone [64] has an unsymmetrical boat-chair-boat (BCB) structure which has four transannular H—H interactions (Anet *et al.*, 1973). At room temperature all ^{13}C chemical shifts except the carbonyl carbon and C-6 are a 1 : 1 average of the shifts of the two equivalent conformations [64a] and [64b]. In cyclodecanone deuteriated at C-6, the symmetry of the conformational equilibration process (58) is not perturbed and hence no shifts arise as a result of unequal weighting; but intrinsic isotope shifts over up to three bonds are observed.

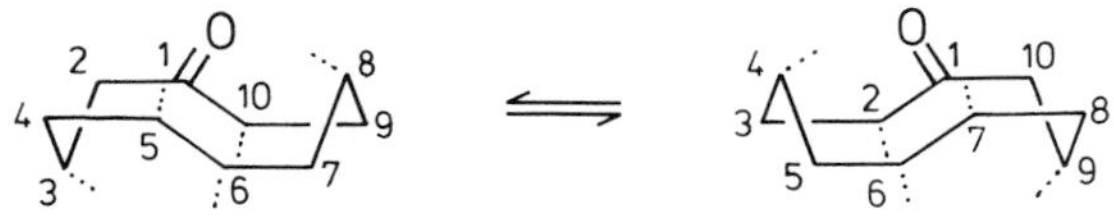

[64a] ⇌ [64b] (58)

In contrast, deuteriation at C-2 lifts the degeneracy of the conformational equilibrium, and, in addition to intrinsic isotope shifts over up to three bonds, equilibrium isotope splittings are observed for the carbons pairs C-2/C-10, C-3/C-9, C-4/C-8 and C-5/C-7 which are averaged in the conformational equilibrium. If corrections for intrinsic isotope shifts are taken into account, the splittings are symmetrical with respect to the unperturbed position in the nonlabelled compound within experimental error. The sign of the equilibrium isotope effect can be extracted from the corrected relative upfield and downfield shifts of the perturbed peak positions in 2-D_1-[64] compared to the corresponding equally averaged position in [64]. Upfield shifts for positions 9 and 4 and downfield shifts of equal magnitude for positions 3 and 8 were observed in 2-D_1-[64] with respect to the corresponding C-9/C-3 and C-4/C-8 resonances in [64]. The C-9 and C-3 and also the C-4 and C-8 peaks in 2-D_1-[64] could be assigned individually on the basis of deuterium coupling to C-3 and C-4. Whipple suggested a reversal of the assignment for C-5 and C-7 in [64] which would also give an upfield shift for C-7 in 2-D_1-[64]. The relative upfield shift for carbons C-4, C-7 and C-9 is in accord with a preference for 2-D_1-[64b] in equilibrium (58). Transannular interaction between protons at positions 2, 6 and 9 and between the protons at positions 4 and 7 result in an upfield steric compression shift for this carbon. The C—D vibrations for deuterium at C-2 are more confined and thus have higher zero point energy in conformation [64b] than in conformation [64a] where transannular interactions occur between positions 3, 6 and 10 and between positions 5 and 8 and do not involve the deuterium at C-2. In accord with this interpretation smaller effects were observed in cyclodecanone deuterated at C-5 which has only one transannular interaction. Similar conformational equilibrium isotope effects are likely to be the cause of the upfield and downfield long range isotope shifts observed in monodeuteriated cyclododecanones (Jeremić *et al.*, 1982).

Cyclohexane

Günther and Aydin (1981) have investigated the conformational equilibrium (59) in d_1-cyclohexane. The two conformational isomers [65a] and [65b] with deuterium in the axial and in the equatorial position can be frozen out in the 100 MHz ^{13}C spectrum below $-80°C$. The deuterated carbons in the two isomers have different intrinsic isotope shifts which are upfield compared to the nondeuteriated carbons. The triplet with the smaller $^1J_{CD}$ coupling constant was assigned to the carbon with axial deuterium and was shifted 0.0482 ppm to higher field than the triplet caused by the carbon with equatorial deuterium.

[65a] ⇌ [65b] (59)

At room temperature the averaged signal for the deuteriated carbons was observed slightly (0.01148 ppm) downfield from the average value expected for an unperturbed equilibrium. This indicates that $K \neq 1$ and that the equilibrium (59) is shifted in favour of conformation [65b] with deuterium in the equatorial position. The small effect was confirmed by two independent experimental observations. In a mixture of D_0-, D_1- and 1,1-D_2-cyclohexane, the deuteriated carbon in D_1-cyclohexane showed an upfield shift which was 0.01689 ppm, less than half the value of the intrinsic upfield shift in

1,1-D_2-cyclohexane. The nonadditivity of the isotope shifts shows that an equilibrium isotope effect is superimposed on the intrinsic isotope shift in D_1-cyclohexane. Integration and line shape analysis of the two signals for the axial and equatorial deuteriums in the 61.4 MHz ^{2}H nmr spectrum of [65] at $-88°$C showed the expected population change and led to the same conclusion. At room temperature the equilibrium constant $K = 1.100$ and $\Delta G^o = -48$ cal mol^{-1} in favour of conformational isomer [65b] with deuterium in the equatorial position.

Vicinal gauche H,H(D) interactions were suggested to be the cause of the observed conformational preference. Equatorial hydrogens in cyclohexane are gauche to both adjacent axial and equatorial hydrogens and the axial hydrogens are gauche to both vicinal equatorial hydrogens but *anti* to two vicinal axial hydrogens. The C—H and C—D vibrations are thus more confined in the equatorial positions. Therefore the equatorial C—H(D) bonds have higher vibrational force constants which in turn result in higher zero point energy for these positions. *A priori*, C—D vibrations have lower zero point energy compared to C—H bonds and the energy decrease is greatest when hydrogen is favoured in the axial and deuterium in the equatorial position.

Recent calculations of the conformational equilibrium isotope effect for [65] using a scaled quantum mechanical 3–21 G force field are in qualitative accord with the experiment and confirm the zero point energy origin of the effect (Williams, 1986).

The smaller force constants for axial C—H than for equatorial C—H bonds are suggested not to be caused by less vicinal H,H steric interactions for axial hydrogens. Instead they may reflect a greater σ-σ^* interaction between an antibonding orbital of an axial C—H bond and the bonding orbitals of a pair of antiperiplanar axial C—H bonds than the interaction between an antibonding orbital of the equatorial C—H bonds and a pair of antiperiplanar C—C bonding orbitals.

The same direction for the isotope effect on the ring inversion equilibrium in *trans*-(1,4-protio)-D_{10}-cyclohexane has been observed using 200 MHz ^{1}H nmr spectroscopy (Anet and Kopelevich, 1986b). The (*aa*)-conformation with the protons at C-1 and C-4 in the axial position is favoured in the equilibrium. At 25°C an upfield shift of 0.00255 ppm was observed relative to the corresponding *cis*-(1,4-protio)-isomer which exhibits no conformational isotope effect ($K_{cis} = 1$) due to symmetry of the *ea*- and *ae*-isomers.

Using the equation $K = (\Delta + 2\delta)/(\Delta - 2\delta)$ with $\Delta = 0.478$ ppm the equilibrium constant $K = 1.022$ is obtained which yields $\Delta G^o = -6.3$ cal mol^{-1} for a single deuterium. These values are significantly smaller than those for [65]. Molecular mechanics model calculations on D_{11}-cyclohexene using stretching force constants for equatorial and axial C—H bonds which were adjusted to reproduce experimental vibrational frequency differences are in agreement with the experimental nmr results.

If vicinal H,H(D) interactions are determining the conformational equilibrium isotope effect, the results can be rationalised in the following way: C—D bonds have lower vibrational amplitude than C—H bonds and therefore it might be anticipated that vicinal gauche interactions of CD_2- and CHD-groups in D_{10}-cyclohexane are smaller, resulting in smaller perturbations than the vicinal gauche interactions between CH_2- and CHD-groups in D_1-cyclohexane [65].

It is difficult, however, to compare small isotope effects obtained by different experimental methods. The data for [65] were obtained over a temperature range of more than 100°C using the peak area method at slow exchange and the chemical shift difference for the averaged intrinsic isotope shift under conditions of fast exchange to determine the equilibrium isotope effect. For trans-(1,4-protio)-D_{10}-cyclohexane the chemical shift difference method was applied but no temperature dependent data were reported; thus an important diagnostic criterion to characterise an equilibrium isotope effect is not accessible. In heavily deuteriated compounds additional problems arise in defining precisely the contribution of intrinsic isotope shifts to the observed isotope splittings.

Cycloheptatriene and cyclopentene

The degeneracy of the rapid conformational equilibrium between the non-planar conformations of cycloheptatriene is lifted in 7-deuteriocycloheptatriene [66]. The equilibrium constant is no longer unity but conformer [66b] with hydrogen *syn* to the ring is present in higher concentration (Jensen and Smith, 1964).

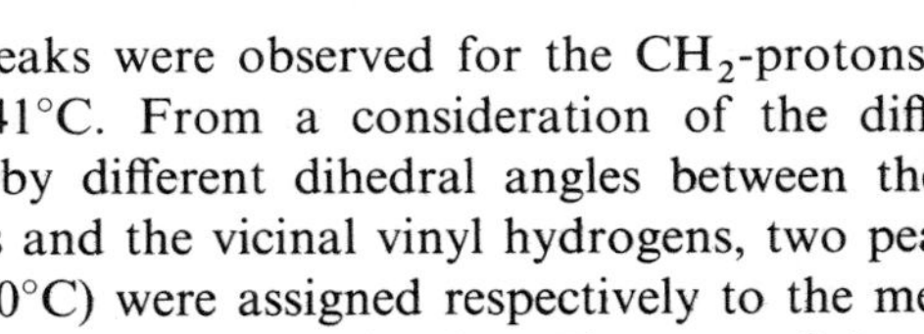

[66a] $\rightleftharpoons$ [66b] (60)

Two separate peaks were observed for the CH_2-protons in cycloheptatriene below −141°C. From a consideration of the different coupling constants caused by different dihedral angles between the two different methylene protons and the vicinal vinyl hydrogens, two peaks at 1.44 and 2.88 ppm (at −170°C) were assigned respectively to the methylene hydrogens in positions *syn* and *anti* to the ring. The area of the peak of the *syn* hydrogen was larger than that for the *anti* hydrogen and the ratio was found to increase in favour of the peak for the *syn*-hydrogen with decreasing temperatures, indicating that the conformer [66b] with *syn*-H and *anti*-D is lower in energy. At fast exchange the shift of the peak for the averaged methylene protons was found to be temperature-dependent, whereas the shift of the vinyl hydrogens was independent of temperature. When the temperature was lowered, the methylene resonance moved upfield which is towards the position of the *syn*-H resonance. The area ratio at slow exchange and the direction of the equilibrium isotope shift at fast exchange led to the same conclusion that conformer [66b] is favoured in equilibrium (60).

The steric effect of the eclipsing interaction between the *anti*-position at the methylene carbon and the two adjacent vinylic hydrogens confines C—H(D) vibrations in this position more than in the *syn*-position. Deuterium is therefore preferred in the *anti*-position and hydrogen in the *syn*-position. The reported values for the isotope effect are $K = 1.41–1.10$ with $\Delta G^o = -72$ to -30 cal mol^{-1} between -168°C and -114°C; $\Delta H^o = -142 \pm 30$ cal mol^{-1} and $\Delta S^o = -0.7 \pm 0.3$ cal mol^{-1} K^{-1}.

The isotope shifts observed in the ^{1}H nmr spectrum of *cis*-3,5-dideuteriocyclopentene [67] (Anet and Leyendecker, 1973) have their origin most probably in conformational equilibrium isotope effects and could be explained in a similar way to the effects in cycloheptatriene, although the size of the effects is different because of different conformational arrangements.

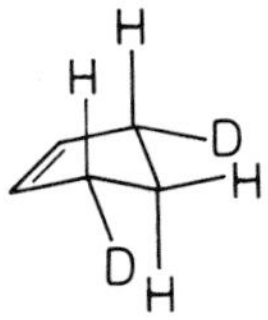

H H D H D H

[67]

BRIDGING AND HYPERCOORDINATION IN TRANSITION-METAL COMPLEXES

Coordinatively unsaturated species can achieve saturation through partial bonding to the hydrogen or carbon atoms of organic ligands. Metal–hydride–metal and metal–hydride–carbon interactions in transition-metal complexes play an important role in catalytic reactions like carbon monoxide reduction, olefin metathesis, alkyne polymerisation and methylene transfer.

Osmium cluster complexes

The nature of the hydrogen bonding in methylated μ-hydridodecacarbonyltriosmium cluster compounds [68] and [69] has been investigated by

Shapley and coauthors using neutron diffraction (Shapley *et al.*, 1979) and nmr spectroscopy in combination with isotope effects (Shapley and Calvert, 1977, 1978; Shapley *et al.*, 1978).

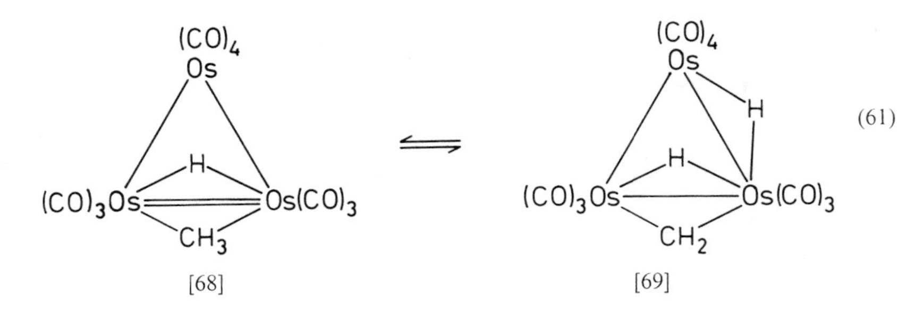

Reaction of di-μ-deuteridodecacarbonyltriosmium with diazomethane provided a mixture of hydridomethyl- and hydridomethylene-tautomers of $Os_3(CO)_4CH_2D_2$, D_2-[68] and D_2-[69], which interconvert slowly in solution. The ^{1}H nmr spectra show a preference for H over D in the metal hydride sites. An isotope effect on the slow exchange equilibrium (61) favours incorporation of deuterium in the methylene- and methyl-positions and hydrogen in the osmium hydride positions. The effect has been interpreted in terms of vibrational zero point energy differences. The C—H bonds have significantly higher vibrational frequencies than the metal hydride bond. Partial substitution by deuterium leads to preferential placement of the lighter isotope in the lower frequency site. Estimation of the corresponding frequencies from vibrational spectra led to a range of calculated equilibrium constants for the H/D incorporation which are consistent with the experimentally observed results. The large disparity in zero point energies found in this case was considered to be generally characteristic of a fully equilibrated H/D distribution among carbon (or nitrogen or oxygen) and metal sites.

Isotopic perturbation of the ^{1}H nmr spectra was found for the shielded (−3.68) methyl signal in partially deuteriated decacarbonylhydridomethyltriosmium (Shapley and Calvert, 1978). The spectrum shows separate CH_2D- and CHD_2-signals displaced significantly to higher field from the CH_3-signal of the nondeuteriated compound [68]. The separations vary strongly with temperature, increasing from 0.34 ppm (CH_2D) and 0.39 ppm (CHD_2) at 35°C to 0.55 and 0.68 ppm at −76°C. These large temperature-dependent shifts are inconsistent with the relatively small intrinsic isotope shifts commonly observed upon geminal substitution of H by D (0.01 ppm) and indicate an isotopic perturbation of a fast equilibrium. No line broadening was observed even at −100°C. This sets an upper limit of approximately 5 kcal mol^{-1} for the barrier of the exchange process.

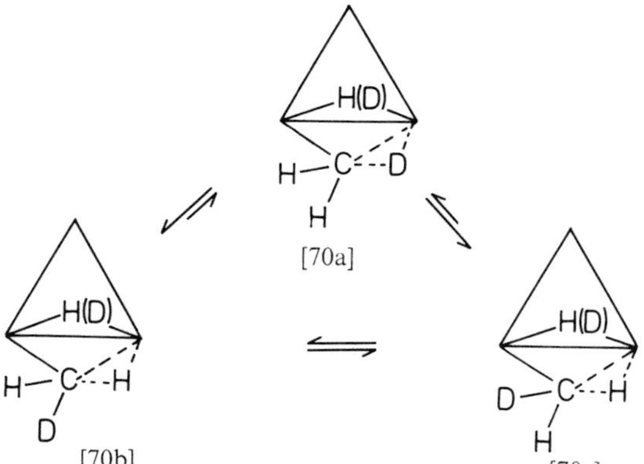

(62)

[70a]

[70b]

[70c]

A model (62) involving C—H—Os interactions can rationalise the observed effect. For the monodeuteriated methyl-compound [70] three structures [70a–c] are possible.[2] An isotope effect on a fast equilibrium between these structures is to be expected. The lower vibrational frequency and hence lower zero point energy of the bridging C—H bond leads to a preference for the lighter nucleus at the bridging site, and both H-bridged forms [70b] and [70c] will be slightly more abundant than the D-bridged form [70a]. Since the bridging atom should resonate at higher field than the terminal hydrogen atom, the nonrandom distribution results in a net upfield shift for the CH_2D-signal relative to the CH_3-signal of the unlabelled compound. As expected for an equilibrium isotope effect, the isotope shifts increase when the temperature is lowered because the equilibrium is shifted in favour of the lower energy (H-bridged) sites, [70b] and [70c]. Neglecting a diastereotopic shift difference of the two terminal hydrogens which is only about 0.2 ppm in analogous static compounds, the exchange process is threefold degenerate, taking place between a doubly populated lowfield site and a singly populated highfield site. Analysis of the temperature dependence of the averaged shifts in the CH_2D- and CD_2H-complexes led to calculated shifts of -15 ± 1 ppm for the C—H—Os bridging hydrogen and 2 ± 1 ppm for the two terminal hydrogens. The energy difference between the D-bridged [70a] and H-bridged [70b] and [70c] was found to be $\Delta H^\circ = -130\ \mathrm{cal\,mol^{-1}}$. Within experimental error identical results were obtained for the mono- and di-deuteriated species.

Support for this interpretation comes from the temperature-dependent isotope effect on the size of the averaged $^1J_{CH}$ coupling constant. Whereas

[2] For clarity, simplified formulae are used in this section for structures [70a–c], [71], [73a,b], [75], [77b,c], [78], [79], [80a,b] and [82].

the averaged value ($^1J_{CH}$ = 121.1 Hz) in the nondeuteriated complex [68] is temperature independent, a change from 118.9 Hz and 116.4 Hz at −27°C to 116.7 Hz and 112.3 Hz at −80°C was observed in the mono- and di-deuteriated complexes respectively. Values of 150 ± 10 Hz and 60 ± 20 Hz for the coupling constants of the terminal and bridging hydrogens were estimated from the temperature dependence.

Alternative structural models which could involve a rapid equilibrium between two different carbon sites were ruled out since only small, temperature-independent, intrinsic upfield isotope shifts (0.2 ppm per D) were observed in the ^{13}C spectrum. Structure [71] with two C—H—Os interactions was also ruled out because for this case different isotope splitting patterns are expected than are observed. Similar equilibrium isotope effects as in [70] have been observed in the homologous ethyl complex (Shapley *et al.*, 1986).

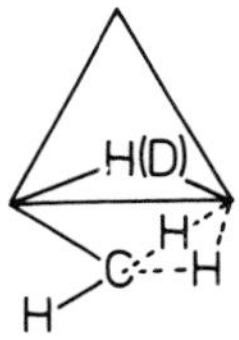

[71]

Alkyl complexes of tungsten and iron

An equilibrium isotope effect has been observed in the tungsten methylene complex [72] (Schrock and Holmes, 1981). The proton spectra show an averaged peak at −0.16 ppm for the methylene protons at room temperature. At −108°C two peaks are observed at about 7.05 ppm and −7.97 ppm.

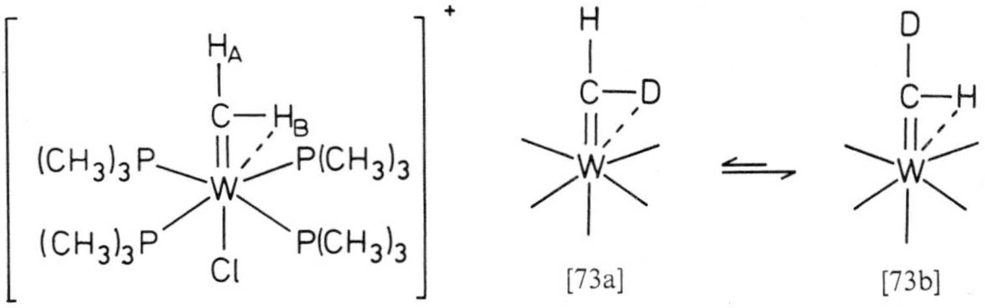

[72] [73a] [73b] (63)

In the monodeuteriated complex $[WCHD(PMe_3)_4Cl]^+$ [73] the averaged proton peak is shifted to −1.40 ppm. The equilibrium isotope effect has been estimated as $K = 1.42$ at 25°C in favour of the proton-bridged form [73b] (ΔG^o = −200 cal mol^{-1}).

The iron complex [74] has a markedly unsymmetrical μ-CH_3-group as was shown by X-ray diffraction studies (Stone *et al.*, 1982). The ^1H and ^{13}C nmr spectra of [74] do not change from 25°C to −80°C and are in accord either with a symmetrical μ-CH_3-bridge like [75] or an unsymmetrical structure [74] undergoing fast dynamic exchange *via* a low barrier. In mono-deuteriated (μ-CH_2D)-[74] the signal for the CH_2D-group is observed at −3.47 ppm in the ^1H nmr and at −1.88 ppm in the ^2H nmr spectrum at 25°C. The temperature dependence is as expected for an equilibrium isotope effect. Compared to the unlabelled methyl group (−2.90 ppm, 25°C), the

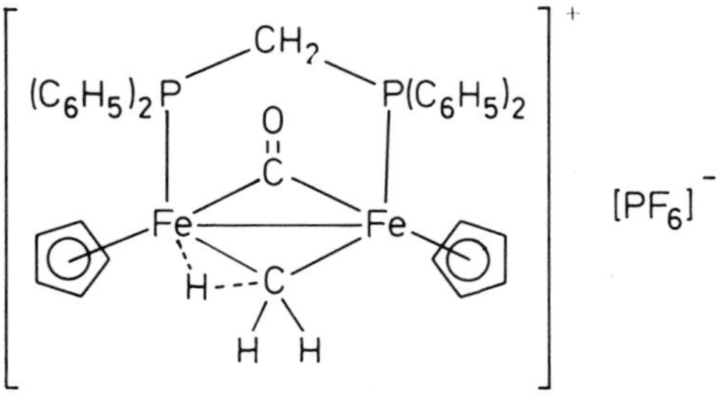

$[PF_6]^-$

[74]

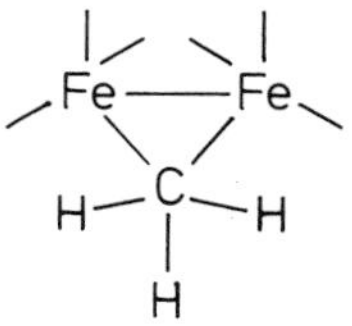

[75]

size of the downfield shift in the deuterium spectrum is approximately twice the size of the upfield shift in the proton spectrum. The lighter isotope is preferred in the bridging position. A structure [74] in which the methyl group is attached to one iron atom by a C—Fe σ-bond and with a η^2-C—H interaction to the other iron is in agreement with the nmr and X-ray data. In solution the fast exchange between the μ-hydrido hydrogen and the two other hydrogens at the bridging methyl group renders all three hydrogens equivalent.

A similar cationic iron complex [76] with a bridging methyl was investigated by Casey *et al.* (1982). The averaged structure of the μ-CH_3 group shows apparent symmetry in the ^1H and ^{13}C nmr spectra, but shift differences between μ-CH_3-[76] (−1.85 ppm) and μ-CH_2D-[76] (−2.62 to

−2.93 ppm between −4°C and −86°C) are observed in the proton spectrum indicating C—H—Fe bridging. The same rationalisation for this equilibrium isotope effect applies as for the complexes mentioned above.

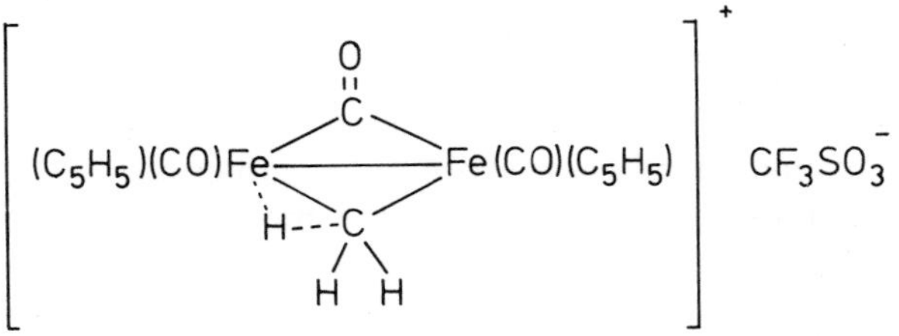

$[(C_5H_5)(CO)Fe(\mu\text{-}CO)(\mu\text{-}CH_3)Fe(CO)(C_5H_5)]^+ \; CF_3SO_3^-$

[76]

Iron alkenyl complexes

Several cationic η[3]-alkenyl-complexes obtained by protonation or deuteriation of tris(phosphite)iron η[4]-diene complexes have been studied by nmr spectroscopy (Ittel *et al.*, 1979). The structures are octahedral, coordinated with phosphite ligands at three of the sites. The η[3]-alkenyl group occupies two sites and a hydrogen atom which is adjacent to the alkenyl-group fills the sixth site giving a noble gas (18-electron) configuration at the iron centre.

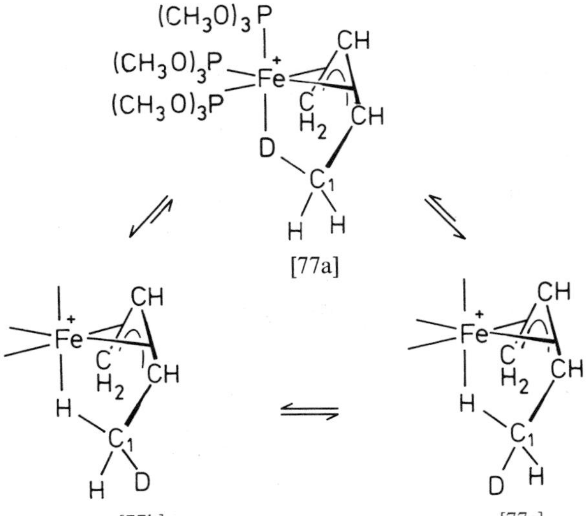

[77a] ⇌ [77b] ⇌ [77c] (64)

In the nonlabelled butenyl complex [77] at room temperature the three hydrogens at C-1 give a single averaged peak (−5.92 ppm) which can be frozen out into three separate resonances at $-90°C$. The signal for the hydrogen bridged to the iron appears at -15.2 ppm and the other two hydrogens at about -2.4 and -2.0 ppm. Under fast exchange conditions the mono-deuteriated complex [77] shows a greatly reduced peak at the original position (-5.92 ppm) arising from some molecules unlabelled at the bridging methyl position. These molecules result from scrambling of one deuterium with four protons among the five positions at the termini of the butenyl ligand *via* a hydrido-diene intermediate [78]. A new peak in the proton spectrum is found shifted more than 1 ppm upfield from the averaged position in the unlabelled complex and a corresponding peak is observed shifted downfield in the deuterium spectrum. These new peaks come from the molecules with the mono-deuteriated methyl group. The fast rotation (64) of the methyl group which interchanges the hydrogens between the bridging and nonbridging positions in [77] is perturbed by the isotope. The direction of the equilibrium isotope effect, i.e. a shift towards [77b] and [77c], shows that the bridging position has the lower zero point energy.

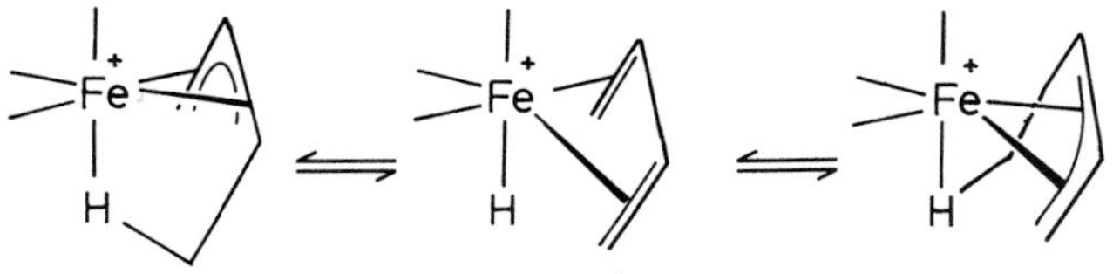

(65)

[78]

In the complexes [79] with cyclic diene ligands, another type of fast averaging process is observed. Only one of the two *endo* hydrogens at the termini of the π-system coordinates to the iron at a given instant. At room temperature a fast exchange of these two hydrogens takes place and gives an averaged signal at -6.5 ppm. For the cyclohexenyl- [79] ($x = 1$) and cycloheptenyl- [79] ($x = 2$) complexes decoalescence into two peaks was

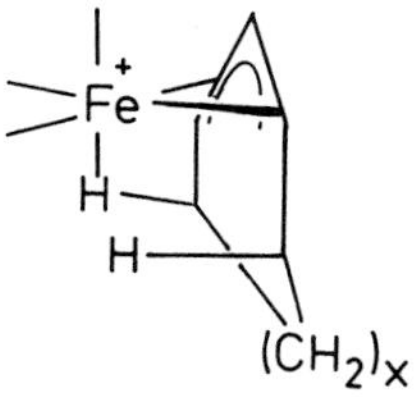

[79]

observed at lower temperatures. One signal appears at -15 ppm and is attributable to the bridged-hydrogen and another signal is found in the normal aliphatic region. The spectra of the deuteriated cyclic enyl complexes [79] indicate that deuterium is exclusively distributed between the *endo* positions and that the *exo* positions are never involved in hydrogen or deuterium transfer to or from the metal. Rotation of the methyl group and subsequent exchange of the termini of the enyl system via a hydrido-diene intermediate, which causes the scrambling in the acyclic enyl complexes, is not possible with the cyclic ligands.

The ^{13}C nmr spectrum of the cyclooctenyl complex [80-H] shows five resonances in the ratio of 1 : 2 : 2 : 2 : 1. If the fast exchange process (66), which renders the two *endo* hydrogen atoms adjacent to the π-allyl unit equivalent, could be frozen out, each of the peaks with the relative intensity 2 would be split into two. In the mono-deuteriated complex [80] the three averaged peaks C-1/C-5, C-2/C-4 and C-6/C-8 show isotopic perturbation.

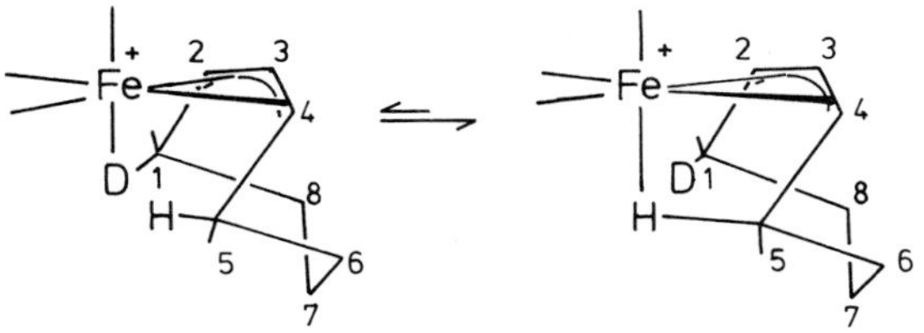

[80a] ⇌ [80b] (66)

Three signals for each averaged peak are observed. A centre peak at the same position as in the unlabelled complex results from the molecules with deuterium at positions where it cannot interact with the metal. This peak is surrounded by a set of two peaks from the molecules with a deuterium at one of either C-1 or C-5 at the termini of the enyl system. For [80a] and [80b] the equilibrium is no longer degenerate. As a result of an isotope effect on the distribution of hydrogen and deuterium between bridging and nonbridging positions, the averaged ^{13}C signals C-1, C-2, C-8 on the deuterium side of the ring are weighted in favour of their nonbridging chemical shifts and are shifted downfield. The averaged signals for C-5, C-4, C-6 on the hydrogen side of the ring are weighted in favour of their bridging chemical shifts and are shifted by about an equal amount upfield. The isotope splittings are largest for C-1 and C-5 which bear the interacting hydrogens, and the size of the splittings are very temperature-dependent.

In the ^{13}C nmr spectrum of the cycloheptenyl complex [79; $x = 2$] the fast exchange can be frozen out and seven separate resonances have been assigned. The small coupling observed for the bridging hydrogen ($J_{CH} = 80$ Hz) compared to the other hydrogen at the same carbon ($J_{CH} = 140$ Hz) shows that the bond to the bridging carbon is considerably weakened. These data confirm the interpretation of the fast exchange spectra for the deuteriated cyclooctenyl complex.

Iridium alkenyl complex

The structure and the fluxional behaviour of the cationic iridium–butenyl–hydride complex [81] has been investigated by nmr spectroscopy (Moore *et al.*, 1981) The allyl hydrogens and the terminal methyl group in the butenyl ligand give singlets at 2.88, 0.48 and −2.23 ppm in the ^{1}H nmr spectrum at −20°C. The shift of the terminal methyl groups indicates C—H—Ir interaction. Rapid methyl rotation averages the bridging and the terminal hydrogens. Line broadening of the methyl peak but no decoalescence was observed at −120°C. In the deuteriated complex, three peaks for the

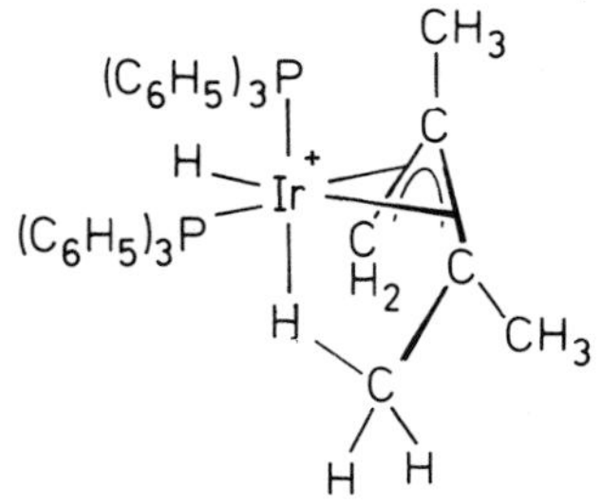

[81]

terminal methyl group were observed at −20°C in the proton spectrum; −2.23 ppm (CH_3), −2.54 ppm (CH_2D) and −2.91 ppm (CHD_2). The upfield shift of the proton signals in the deuteriated methyl group are complemented by a corresponding downfield shift in the ^{2}H nmr spectrum. This equilibrium isotope effect on the averaged proton and deuterium signals of the rapidly rotating terminal methyl group can be explained analogously as the result of zero point energy differences favouring hydrogen rather than deuterium at the bridging position.

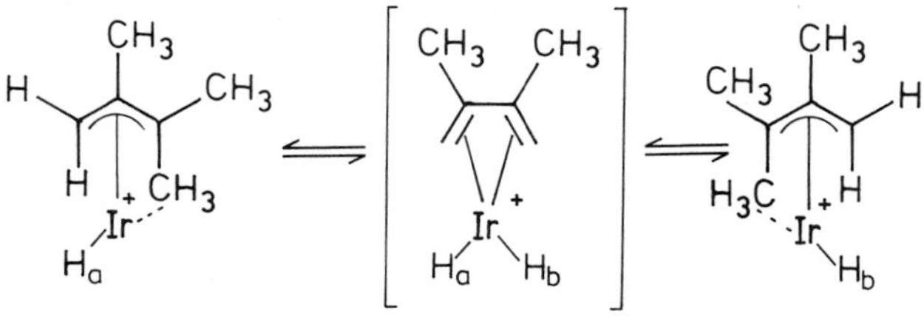

(67)

[82]

Another fluxional process (67) is indicated by dynamic nmr and magnetisation transfer experiments. H- or D-atoms at the terminal carbons are interchanged *via* a diene-dihydride intermediate [82]. The central methyl groups are not involved. The rapid rotation of the terminal methyl group allows transfer of either hydrogen from the diene-dihydride intermediate to the terminal carbons, thus explaining incorporation of two deuteriums at the same carbon and the occurrence of the CHD_2-signal. This process is slow up to +25°C when the proton signals of the inner methyl groups and the two phosphorus signals also coalesce (20°C) to give the same but time-averaged symmetry as [82]. Other rearrangements involving the diene-dihydride intermediate could not be ruled out.

Manganese alkenyl complexes

The C—H—Mn interaction in η^3-cyclohexenyl-manganese tricarbonyl was investigated by X-ray and ^{1}H nmr spectroscopy (Brookhart *et al.*, 1982). Alternative metal-coordination of the two *endo* protons adjacent to the π-allyl unit is a fast dynamic process at 15°C, which averages the bridging proton (−12.8 ppm at −99°C) and the terminal *endo* proton (1.4 ppm at −99°C) to give a single peak at −5.8 ppm. The deuteriated complex [83] shows one peak at the same position as the averaged signal in the unlabelled complex. This resonance results from the rapid exchange of the *endo* hydrogens at the methylene groups adjacent to the allyl moiety in those molecules which contain only hydrogens in the *endo* positions. Another signal is found upfield (−7.3 ppm to −6.9 ppm between −41°C and +44°C). This resonance arises from those molecules with one hydrogen and one deuterium at the exchanging *endo* positions. The chemical shift of this peak is determined by the weighted average shifts of the terminal and bridging protons and is a function of the isotope effect on equilibrium (68).

$(CO)_3Mn$ … H, H(D), D $\rightleftharpoons$ $(CO)_3Mn$ … H, H(D), D (68)

[83a] [83b]

The upfield shift compared to the unlabelled species is rationalised in terms of relative differences in the zero point energy of the C—H and C—D bonds at the two different sites which shift the equilibrium in favour of the ^{1}H-bridged form [83b].

The equilibrium constant has been calculated from the temperature dependence of the chemical shift of the upfield peak and varies from $K = 1.54$ to $K = 1.35$ between −41°C and +44°C. The thermodynamic parameters were determined as $\Delta H° = -220 \text{ cal mol}^{-1}$ and

$\Delta S^{\circ} = -0.07\ \mathrm{cal\ mol^{-1}\ K^{-1}}$. The small value for ΔS° shows that the isotopic site preference in the labelled complex is determined almost exclusively by ΔH°.

Two other dynamic processes with higher energy barriers were also observed in this complex: 1,2-metal migration (69) *via* a diene-hydride intermediate [84] permits a degenerate metal migration around the ring; rotation of the allyl unit at higher temperatures leads to scrambling of the carbonyl ligands.

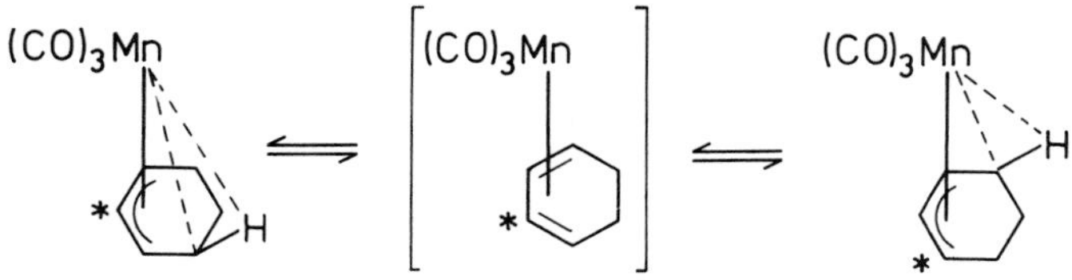

$(CO)_3Mn$ ⇌ [$(CO)_3Mn$] ⇌ $(CO)_3Mn$ (69)

[84]

The interpretations of all data in the various transition metal complexes are in agreement concerning the hypercoordination of hydrogen in these molecules. Slightly different interpretations have been presented to describe the exact nature of the two-electron-three-centre bond in the iron and manganese alkenyl complexes. It is clear from a neutron diffraction and X-ray study of [80] (Williams *et al.*, 1980) that the Fe—H—C bond is not collinear but has a triangular arrangement. This was conceived to be a fortuitous result of the structural constraints dictated by the metal-hydrogen bonding and the internal ring-bonding of the alicyclic ligand, and an open three-centre Fe—H—C bond [85] was suggested. For the manganese complexes the closed three-centre bond description [86], in which substantial bonding interaction exists between the metal and both the carbon and the attached hydrogen, was preferred.

C···H···M [85] C···H···M, C---M [86]

HYDRIDO-BRIDGED CARBOCATIONS

Medium ring compounds commonly show enhanced solvolysis rates and solvolysis products derived from transannular hydride transfer. Ionisation in these systems may take place with some degree of transannular C—H bond participation, the extent of bridging in the transition state and competitive nucleophilic solvent assistance (k_s-process) being broadly and independently variable.

Sorensen and coworkers (Sorensen and Kirchen, 1978, 1979; Sorensen *et al.*, 1978, 1981a,b,c) have studied medium sized cycloalkyl cations under conditions where their lifetime is long. For more than ten differently substituted C-8 to C-11 cycloalkyl cations, compelling nmr spectroscopic evidence for 1,5- and 1,6-μ-hydrido-bridged structures [87] for these cations has been obtained. Nmr shift and coupling constant data have been reviewed (Ahlberg *et al.*, 1983a). Only isotope effect studies are discussed here and compared with isotope effect data of analogous alicyclic carbocations.

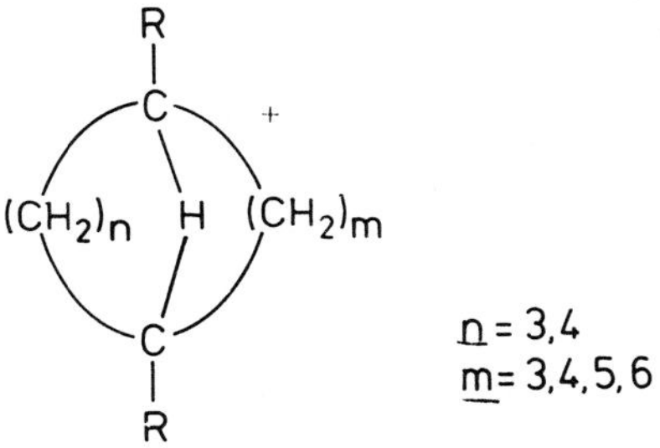

R
C +
$(CH_2)_n$ H $(CH_2)_m$
C
R
[87]

n = 3,4
m = 3,4,5,6

Evidence for a single energy minimum

The μ-hydrido bridged cation [90] is equivalent to a resonance system such as exists in allyl cations (single energy minimum) and can be distinguished from a set of two degenerate cycloalkyl cations [88] undergoing a fast hydride shift (double energy minimum) as shown in (70) by comparison with suitable model cations.

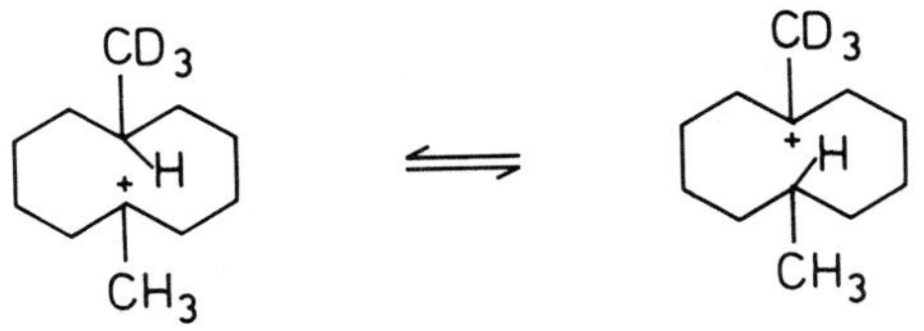

CD_3 + H CH_3 ⇌ CD_3 + H CH_3 (70)

[88]

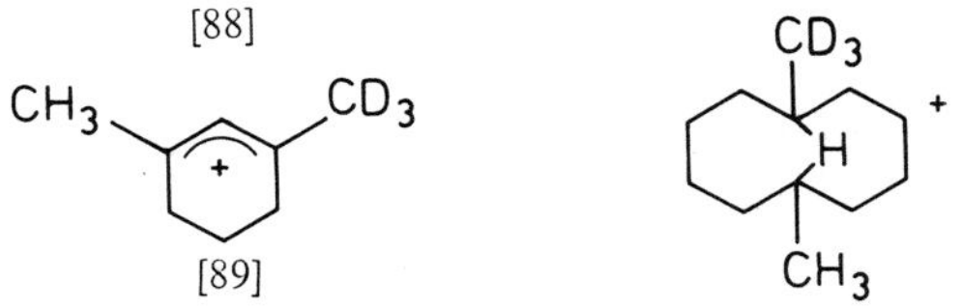

CH_3 + CD_3

[89]

CD_3 H CH_3 +

[90]

The 1,3-dimethylcyclohexenyl cation serves as a model for a symmetrical resonance system. When one methyl group is replaced by a CD_3-group in [89], the symmetry of the cation is perturbed and the averaged allylic carbons show a small splitting of 0.3 ppm with no measurable temperature dependence (Sorensen *et al.*, 1981a). In the 1-methyl-6-trideuteriomethylcyclodecyl cation [90] a splitting of similar size, 0.6 ppm at −110°C, is observed for the carbon signal at 142 ppm which is caused by the C-1 and C-6 carbons (Sorensen *et al.*, 1981a). No measurable temperature dependence was observed in accord with a single minimum resonance structure. 1-CH_3-5-CD_3-Cyclononyl cation [91] (δ = 1.5 ppm) (Sorensen *et al.*, 1981c) and 1-CH_3-5-CD_3-cyclodecyl cation [92] (Sorensen *et al.*, 1981a) (three different conformers: δ = 2.75; 1.1; 1.1 ppm) also show no temperature dependence of the isotope splittings although the size of the splitting is significantly larger than in the allyl cation model.

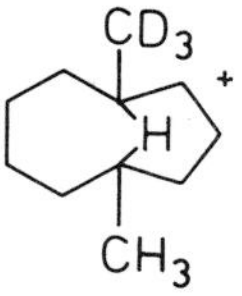

CD_3 H CH_3 [91] CD_3 H CH_3 [92]

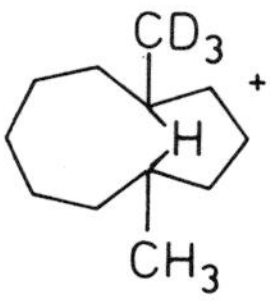

The small temperature dependence of the isotope effect splittings observed in the 1-CH_3-5-CD_3-cyclooctyl cation [93] (δ = 1.43–2.7 ppm between −127°C and −75°C) is in the opposite sense compared to equilibrating cations and thus cannot be attributed to an equilibrium isotope effect for two cations undergoing a fast degenerate 1,5-hydride shift. Instead it has been suggested to be caused most probably by conformational interchange processes (71) in μ-hydrido-bridged cations [93a]–[93c] (Sorensen *et al.*, 1981c).

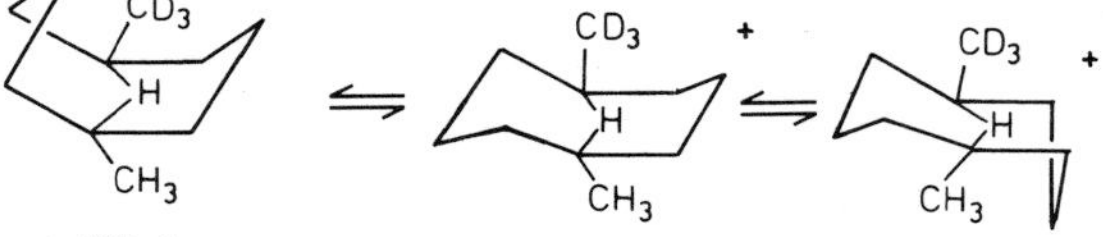

CD_3 H CH_3 [93a] ⇋ CD_3 H CH_3 [93b] ⇋ CD_3 H CH_3 [93c] (71)

The isotopic perturbation technique has also been used to confirm the 1,6-μ-hydrido-bridged structure of the bicyclo[4.4.4]tetradecyl cation [94] which was generated by protonation of *in*-bicyclo[4.4.4]-1-tetradecene (McMurry and Hodge, 1984). When deuteriated fluorosulfuric acid was used the peak at 136.7 ppm for the bridgehead C-1 and C-6 carbons showed only a small isotope splitting of 0.8 ppm. This excludes a pair of rapidly equilibrating degenerate secondary cations for which much larger splittings would be expected.

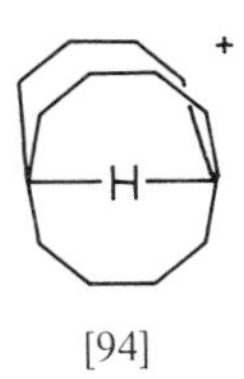

[94]

The isotope effects observed in cycloalkyl cations may be compared with the isotope effect in analogous acyclic cations. The 2,6-dimethylheptyl cation [95] can be regarded as an open chain counterpart of 1,5-dimethyl-substituted cycloalkyl cations [91]–[93]. Cation [95] undergoes a fast degenerate 1,5-hydride shift (72) with a low energy barrier ($E_a = 4.9\,\text{kcal mol}^{-1}$), reflecting the low energy of the six-membered ring hydrido-bridged transition state [96] (Saunders and Siehl, 1985). In cation [95] which was deuteriated unsymmetrically in the methylene or methyl positions large equilibrium isotope splittings were observed using ^{1}H and ^{13}C nmr spectroscopy. The equilibrium constants decrease with temperature in the normal direction expected for an equilibrium isotope effect.

(72)

[95a] [96] [95b]

Comparison of the equilibrium isotope effects in 2-CD_3-[95] with the small isotope effects in 1-CD_3-5-methylcycloalkyl cations [91]–[93] shows the difference between these two types of cation structure. In the high field ^{13}C spectrum of [95] the averaged signal for the C^+/C-6-methine carbons is merged into the baseline due to extensive kinetic line-broadening. The expected isotope splitting ($\delta = 78$ ppm, at $-120°$C) for these carbons can be calculated from the equilibrium constant $K(CD_3) = 1.69$ ($-120°$C). The large effect in [95] clearly indicates a double minimum energy surface for this carbocation and strongly supports the static, symmetrical single energy minimum μ-hydrido-bridged structure for the cyclodecyl cations [90]–[94].

Structural changes of the acyclic cation [95], such as introduction of substituents could in principle lower the barrier, perhaps even enough to make the μ-hydrido-bridged cation [96] the minimum energy structure. It has been shown, however, that the energy barrier ($\Delta G^{\ddagger} = 5.2\,\text{kcal mol}^{-1}$ at $-130°$C) in the more substituted 2,4,4,6-tetramethylheptyl cation [97] is very similar to the barrier in the less substituted cation [95] (Siehl and Walter, 1985).

[97a] ⇌ [97b] (73)

C—H bond force constants and structure

In the 1,5-μ-hydrido-1-alkylcyclooctyl cations [98] (R = t-butyl, isopropyl, ethyl or methyl) a fast exchange is observed between the terminal hydrogen (H_T) at C-5 and the bridging hydrogen (H_μ). When the cation is generated from a precursor which is monodeuteriated at the C-5 position, two isotopic isomers with deuterium at the terminal or at the bridging position interconvert rapidly, and a primary isotope effect on this equilibrium (74) is observed (Sorensen *et al.*, 1981b). For cation [98] (R = t-butyl) the equilibrium constant was measured under conditions of slow exchange by direct integration of the signals for the terminal and the bridging hydrogen. The intensity ratios at $-130°C$ are $H_T : H_\mu = 1 : 4.5$ and $D_T : D_\mu = 4.35 : 1$ in the ^1H and ^2H spectra respectively. This large primary isotope effect shows that cation [98b] with deuterium in the terminal position and hydrogen in the bridging position is strongly favoured in the equilibrium.

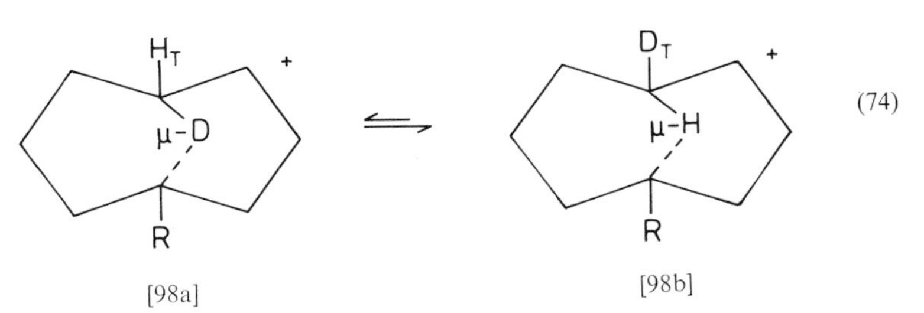

[98a] [98b]

The equilibrium constants for the other cations were obtained under conditions of fast exchange from the population-dependent averaged chemical shift of the H_T-H_μ peak in the deuteriated cation, using the equation $K = (\Delta + 2\delta)/(\Delta - 2\delta)$, where Δ is the shift difference between H_T and H_μ in the static cations and δ the equilibrium isotope shift difference between the averaged proton shift in the unlabelled and the deuteriated cations. The equilibrium constants for the substituted cyclooctyl cations [98] at $-70°C$ are $K = 3.0$; (R = t-butyl); 1.40 (R = isopropyl); 1.20 (R = ethyl); and 1.15 (R = methyl). The observation of an equilibrium isotope effect between the two hydrogens at C-5 and particularly the dependence of this effect on the C-1 substituent is direct evidence for the μ-hydrido-C-1 bonding interaction.

The direction of the isotope effect shows that the C-5-H_μ bond has lower vibrational force constants, i.e. is less stiff, than the C-5-H_T bond. The electron deficiency of the bridging bond (two electrons connecting the three centres C-5, μ-H, C-1) can give only fractional bonds to the hypercoordinated hydrogen. The decrease of the equilibrium isotope effect in [98] with smaller substituents at the tertiary carbon C-1 indicates a gradual loss of transannular bonding interaction. The isotope effect approaches unity in the 1-methylcyclooctyl cation.

In close analogy to unsymmetrical substituted allyl cations (unsymmetrical, three-centre, two-electron π-bonding) unsymmetrical substituted (tertiary/secondary) μ-hydrido bridged cycloalkyl cations [98] (unequal three-centre, two electron σ-bonding) have a single but unsymmetrical potential energy minimum which may gradually change to an unsymmetrical double energy minimum depending on the substituent at the cationic centre.

CARBON HYPERCOORDINATED CARBOCATIONS

2-Norbornyl cation

The protracted and controversial discussions on the structure and dynamics of the 2-norbornyl cation have been recently summarised (Walling, 1983; Olah *et al.*, 1983; Brown, 1983; Grob, 1983). The question has been whether the 2-norbornyl cation has a symmetrically bridged hypercoordinated structure [99] with a pentacoordinated carbon atom or whether it is a rapidly equilibrating pair of trivalent cations [100]. In the context of this review only the evidence from isotope effects on the nmr-spectra of the 2-norbornyl and related cations is discussed.

H H ⇌ 7 4 5 H 6 H 3 + 1 2 [99] ⇌ H H (75)

The ^{1}H and ^{13}C nmr spectra of the 2-norbornyl cation are complicated by the fact that the 3,2-hydride shift and the 6,2,(1)-hydride shift lead to averaged carbon and proton signals. At $-80°C$ the 3,2-hydride shift is frozen out and three signals for the averaged C-1, C-2, C-6 carbons, the C-4 carbon and the averaged C-3, C-5, C-7 carbons are observed. Cooling the solution to $-159°C$ leaves the peak for C-4 unchanged but the averaged peak for C-1, C-2, C-6 decoalesces into one peak for C-1, C-2 and another for C-6 and also the averaged peak for C-3, C-5, C-7 splits into one peak for C-3, C-7 and another peak for C-5. Since the 6,2,(1)-hydride shift has a barrier of only 5.9 kcal mol^{-1}, a certain amount of line-broadening (2–2.5 ppm) caused by slow exchange is observed in the ^{13}C spectrum even at $-159°C$.

[100a] W.M. [100b]

6,2-H (76)

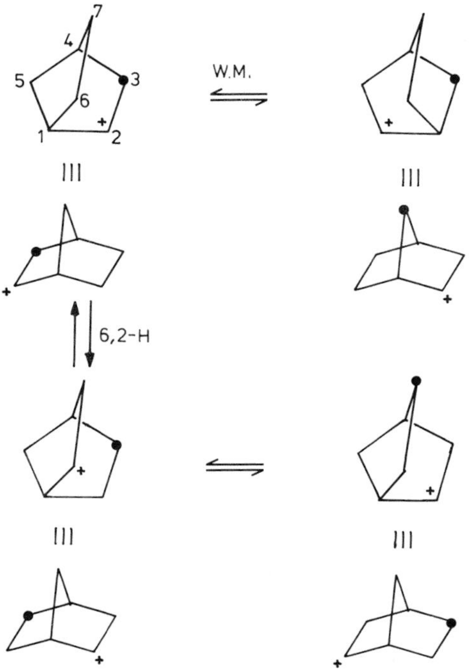

[101a] [101b]

Evidence for a single energy minimum. Saunders and Kates (1980) have measured deuterium isotope effects on the ^{13}C nmr spectra of deuteriated 2-norbornyl cations. No additional isotope splitting or broadening was observed in the spectra of either 2-D_1- or 3,3-D_2-norbornyl cations. The upper limit for an isotope splitting thus cannot be more than 2–2.5 ppm. This result is in marked contrast to the large splittings observed in the model cations [104] and [105].

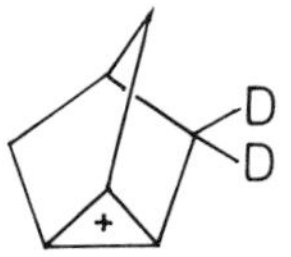

[102]

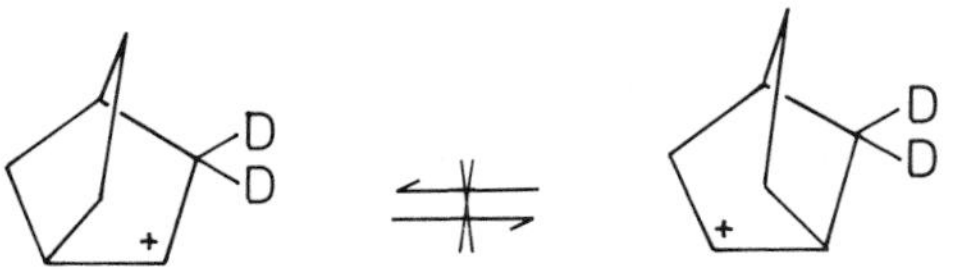

[103]

(77)

[104]

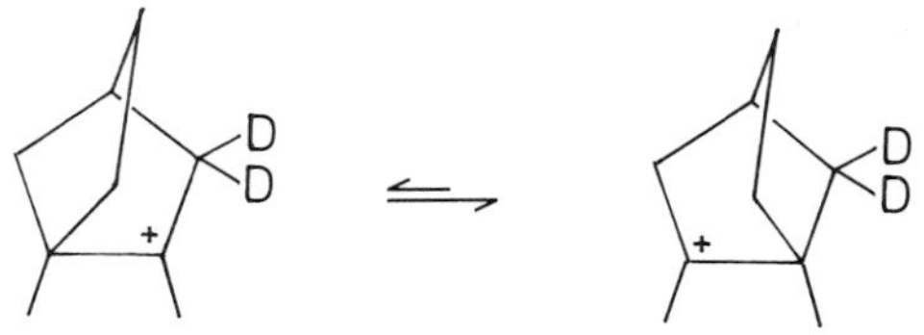

(78)

[105]

The 1,2-dimethylcyclopentyl cation is known to be a rapidly equilibrating cation (77) at all accessible temperatures in solution. The averaged peak for the C^+/CH carbons at 203 ppm gives two peaks at 329 ppm and 75 ppm for the C^+ and CH carbon in the solid state nmr spectrum at $-145°C$ indicating a static trivalent cation (Myhre *et al.*, 1984). In the 3,3-dideuteriated cation [104] (Saunders *et al.*, 1977a), large temperature-dependent isotope splittings $\delta = 105$ ppm at $-130°C$ were observed for the averaged C-1/C-2 signal.

The 1,2-dimethylnorbornyl cation is a structurally even more closely related cation. The fast Wagner-Meerwein shift that averages C-1 and C-2 at 168 ppm can be slowed down in the solid state and at $-196°C$ a near limiting spectrum of the static cation was obtained (Yannoni *et al.*, 1985) which showed two peaks for the C^+-carbon (250 ppm) and the adjacent C-2 quaternary carbon (93 ppm). The equilibrium isotope splittings observed for the averaged C-1/C-2 carbons ($\delta = 24$ ppm at $-110°C$) in the 3,3-D_2-1,2-dimethyl-2-norbornyl cation [105] (Saunders *et al.*, 1977a) are significantly smaller than in the dimethylcyclopentyl cation [104] but are still an order of magnitude larger than any maximum isotope effect estimated from the linewidth of the C-1/C-2 peak in the parent 2-norbornyl cation. The smaller shift difference in 1,2-dimethylnorbornyl cation and the smaller equilibrium isotope splittings as compared to the dimethylcyclopentyl cation support the current view that the 1,2-dimethylnorbornyl cation is a partially σ-delocalised cation (Saunders *et al.*, 1977b) equilibrating over a very low barrier (Sorensen *et al.*, 1975).

The postulated Wagner-Meerwein rearrangement in the 2-norbornyl cation [100a]/[100b] is not consistent with the results obtained. For a set of rapidly equilibrating secondary cations one should observe large isotope splittings for the isotopically labelled cation [103] resulting from perturbation of the equilibrium by a differential β- *vs* γ-equilibrium isotope effect.

This is true even if a slow 6,2-hydride shift in secondary cations converts part of [103] (i.e. 3,3-D_2-[100a,b]) into 5,5-D_2-[101]. Cation 5,5-D_2-[101] would lack an equilibrium isotope effect because of the symmetry of an assumed Wagner-Meerwein rearrangement between D_2-[101a] and D_2-[101b]. This does not vitiate the validity of the isotopic perturbation test, which shows that a double energy minimum of secondary cations [103] is not present. A static symmetrical hypercoordinated structure [102] is in accord with the observed results. If the 6,2,1-hydride shift in [102] was not frozen out on the timescale of the experiment, the ^{13}C spectrum of [102] measured below $-150°C$ is still conclusive, although it results from an equal mixture of isomers of [102] labelled at C-3, C-5 and C-7.

It has been suggested, however, that the ^{13}C nmr data on deuteriated 2-norbornyl cations do not rule out an equilibrium between two unsymetrically hypercoordinated cations [106] which are separated by a very small barrier and might have reduced isotope effects (Walling, 1983). Solid state

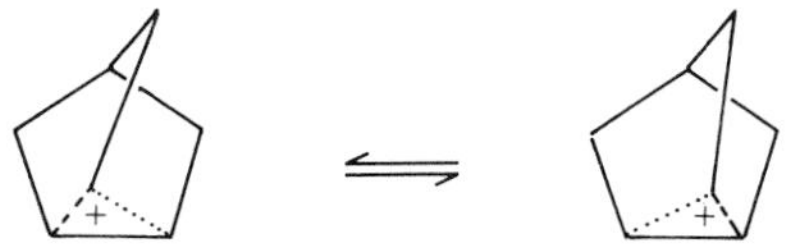

[106a] ⇌ [106b] (79)

nmr studies at temperatures as low as 5 K restrict the activation energy for an assumed equilibrium to no greater than 0.2 kcal mol^{-1} (Yannoni *et al.*, 1982).

C—H bond force constants and structure. At temperatures above $-130°C$, the ^{1}H nmr spectrum of the 2-norbornyl cation has only three peaks. A rapid process averages the environments of four protons at C-1, C-2 and C-6 and also the six protons at C-3, C-5 and C-7 and leaves the proton at C-4 unaffected. In the hypercoordinated structure [99] this process is a corner-to-corner proton shift between the C-1, C-2 and C-6 positions of the cyclopropyl ring. The exchange process takes place between one doubly populated site (the protonated corner of the cyclopropane) and two equivalent singly populated sites (the two other corners of the cyclopropane). The isotopic perturbation of this rapid equilibrium has been investigated by Saunders and Kates (1983) and by Sorensen and Ranganayakulu (1984). If the 6,2,1-shift is stereospecific and, if one of the four protons at the C-1,2,6 positions in [99] is substituted by deuterium as in [108], only four possibilities, two with deuterium at the corner, [107b] and [107c], and two with deuterium at the base of the protonated cyclopropane, [107a] and [107d], are accessible on the nmr timescale (Sorensen and Ranganayakulu, 1984). Two of the remaining three protons (A, B, C) will never get to share the same site;

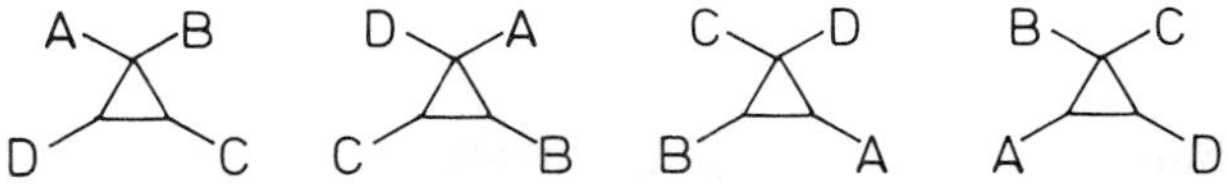

[107a] [107b] [107c] [107d]

thus, independent of any isotope effect on the relative populations of corner-deuteriated and corner-protonated cyclopropane structures, two protons (A and C) will always average at the position identical with that in the unperturbed "normal" position. The third proton (B) has the same relationship with one deuterium (D). Proton B and the deuterium D are averaged between two different sites and therefore the averaged chemical shift of this proton and deuterium are influenced by an equilibrium isotope effect on the 6,2,1-hydride shift in [108a–c].

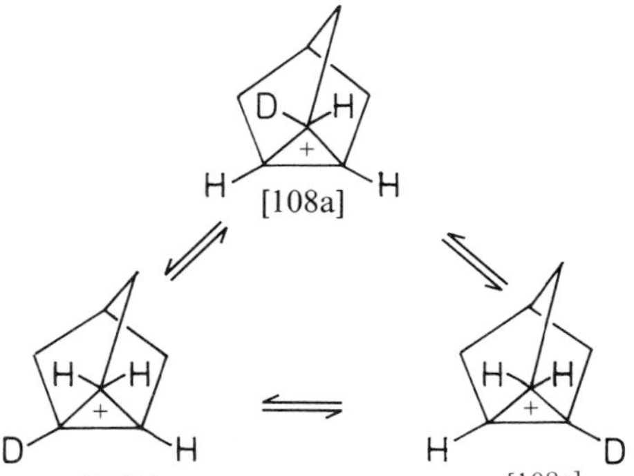

[108a] [108b] [108c] (80)

One cannot fix isotopic labels in the 2-norbornyl cation because at temperatures above −130°C any hydrogen or deuterium is equilibrated among all 11 positions in a few minutes due to the slow (at these temperatures) 3,2-hydride shift. The deuteriated cation will be a mixture of the two types of labelled cations. Neglecting an isotope effect on the 2,3-hydride shift, 7/11 of the cations will have one deuterium located at C-3,5,7 or C-4. In these cations the 6,2,1-shift is not perturbed and a peak at the normal unperturbed position is observed. The remainder (4/11) of the cations will have one deuterium located at one or other of the C-1,2,6 positions.

The intensity of the unperturbed and perturbed proton peaks in a monodeuteriated norbornyl cation is predicted to be $7/11 \times 4\,H + 2/3(4/11 \times 3\,H)$ for the "normal" peak and $1/3(4/11 \times 3\,H) = 9:1$ for the perturbed peak. The factor 1/3 arises from the fact that only one out of three protons at the C-1, C-2 and C-6 positions is perturbed by deuterium. The experimentally observed intensity ratio of the "normal" peak to the perturbed upfield peak is $10 \pm 0.5:1$. The slightly higher ratio than predicted indicates a preference for deuterium at positions 3,4,5 and 7 over the 1,2 and 6 positions. This is the same direction as found in other systems where an equilibrium exists between H(D) bonded to a cyclopropyl carbon or to an sp^3 hybridised carbon position (Schleyer *et al.*, 1971; Kalinowski, 1984).

The upfield shift of the peak of the perturbed protons is temperature-dependent and varies between 0.215 ppm at $-105°C$ and 0.146 at $-43°C$ (Saunders and Kates, 1983). The equilibrium constant for the stereospecific hydride shift ($K = 1.28$–1.18 between $-105°C$ and $-43°C$) can be calculated using the modified equation $K = (\Delta + 2\delta)/(\Delta - 2\delta)$, (Sorensen and Ranganayakulu, 1984). The same equilibrium constants within experimental error were obtained for the complementary isotope effect on the proton chemical shift in the decadeuteriated cation ($C_7D_{10}H^+$) and for the isotope effect on the deuterium chemical shift in the monodeuteriated cation ($C_7H_{10}D^+$).

The significance of these smaller values of the equilibrium constant for the interpretation of the structure of the 2-norbornyl cation has been questioned by Sorensen and Ranganayakulu (1984). It is, however, not so much the size but the direction of the equilibrium isotope effect which is the significant probe of the structure of the cation. The size of the isotope effect depends on absolute magnitudes of force constants for the bending and stretching frequencies at the exchanging sites. The magnitude of these force constants is unknown in many cases and difficult to predict, especially when no data from adequate model compounds are available. The direction of the isotope effect is often the more useful diagnostic tool because it depends only on the relative differences of the force constants at the exchanging sites.

The upfield shift of the perturbed proton peak (or the downfield shift of the perturbed peak in the deuterium spectrum) has been rationalised by Saunders and Kates (1983).

In the ^1H nmr spectrum of [99] at $-158°C$, the protons at the C-6 carbon were assigned to a peak at higher field. The observed upfield (downfield) shift for the perturbed peak in the proton (deuterium) spectrum of the D_1-2-norbornyl cation [108] shows that protons are preferred over deuterium at C-6 and thus [108b] and [108c] are favoured in the equilibria (80). The C—H(D) bonds at C-6 must therefore have lower force constants for bending and stretching vibrations and in turn lower zero point energy than the H(D) bonds to C-1 and C-2. The lower force constants for the C—H bonds at C-6 are a consequence of the pentacoordination of that carbon. There are not enough electrons at C-6 to make five full single bonds; the bonds are therefore expected to be longer and looser.

Below $-140°C$ the shifts of the protons at C-6 or at C-1 and C-2 are virtually identical in the monodeuteriated and in the unlabelled cations [108] and [99]. This shows that the fast 6,2,1-hydride shift which is perturbed by deuterium in [108] is frozen out and consequently the isotope effect disappears. This indicates also that there can be no equilibrium between the postulated unsymmetrical bridged cations [106a] and [106b]. If present such an equilibrium should show isotopic perturbation at these low temperatures.

The postulated fast Wagner-Meerwein σ-bond shift (79) in unsymmetrically bridged species [106] would average H-6 *exo* and H-6 *endo* and also H-1 and H-2. Cation [106] would have different bonding situations at C-1 and C-2, which means also different force constants for the C—H bonds at C-1 and C-2. As long as there are force constant differences between exchanging sites, the isotopic equilibrium constant will deviate from unity and equilibrium isotope splittings in the nmr spectra should be visible if there are shift differences between the exchanging sites. It can be anticipated from comparison with the 2-methylnorbornyl cation that the shift differences in an unsymmetrically bridged 2-norbornyl cation [106] are substantial (Sorensen and Ranganayakulu, 1984). Assuming no shift difference is equivalent to saying that the species is symmetrical and has a single minimum energy surface.

When the central barrier in a double energy minimum decreases below the lowest zero point vibrational level, the situation is equivalent to a true single minimum case (Forsén *et al.*, 1978). A carbon tunnelling mechanism suggested to explain the dynamic behaviour in 2-norbornyl cations, (Fong, 1974; Dewar and Merz, 1986) has been shown to be unimportant (Yannoni *et al.*, 1985).

All results obtained on equilibrium isotope effects are consistent and support conclusively that the 2-norbornyl cation has a single energy minimum and the symmetrical protonated cyclopropane structure [99].

2-Methylnorbornyl cation

Equilibrium isotope effects in 2-methyl-2-norbornyl cation have been investigated and it was inferred from the direction and the small size of temperature-dependent isotope effects on the carbon chemical shifts of 2-CD_3-2-norbornyl cation (Lloyd, 1978) that the carbon spectrum is actually an averaged spectrum of three rapidly equilibrating isomeric cations. One tertiary cation [109] and two hypercoordinated cations [110] and [111] were estimated to have about the same energy and to be present in a ratio of about 70 : 25 : 5 at $-115°C$. The equilibrium is nondegenerate and K_H is very large in favour of the tertiary cation [109]; only small isotope effects are therefore observed.

The deuterium isotope effect on the proton and deuterium chemical shift in methyl and/or ring-deuteriated 2-methyl-2-norbornyl cations have also

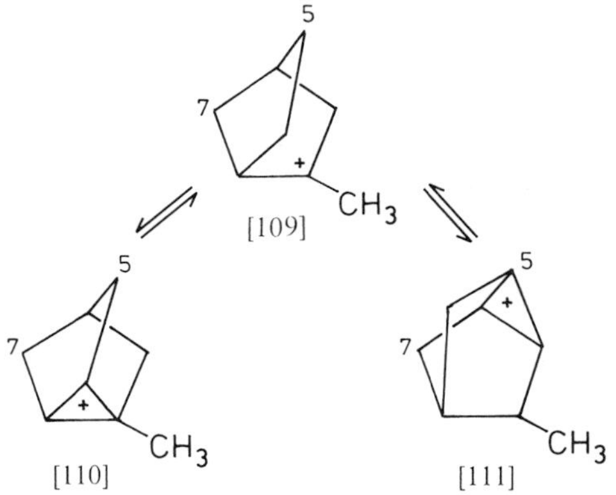

(81)

been investigated (Sorensen and Ranganayakulu, 1984). The largest equilibrium isotope effect was observed between the H-5 and H-7 protons. For one H-5/H-7 pair $K = 1.3$ ($-5°C$) and for the corresponding D-5/D-7 pair $K =$ *ca* 1.2 ($-20°C$). These isotope effects could not be explained in terms of structural or electronic reasons if a degenerate rearrangement mechanism were assumed.

In fact, if the equilibration occurred only by a consecutive Wagner-Meerwein shift/6,2-hydride shift/reverse Wagner-Meerwein shift (82) between [109] and [112] (Sorensen *et al.*, 1972), the C-5,C-7 positions would be symmetrical in the degenerate cations. Deuterium at either the C-5 or C-7 position should lack an equilibrium isotope effect in the equilibration process.

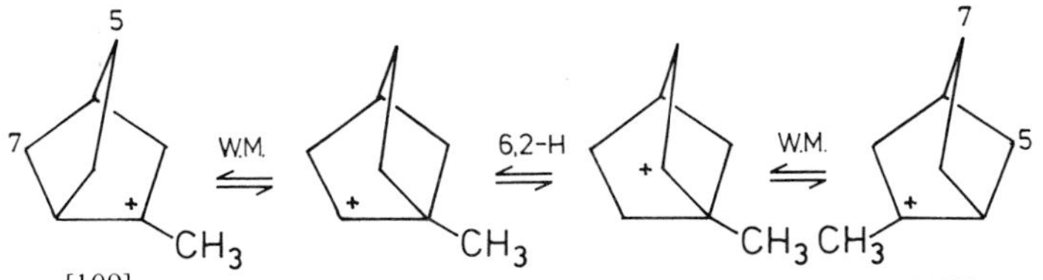

(82)

In the nondegenerate equilibration scheme (81), the cation [111], with C-5 pentacoordinated, is accessible *via* a 2,3-hydride shift. The C—H bonds of a pentacoordinated C-5 have lower vibrational force constants and, in an equilibrium where H-5 and H-7 are interchanged, protons are predicted to be favoured at the H-5 position. This cannot be verified from the data available because the direction of the isotope effect between H-7/H-5 has not been reported. Absolute assignment of the separate H-5 and H-7 protons has not been achieved so far but might be possible considering the direction of the isotope shift.

2-Bicyclo[2.1.1]hexyl cations

The 2-bicyclo[2.1.1]hexyl cation [113] or [114] shows only three carbon signals, one for C-4 and two averaged signals for C-1/C-2 and C-3/C-5/C-6 respectively. The fast dynamic process leading to the observed symmetry could be either a Wagner-Meerwein shift between trivalent cations (83) or a three-fold degenerate rearrangement of hypercoordinated cations [114] as shown in (84).

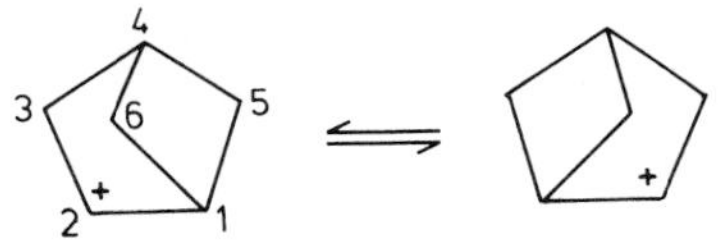

(83)

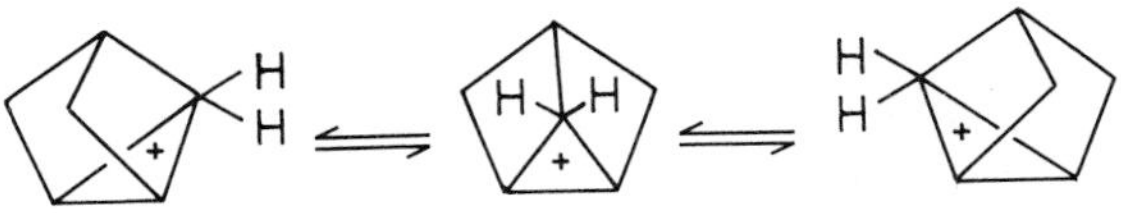

[113]

(84)

[114]

An unsymmetrical bridged structure (C1–C6 > C2–C6) which undergoes a very rapid C2–C6 → C2–C5 bridging flip and also a degenerate Wagner-Meerwein shift, whereby the C1–C6 and C2–C6 bonds interchange, has also been suggested (Sorensen and Schmitz, 1980). In the ^{13}C nmr spectrum of 2-deuterio-2-bicyclo[2.1.1]hexyl cation, the averaged C-1/C-2 carbons give two peaks separated by 1.63 ppm. The triplet for the deuteriated carbon C-2 is shifted upfield from the singlet peak of C-1. In a mixture with nonlabelled ion, this singlet is found 0.49 ppm downfield from the averaged C-1/C-2 peak (Saunders *et al.*, 1977b). The small relative size of the splitting is strong evidence that no isotope effect on an equilibrium of trivalent bicyclohexyl cations is being observed but that the structure is extensively σ-delocalised.

The proton spectrum (100 MHz, −125°C) of the monodeuteriated cation is in accord with this conclusion. The same spectrum was observed as for the unlabelled cation except for the reduced intensity of the peak for the averaged hydrogens on C-1/C-2 (Saunders *et al.*, 1973). In the ^{13}C nmr spectrum of the cation with one dideuteriomethylene group, the undeuteriated methylene carbon peak was found shifted 0.94 ppm downfield from the methylene peak in the nonlabelled cation. This shift must be caused by an equilibrium isotope effect for either one of the proposed structures [113] or [114] but the direction of this equilibrium isotope effect could not be correlated to the structures because the shift difference Δ of the averaged methylene groups is not accessible from nmr spectra in solution.

The 1-methyl-2-trideuteriomethyl-2-bicyclo[2.1.1]hexyl cation [116] has a definite equilibrium isotope effect. The averaged C-1 and C-2 carbons show an isotope splitting of 46.5 ppm at $-128°C$. The carbon bonded to the CD_3-group is shifted to higher field. The enthalpy difference ($\Delta H° = -50\ \mathrm{cal\ mol^{-1}}$ per D) for the isotopic equilibrium was calculated from the temperature dependence of the equilibrium constant between $-147°C$ and $-102°C$ using a static shift difference of $\Delta = 225$ ppm (Sorensen and Schmitz, 1980).

A comparison of the equilibrium isotope effects in the trivalent cation [115] [$\Delta = 261$ ppm, $K(CD_3) = 1.77$ at $-126°C$, $\Delta H° = -60\ \mathrm{cal\ mol^{-1}}$ per D] on the one hand with the partially σ-delocalised cation [117] [$\Delta = 202$ ppm, $K(CD_3) = 1.23$ at $-127°C$, $\Delta H° = -20\ \mathrm{cal\ mol^{-1}}$ per D; using $\Delta = 157$ ppm (Yannoni *et al.*, 1985), one obtains $K(CD_3) = 1.30$ at $-127°C$; $\Delta H° = -26\ \mathrm{cal\ mol^{-1}}$ per D] on the other shows that the dimethylbicyclo[2.1.1]hexyl cation falls in the intermediate range. The decreasing size of the CD_3-isotope effect on (85)–(87) indicates a decreasing demand for C—H hyperconjugative stabilisation due to increasing σ-stabilisation in

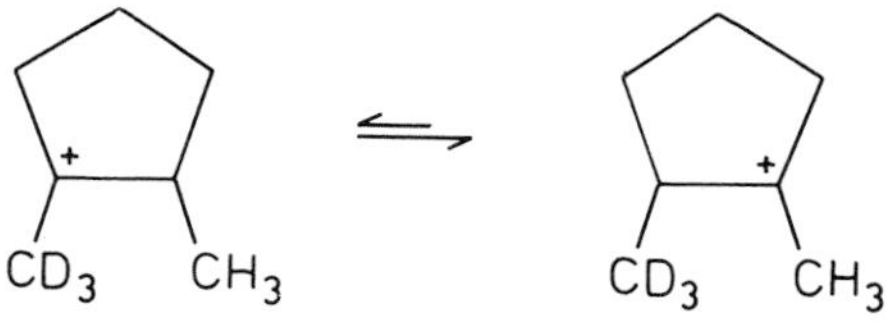

CD_3 CH_3 ⇌ CD_3 CH_3 (85)

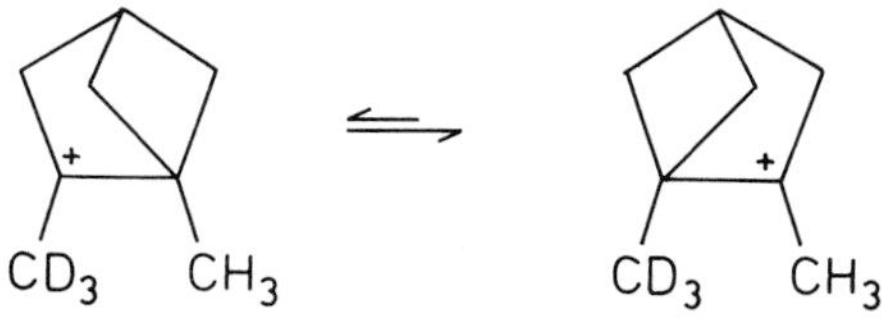

[115]

CD_3 CH_3 ⇌ CD_3 CH_3 (86)

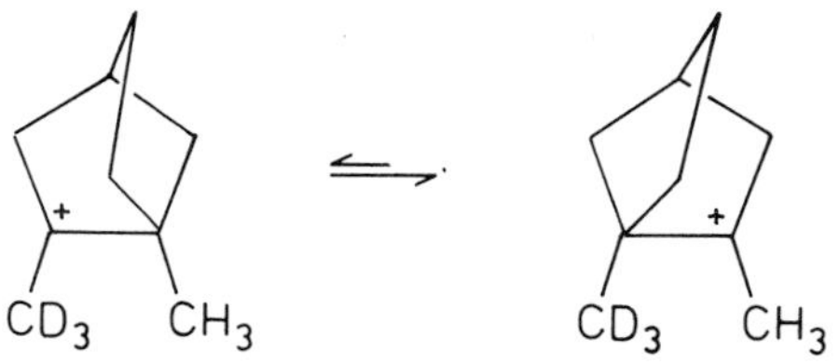

[116]

CD_3 CH_3 ⇌ CD_3 CH_3 (87)

[117]

cations [115]–[117]. The change of geometry from 2-norbornyl- to 2-bicyclo-[2.1.1]hexyl cation is important in determining the degree of σ-delocalisation. Comparison of the parent cations [99] and [114] with the dimethyl-substituted systems [117] and [116] demonstrates the facile variation of the shallow energy surfaces of carbocations by substituents. The minimum energy structure is easily changed and can have a single or multiple minima and various degrees of σ-delocalisation.

Pentacyclononyl and trishomocyclopropenyl cations and pyramidal dications

The ^{13}C nmr spectrum of the 9-pentacyclo[4.3.0.^{2,1}0^{5,7}]nonyl cation [118] (Coates and Fretz, 1975) shows only three peaks for the averaged C-6, 7, 9, C-1, 5, 8 and C-2, 3, 4 carbons, independent of temperature. Deuterium at C-9 is scrambled only among the three carbons C-6, 7, 9. This is consistent only with a static trishomocyclopropyl cation type structure. In the spectrum of the 9-deuteriated cation, the peak corresponding to C-6, 7, 9 is shifted less than 0.1 ppm, confirming the static symmetrical structure of the cation (Saunders and Kates, 1980).

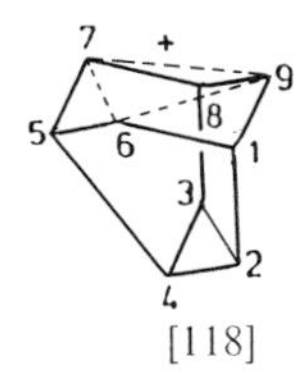

[118]

The isotope effect in the trishomocyclopropenyl cation [119] has been investigated (Olah *et al.*, 1985b). The virtual identity of the ^{13}C nmr spectrum of 3-deuteriobicyclo[3.1.0]hexyl cation D_1-[119] with that of the nonlabelled cation, except for a small (0.2 ppm; −80°C) upfield shift of the deuteriated methine triplet, is in accord with the symmetrical single minimum hypercoordinated structure of cation [119].

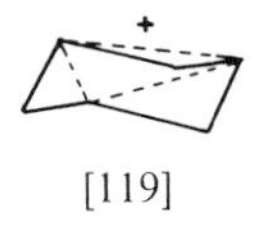

[119]

The ^{1}H and ^{13}C nmr spectra of the pyramidal dication of Hogeveen shows a five-fold symmetry with no significant line-broadening at temperatures as low as −150°C. The alternative structures could be either a hypercoordinated five-fold symmetrical pyramidal dication [120] or a set of rapidly equilibrating carbodications [121] with the same time-averaged symmetry, as indicated in (88).

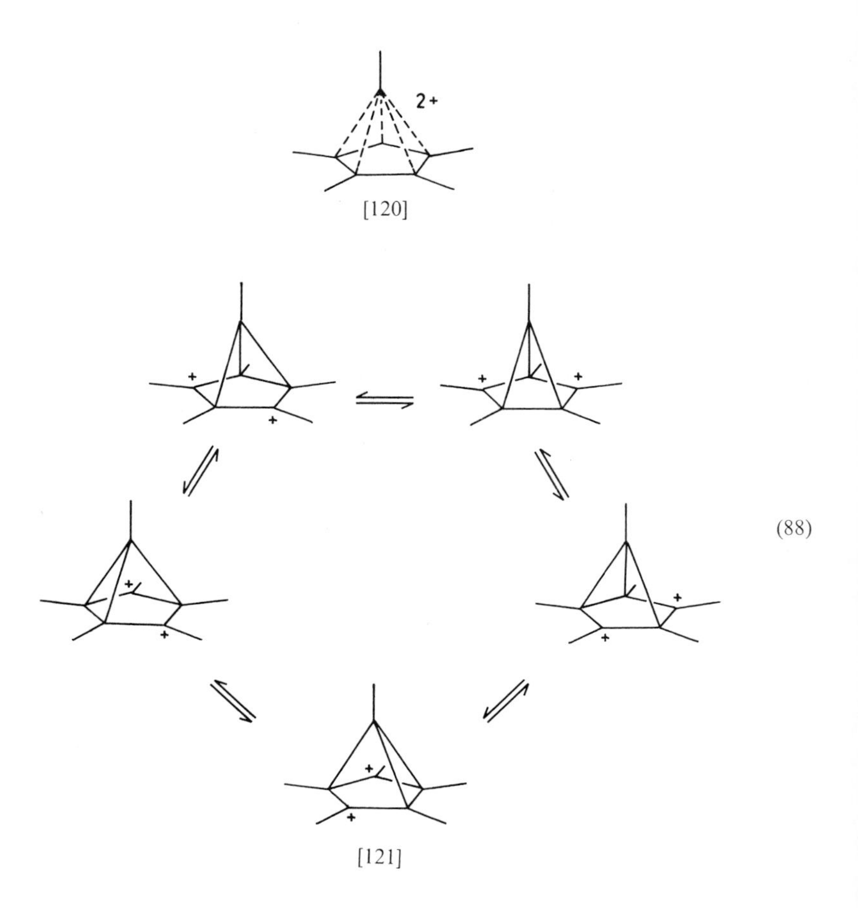

Decisive evidence for the single energy minimum structure comes from isotopic perturbation studies (Hogeveen and van Kruchten, 1981). The dication with one CD_3-group bonded at the basal carbons showed only very small splittings in the ^{13}C nmr spectrum. The basal carbons bonded to the CH_3-groups appear 0.14 ppm downfield and the basal carbon bonded to the CD_3-group are shifted 0.27 ppm upfield compared to the unlabelled dication. Nearly identical shifts for the basal carbons were observed in the cation which has two CD_3-groups at the basal carbons. The temperature-independence and the small magnitude of these shifts are not in agreement with a mixture of rapidly equilibrating trivalent dications [121], and the magnitude of the observed effect for [120] corresponds well to the small effects found in other single minimum energy structures.

Cyclopropylmethyl, cyclobutyl and bicyclobutonium cations

The fast rate of solvolysis and the rapid interconversion of cyclopropylmethyl and cyclobutyl derivatives have been extensively investigated. The dominant structure in the dynamic equilibria of the $C_4H_7^+$ cation [122] and its methyl analogue $C_4H_6CH_3^+$ [123] have been in dispute for a long time (Brown and Schleyer, 1977). The structures envisaged include cyclopropylmethyl cations [124], cyclobutyl cations [125] or bicyclobutonium ions [126].

$[C_4H_7]^+$

[122]

$[C_4H_6CH_3]^+$

[123]

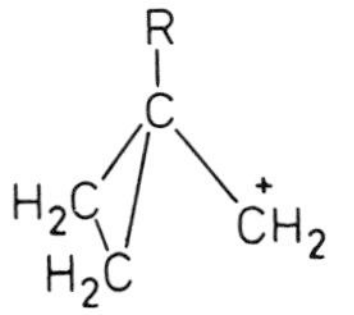

R–C, H_2C, H_2C, $\overset{+}{C}H_2$

[124]

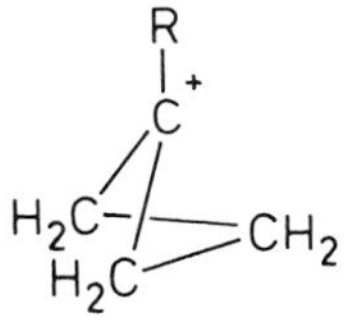

R–$\overset{+}{C}$, H_2C, CH_2, H_2C

[125]

R–C, H_2C, +, CH_2, H_2C

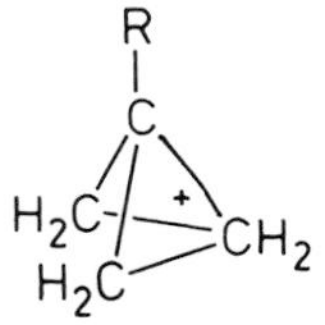

[126]

Different structures and various equilibria have been suggested to account for the observed nmr spectra. The scheme in (89) shows the relation and principal interconversion cycle of cations [124]–[126] around a tricyclobutonium ion structure [135] which is energetically disfavoured according to the Jahn–Teller theorem.

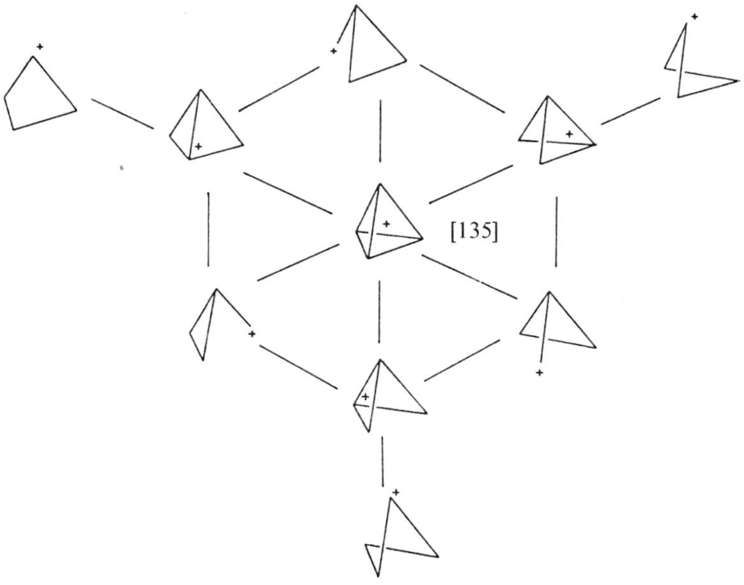

[135]

(89)

The application of the isotopic perturbation method to a model cation [127] is discussed first followed by the results obtained on the $C_4H_6CH_3^+$ and $C_4H_7^+$ cations. This gives a conclusive line of argument and shows that only noncontroversial basic nmr data taken together with the isotope effect data are sufficient to unravel structural characteristics of the $C_4H_7^+$ cation [122] and its methyl analogue [123] which have been enigmatic for so long.

1-(2,3-Dimethylcyclopropyl)ethyl cation. The 1-(2,3-dimethylcyclopropyl)-ethyl cation [127] undergoes a fast triply degenerate rearrangement (Olah *et al.*, 1982). The structure is not controversial and therefore this system serves as a model cation for the investigation of equilibrium isotope effects in cyclopropylmethyl-type cations (Siehl and Koch, 1985).

H_3C H_3C CH_3 +

[127]

The ^{13}C nmr spectrum of [127] measured at $-79°C$ shows averaged signals for the C^+ and the two β-methine carbons and also for the three methyl groups. The slow exchange spectrum can be obtained at $-154°C$. The signal for the three methine carbons decoalesces into one peak for the C^+-carbon at 232 ppm, and another peak of double intensity for the two equivalent β-carbons at 88 ppm. Accordingly the peak for the three methyl groups gives one lowfield signal for the $C^+—CH_3$ and one upfield signal of twice the intensity for the two β-CH_3-groups. The ^{13}C spectrum of cation [9] (p. 78) which has a deuterium at one methine position showed at $-70°C$ typical isotope splittings of those peaks which are averaged in the unlabelled ion [127], whereas the signal for the α-cyclopropyl carbon which is not averaged was not affected. The splittings are due to the lifting of degeneracy of the dynamic process in the deuteriated ion. The two site fast exchange taking place between a single populated lowfield site (the C^+- or the C^+-CH_3 position) and a doubly populated highfield site (the two equivalent β-CH or β-CH-CH_3 positions) is perturbed by deuterium. This leads to a relative shift ratio of 1 : 2 for the downfield and upfield signals in both sets with respect to the shift of the unlabelled cation (when corrections for intrinsic isotope shifts are taken into account). As expected, the isotope splittings disappear when the dynamic process is frozen out at $-154°C$.

The deuteriated methine carbon was shifted upfield and the intensity ratio observed for the downfield to upfield methyl peaks was 2 : 1. This ratio results from the fact that only one out of three exchanging methine positions is deuteriated. The upfield shift of the deuteriated methine peak towards the position of the β-cyclopropyl methine-carbon in the static ion and the corresponding upfield shift of the methyl group which is attached to the C—D-carbon indicate that the ring-deuteriated cations [9b] and [9c] are preferred in the equilibrium (23). This proves that the C—H(D) bond vibrations are more confined at the cyclopropane ring position than at the C^+-carbon position. The zero point energy for C—H and C—D bond vibrations and the vibrational force constants are therefore higher for bonds at the cyclopropane position compared to the sp^2-carbon position. The equilibrium constant $K = 1.299$–1.205 between $-110°C$ and $-65°C$ was calculated using the equation $K = (\Delta + \delta)/(\Delta - 2\delta)$ where Δ is the shift difference of the exchanging positions in the static cation and δ are the observed isotope splittings of the signal pairs, corrected for intrinsic isotope shifts. Thermodynamic parameters were obtained as $\Delta H° = -115.5$ cal mol^{-1} and $\Delta S° = -0.18$ cal mol^{-1} K^{-1}.

The equilibrium isotope effect in the cation is a differential α- *vs* γ-effect. Kinetic investigations have shown, however, that γ-deuterium isotope effects in cyclopropylmethyl cations are small (Sunko *et al.*, 1962); therefore neglect of the γ-effect and comparison with calculated α-effects seems justified. The fractionation factor $K_f = 1.11$ (25°C) for an isotopic exchange equilibrium between D_1-cyclopropane and D_1-ethene (Shiner and Hartshorn, 1972) is a model value for the maximum expected α-deuterium isotope effect and is in agreement with the experimental value in cation [9] extrapolated to 25°C ($K = 1.11$).

The α- *vs* γ-equilibrium isotope effect observed in the Cope rearrangement of 1,5-dimethyl-2-deuteriosemibullvalene [46] (Askani *et al.*, 1982), in which a comparable exchange situation exists for deuterium bonded either to an sp^2-hybridised carbon or to a cyclopropane ring, also shows very good agreement with the observed values for the equilibrium constant and the thermodynamic parameter in cation [9].

The analogous equilibrium isotope effects in two-fold degenerate secondary cyclopropylmethyl cations [8] (Koch, 1985) and [128] (Schneider, 1985) have also been investigated by nmr spectroscopy. The results are similar and confirm the interpretation of the isotope effect in cation [9].

CH3 CH3 CH3 H D ⇌ CH3 CH3 CH3 H D (90)

[128]

Methylbicyclobutonium cation. The ^{13}C nmr spectrum of the $C_4H_6CH_3^+$ cation [123] at −70°C shows only three peaks, one for the quaternary carbon, one for the methyl group and another one for the three methylene carbons indicating three-fold time-averaged symmetry. Kinetic line-broadening shows up at lower temperatures, and at −154°C two different methylene peaks are observed (Sorensen and Kirchen, 1977). One peak at 71.3 ppm corresponds to two methylene carbons and another peak for one methylene carbon gives a signal at an extraordinary position of −2.8 ppm.

The ^1H- and ^{13}C-nmr spectra of a mixture of $C_4H_6CH_3^+$ and CD_2-labelled $C_4H_4D_2CH_3^+$ show isotopic perturbation for the averaged signal of methylene groups, whereas the other peaks are unaffected (Siehl, 1985). The peak for the CH_2-groups in the deuteriated cation are shifted upfield compared to the protio-cation between 1.45 (−80°C) and 1.22 ppm (−56°C) in the ^{13}C-spectrum, whereas the peak for the deuteriated carbon was found shifted downfield.

Small but significant upfield shifts for the methylene protons were observed in $C_4H_4D_2CH_3^+$ at 400 MHz (0.043–0.041 ppm between −110°C and −90°C). In an earlier investigation of the $C_4H_2D_4CH_3^+$-cation (Telkowski, 1975) an upfield shift of 0.08 ppm (100 MHz, −70°C) was reported. No isotopic perturbation was found, however, in $C_4H_6CD_3^+$.

The fact that the shifts of the quaternary carbon and the attached methyl group are independent of temperature (Sorensen and Kirchen, 1977) together with the observation that deuterium perturbs only the process which effects the equivalence of the methylene groups and does not change the environment of the other carbons, is evidence that the equilibrium of $C_4H_6CH_3^+$-cations is degenerate.

In the monodeuteriated cation $C_4H_5DCH_3^+$ a set of two new peaks at both sides of the averaged methylene position for the protio-cation was observed in the proton and the carbon spectra. The signal for the proton geminal to deuterium and the signal for the deuteriated carbon triplet (J_{CD} = 25.3 Hz) move downfield (^1H spectrum: 0.099–0.063 ppm, between −110°C and −50°C; ^{13}C-spectrum: 0.898–0.670 ppm, between −80°C and −46°C), whereas the remaining methylene protons show a small upfield shift. The two CH_2-carbons are shielded between 0.607 (−80°C) and 0.536 ppm (−46°C) relative to the unlabelled cation. The relative upfield and downfield ^{13}C-shift ratio of 1 : 2 (corrected for an intrinsic upfield shift) confirms that a fast two-site exchange is taking place between a singly populated highfield site and a doubly populated lowfield site.

The slow exchange spectra of the mono- and di-deuteriated cations at −153°C have essentially the same appearance as the spectra of the protio-ion except for intrinsic deuterium isotope shifts. This shows that the low temperature species is a static cation.

The direction of the equilibrium isotope shifts in the averaged spectra can be correlated with the shift and intensity ratio of the two types of methylene in the frozen out spectra. The upfield shift of the averaged CH_2-carbons in the deuteriated cations towards the one methylene group at -2.8 ppm and the corresponding downfield shift of the deuteriated methylene carbon shows that the mean of the force constants of the two C—H bonds at the high field carbon (-2.8 ppm) must be lower than those of the C—H bonds at the two low field methylene carbons.

Two of the three types of structure, [124], [125] and [126], considered for the $C_4H_6CH_3^+$ can be excluded by a comparison of the relative chemical shift, intensity ratio and C—H bond force constants of the methylene signals in [123] with those in model cations. The different isotope effect observed in [9] rules out a cyclopropylmethyl-type cation structure for $C_4H_6CH_3^+$.

In a degenerate rearrangement (91) of D_1-cyclobutyl cation [129] deuterium would be preferred in the γ-position (upfield) and protons in the β-position (low field) (Siehl, 1985, 1986a). The opposite direction observed in deuteriated $C_4H_6CH_3^+$ excludes a cyclobutyl-type structure for this cation.

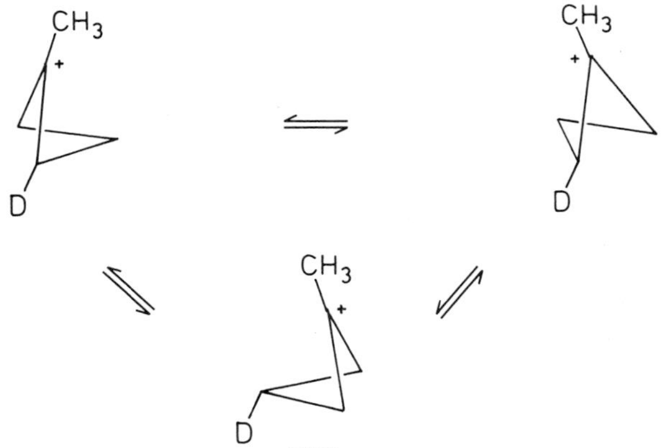

(91)

[129]

The experimental results for [123] are in agreement with the methylbicyclobutonium structure [130] which has one methylene group at higher field with a pentacoordinated carbon expected to have lower bond force constants. The isotope effects observed in $C_4H_6CH_3^+$ are fundamentally coincident with those in the $C_4H_7^+$ cation. The direction of the isotope shifts are the same as were observed in the CD_2-deuteriated parent cation, where the isotope effect is an average of a large effect preferring deuterium in the low field position and a smaller one preferring deuterium in the upfield position.

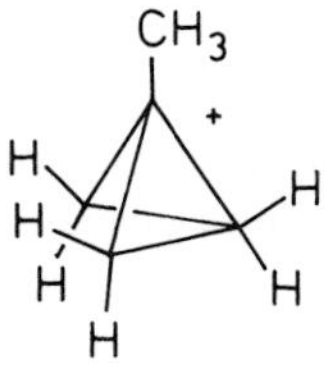

CH_3 H H H H H H +

[130]

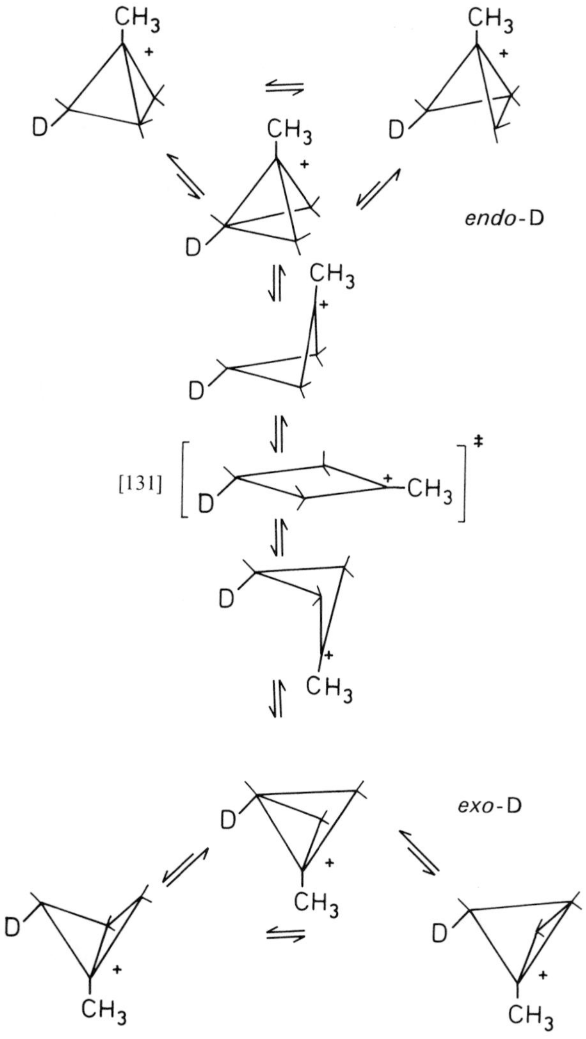

CH_3 + D ⇌ CH_3 + D

CH_3 + D *endo*-D

CH_3 + D

[131] [D — + CH_3]‡ (92)

D + CH_3

D *exo*-D + CH_3

D + CH_3 ⇌ D + CH_3

Contrary to the parent ion [139] the geminal protons at the averaged methylene carbons in [130] are not distinct. Averaging of the geminal protons occurs *via* a bent methylcyclobutyl cation with sufficient lifetime to allow inversion by way of a planar cyclobutyl cation transition state [131]. The two isotope effects for *exo*- and *endo*-D_1-[139] are different in size and sign, but are averaged and thus smaller in D_1-[130]. The size of the isotope splitting observed in the ^{13}C nmr spectrum of D_1-[130] corresponds to the arithmetic mean of the two isotope effects observed in *exo*- and *endo*-D_1-[139].

[132a] ⇋ [133] ⇋ [132b] (93)

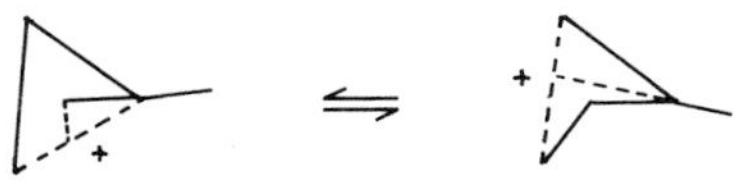

[134a] ⇋ [134b] (94)

The σ-delocalised bicyclobutonium ion structure was suggested to be a set of unsymmetrical cations [132] by Olah *et al.* (1970). The possibility of a symmetrical single minimum structure [133] (i.e. [130] or [139] respectively) was also considered but claimed to be indistinguishable from a double potential energy minimum with a very low barrier (Olah *et al.*, 1978a). The unsymmetrical structures [132] which are still depicted as in (94) (Olah *et al.*, 1985c), imply three different CH_2-groups (Olah *et al.*, 1972); in fact three different chemical shifts for the three CH_2-groups were used to calculate the anticipated averaged position in the bicyclobutonium ion [139] (Olah *et al.*, 1978b). This would require a three-site, singly populated, fast exchange at intermediate temperatures which is not consistent with the observed pattern for the isotope splittings in the $C_4H_5DCH_3^+$ cation.

Equilibrium isotope effects depend only on force constant differences of C—H bonds, which are to be expected for the CH_2-groups in the envisaged exchanging structures [134a] and [134b] and are independent of the energy barrier between these two structures. The absence of equilibrium isotope effects in the very low temperature spectra of the methylene-deuteriated bicyclobutonium ions rules out any equilibrium between the unsymmetrical structures [134a] and [134b] and provides convincing evidence that the symmetrical methylbicyclobutonium ion [130] is the minimum energy structure for cation [123]. Similar conclusions have been drawn recently by Olah *et al.* (1985b), who have also observed isotope effects in CD_2-labelled $C_4H_6CH_3^+$.

Bicyclobutonium cation

Evidence for equilibrating cations and stereospecific isotope effects
Even at the lowest temperatures the ^{13}C nmr spectrum of the $C_4H_7^+$ cation, unlike that of its substituted homologue $C_4H_5CH_3^+$, has only a single peak for the three methylene carbons, indicating real or time-averaged three-fold symmetry.

The CD_2-labelled cation $C_4H_5D_2^+$ shows a temperature-dependent upfield shift for the nondeuteriated methylene carbons of 1.77 ppm (−135°C) to 1.24 ppm (−107°C) (compared with the protio-ion), indicating a definite equilibrium isotope effect. A three-fold symmetrical static structure like the tricyclobutonium [135] is not consistent with this result and cannot be the main species present (Saunders and Siehl, 1980).

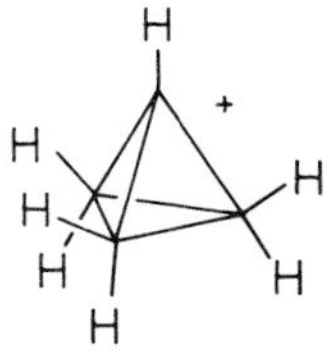

[135]

The $C_4H_7^+$-cation has two different sets of three averaged methylene protons. Mono- and trideuteriated cations prepared from 1-D_1-cyclopropylmethanol [136] (Saunders and Siehl, 1980) or from 1-D_1-(2′,2′-D_2-cyclopropyl)methanol [137] (Roberts *et al.*, 1984) are a mixture of stereoisomeric cations with the CHD-deuterium in the low field and in the high field methylene positions.

When (2′*E*, 2′*Z*, 3′*E*)-D_3-cyclopropylmethanol [138] is used as a precursor (Roberts *et al.*, 1984), only one stereoisomeric $C_4H_4D_3^+$ is observed. The deuterium at the monodeuteriated methylene carbon is in the *endo* position, i.e. *trans* to the methine proton. From the intensity ratio of the signals for the remaining two *exo*-protons and the single *endo*-proton, the lowfield signals could be assigned to the *endo*- and the highfield signals to the *exo*-protons. The interconversion of *exo*- and *endo*-stereoisomers [139] is slow on the nmr timescale at −90°C. This indicates that inversion *via* [131], which is a tertiary cation for [130], is unfavourable for [139] because a secondary cyclobutyl cation would be involved. The alternative pathway *via* rotation about a methylene to methine carbon bond also has a high activation energy.

CHDOH ⟶ $C_4H_6D^+$ *exo* and *endo*

[136]

D, D, CHDOH ⟶ $C_4H_4D_3^+$ *exo* and *endo*

[137]

D, D, H, D, CH_2OH ⟶ $C_4H_6D_3^+$ *endo*

[138]

The proton spectra of a mixture of stereoisomeric D_1- or D_3-cations show different equilibrium isotope effects for the *endo*- and *exo*-deuteriated cations. When the deuterium at the CHD-methylene group is in the *endo*-(lowfield) position, large isotope splittings between 0.388 ppm (−145°C) and 0.142 ppm (−55°C) in $C_4H_6D^+$ and 0.29 ppm (−121°C) in $C_4H_4D_3{}^+$ were observed for the *exo*-(upfield) protons but there was no measurable effect on the remaining *endo*-protons. The stereoisomeric $C_4H_4D_3{}^+$-cation with the CHD-deuterium in the *exo*-position showed no significant isotope effect and, *exo*-D-$C_4H_6D^+$, only a small temperature-dependent downfield shift was observed for the *exo*-protons and there was no effect on the *endo*-protons. In the proton spectra of the geminally dideuteriated cation $C_4H_5D_2{}^+$, the upfield methylene protons were found to be shifted to higher field (relative to the protio-ion) by between 0.087 (−130°C) and 0.057 ppm (−80°C) and the downfield methylene peak was unaffected.

This observation, that the equilibrium isotope shifts for *endo*- or *exo*-deuterium are sizeable for the high field averaged *exo*-protons only, but negligible for the low-field averaged *endo*-protons can only be rationalised by assuming that equilibrium among nonequivalent methylenes is perturbed by deuterium, but that the chemical shift difference between the rapidly equilibrating hydrogens, if the fast interconversion process could be stopped, is much larger for the *exo*-protons than for the *endo*-protons.

The isotope effects in the two $C_4H_6D^+$-cations give additional information on the dynamics and structure of the $C_4H_7{}^+$-cation (Saunders and Siehl, 1980). The intensity pattern observed in the proton spectrum shows the multiplicity of the exchange and the relative populations of the exchanging sites. The direction of the isotope shifts can be assigned to the *exo*- and *endo*-deuteriated cations using the *exo*/*endo*-proton assignment of Roberts *et al.* (1984). The sign of the equilibrium isotope effect and the relative C—H bond force constants at the exchanging sites can be determined.

In the *endo*-D_1-cation, a signal of intensity 1 for the CHD-*exo*-proton moves about twice as much to lower field as a signal of intensity 2 for the two CH_2-*exo* hydrogens moves upfield compared to the protio-ion and corrected for geminal intrinsic isotope shift. The relative size of the shifts indicates that a two-site exchange between a doubly populated low field site and a singly populated high field site is perturbed by deuterium. The intensity ratio of the two peaks for the *exo* protons is 1 : 2, one *exo*-CHD to lowfield and two *exo*-CH_2 to high field. This shows that CH_2-*endo*-protons are preferred at the upfield site. The upfield *endo*-protons have lower vibrational bond force constants than the low field *endo*-protons.

In the stereoisomeric D_1-*exo*-cation, the two remaining *exo*-protons give a peak of intensity 2 which is close to the unlabelled position but shifted downfield. This indicates that the force constants for the *exo*-protons at the downfield site are lower than those at the upfield site.

The ^{13}C nmr spectrum of the mixture of *exo*- and *endo*-$C_4H_6D^+$ cations convincingly confirms the observation of two different isotope effects. As in the proton spectrum, the methylene resonance in the *endo*-D-cation showed the larger isotopic perturbation. With respect to the protio-ion, the CHD-carbon moved downfield and two CH_2-carbons moved upfield. The shift between them varies with temperature from 7.05 ppm (−133°C) to 3.82 ppm (−87°C). The *exo*-D-cation showed smaller isotope splittings between 3.16 ppm (−118°C) and 2.78 ppm (−87°C) in the opposite direction, with the CHD-carbon moved upfield and the CH_2-carbons moved downfield.

Analogous ^{13}C-shift patterns, although with different intrinsic isotope shifts superimposed, were observed by Roberts *et al.* (1984) in the $C_4H_4D_3{}^+$-cations. The isotopic perturbation for the CHD- and CH_2-resonances at −95°C are 5.34 ppm for the CHD-*endo*-cation and 3.17 ppm in opposite direction for the CHD-*exo*-cation.

Additional support for the opposite sign of the equilibrium isotope effects in *exo*- and *endo*-d-$C_4H_6D^+$ comes from the observation that the upfield shift of the CH_2-carbons in CD_2-labelled $C_4H_5D_2{}^+$ (compared to the protio-ion) is the algebraic sum of the upfield and downfield shifts of the CH_2-carbons in the two monodeuteriated cations.

C—H bond force constants and structure

Despite the fact that the fast averaging process which renders the methylene carbons equivalent cannot be frozen out, the isotope shifts give information on the relative stiffness of C—H bond force constants in a frozen-out structure. The *endo*-protons at the single methylene carbon which would give an upfield peak in the spectrum of a static structure have much weaker vibrational force constants than the *endo*-protons at the two methylene carbons which would give a downfield peak. The *exo*-protons at the high field methylene carbon have moderately stiffer bond force constants than the *exo*-protons at the two methylene carbons which would cause the downfield peak in the spectrum of the frozen-out cation.

The much larger shift difference for the averaged *exo*-methylene protons than for the *endo*-protons suggests that the two protons at the upfield methylene carbon have very different chemical shifts and that the *exo*-proton at this carbon is located in a unique environment compared to the other protons.

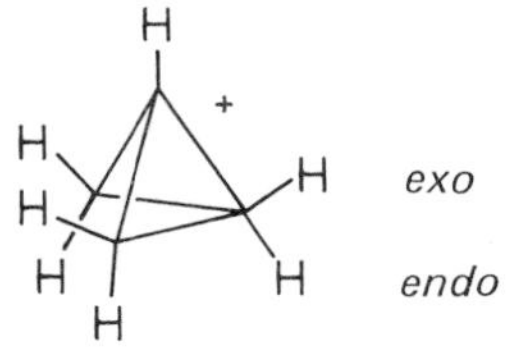

[139]

These two observations exclude cyclopropylmethyl and cyclobutyl as the dominant structures in the equilibrium of $C_4H_7^+$-cations. In a bicyclobutonium ion [139], the pentacoordinated carbon is likely to be upfield as for the methyl-homologue. The vibrations for the *endo*-C—H bond at the pentacoordinated carbon are less confined compared to the other *endo*-C—H bonds because bridging could drain electron density out of this bond. MINDO/3 calculations of Dewar and Reynolds (1984) have shown a larger bond length and some involvement in multicentre bonding only for this *endo*-C—H bond.

The reverse but smaller isotope effect observed for the *exo*-protons at the pentacoordinated carbon must be caused by more hindrance to vibration compared to the other *exo*-hydrogens. This may be due to greater crowding at the pentacoordinated carbon and 1,3-hydrogen interaction between the 3-*exo*- and the 1-methine-protons. The unique shift of the *exo*-proton was suggested by Roberts to be caused by its apical position at the pentacoordinated carbon which could possess a configuration with the elements of square-pyramidal geometry.

Both directions for the equilibrium isotope effects have their counterparts in kinetic isotope effects. The situation for the *endo*-proton resembles the transition state in an S_E2 reaction with an α-primary kinetic isotope effect ($k_H/k_D > 1$). The *endo*-proton would be the leaving group with a loose bond, and the bridging cationic carbon an incoming electrophile. The confined situation for the *exo*-proton can be compared with an α-secondary kinetic isotope effect ($k_H/k_D < 1$) which results from increased hindrance to bending vibration in the pentacoordinated transition state.

Olah and Roberts (Olah *et al.*, 1978b) interpreted a small temperature dependence of the ^{13}C nmr spectrum of ^{13}C-labelled $C_4H_7^+$ as indicating a minor species like [124] in rapid equilibrium with the major ion [139]. A splitting of 0.4 ppm between the methine peaks of two isomeric deuteriated cations observed by Saunders and Siehl (1980) and also by Roberts *et al.* (1984) might indicate an isotope effect on a nondegenerate equilibrium in support of this idea.

HYPERCONJUGATION IN CARBOCATIONS

C—H Hyperconjugation and deuterium isotope effects

The origin of kinetic β-deuterium isotope effects in solvolysis reactions proceeding through carbocation-like transition states has generally been attributed to C—H hyperconjugation (Sunko and Hehre, 1983). Drainage of electron density from the valence orbitals of C—H bonds into the vacant orbital at the carbocation centre leads to a weakening of all force constants for C—H vibrations and not just those associated with C—H stretching as suggested by a simple hyperconjugative model (Hehre *et al.*, 1983). Vibrational frequencies and zero point energies of β-C—H bonds involved in hyperconjugation are therefore lower than for C—H bonds which are either not involved or are at more remote positions.

In an isotopic equilibrium of otherwise degenerate carbocations, where an exchange situation for H and D exists between a hyperconjugating β-position to the positive charge and another position, the equilibrium is shifted to the side where the hydrogens are preferentially bonded at the β-carbon and deuteriums at the more remote carbon position.

Saunders and coworkers have applied the isotopic perturbation technique to a number of persistent carbocations and have demonstrated that C—H hyperconjugation is a principal cause of β-secondary deuterium equilibrium isotope effects in stable carbocations. The results fully support the interpretation of secondary kinetic isotope effects in solvolytic substitution reactions.

2,3-Dimethyl- and 2,2,3-trimethyl-2-butyl cations. Saunders and coworkers (Saunders *et al.*, 1971; Saunders and Vogel, 1971b) have studied the influence of methyl deuteriation in the 2,3-dimethyl-2-butyl cation using ^1H-nmr spectroscopy. This cation undergoes a rapid degenerate hydride shift leading to a single time-averaged doublet for the methyl protons. Substitution of the methyl protons on one side of the molecule by 1 to 6 deuteriums lifts the degeneracy between the two isomers. In the monodeuteriated cation [140] two new time-averaged methyl doublets are observed. The low field doublet results from averaging of six protons of the methyls attached to the C^+-carbon in [140b] with the six protons of the methyl groups attached to the methine—carbon in [140a]. The doublet at higher field has the lower intensity and is due to the averaging of five protons of the methyl groups attached to the methine carbon in [140b] with five protons of the methyls on the C^+-carbon in [140a]. The six-proton doublet being located at lower field than the five-proton doublet indicates that the carbocation prefers to be substituted by unlabelled methyl groups, i.e. the equilibrium is shifted towards [140b].

$$(CH_3)(CH_2D)\overset{+}{C}{-}CH(CH_3)_2 \rightleftharpoons (CH_3)(CH_2D)CH{-}\overset{+}{C}(CH_3)_2$$

[140a] [140b]

The hydride shift in [140] is fast even at low temperatures; the shift difference for the two methyl-positions was therefore estimated to be 2.1 ppm using the 2-methyl-2-butyl cation as a model. The isotope splittings δ for the methyl peaks increase with the number of deuteriums from 0.145 ppm, 0.30 ppm and 0.46 ppm at $-76°C$ in the D_1-D_3-labelled cations to 0.623 ppm, 0.779 ppm and 0.928 ppm at $-79°C$ in the D_4-D_6-labelled cations.

The equilibrium constants were calculated [using $K = (\Delta + \delta)/(\Delta - \delta)$] as $K = 1.148, 1.33, 1.561$ at $-76°C$ for the D_1-D_3-cations [140] and 1.844, 2.18, 2.585 at $-79°C$ for the D_4-D_6-[140]. The corresponding enthalpy differences $\Delta H°$ are 54.2, 56.2, 58.3, 58.5, 59.6 and 61.3 cal mol^{-1} per D.

The symmetrically hexadeuteriated cation [1] showed no isotope effect because the two isomers related by the hydride shift (95) are identical and the equilibrium constant is exactly unity.

$$(CH_3)(CD_3)\overset{+}{C}{-}CH(CH_3)(CD_3) \rightleftharpoons (CH_3)(CD_3)CH{-}\overset{+}{C}(CH_3)(CD_3) \qquad (95)$$

[1]

$$\upharpoonleft\!\downharpoonright$$

$$(CD_3)_2\overset{+}{C}{-}CH(CH_3)_2 \rightleftharpoons (CD_3)_2CH{-}\overset{+}{C}(CH_3)_2 \qquad (96)$$

[141a] [141b]

A slow methyl shift equilibrates the unsymmetrical deuteriated ion [141] with the symmetrical deuteriated cation [1]. This process is affected by the equilibrium isotope effect on the fast hydride shift (96) in the unsymmetrical deuteriated cations [141a] and [141b]. Cation [141b] with six deuteriums away from the positive charge is favoured over the symmetrical hexadeuteriated cation [1]. An isotope effect of $K = 0.95$–0.85 for the slow exchange process from [141] to [1] has been determined from peak intensities of an equilibrated mixture of these cations.

The ^{13}C nmr spectrum of cation [141] shows a very large isotope splitting of $\delta = 167.7$ ppm at -138°C for the C^+-/methine-carbon pair. The equilibrium constant $K = 1.264$–1.158 between -138°C and -71°C and $\Delta H^o = -68 \pm 2\ \text{cal mol}^{-1}$ per D were calculated using $\Delta = 277$ ppm as the shift for the C^+ and C—H carbons (Kates, 1978). These data are probably more accurate than those determined from proton spectra because a larger temperature range was investigated and the assumed Δ agrees within 1 ppm with the shift difference obtained recently from the solid-state nmr spectrum of the unlabelled cation (Myhre *et al.*, 1984).

The same direction of the equilibrium isotope effect was observed in the nondegenerate 1,2-hydride shift equilibrium of 2-(4′-trifluoromethylphenyl)-3-methyl-2-butyl cation [142] with one trideuteriomethyl group at C-2 or C-3 respectively (Forsyth and Pan, 1985). The isotope shifts in the ^{13}C spectrum are much smaller (1.3 ppm–1.45 ppm) than in degenerate cations like [141] because K_H is very much in favour of the benzylic cation structure for [142].

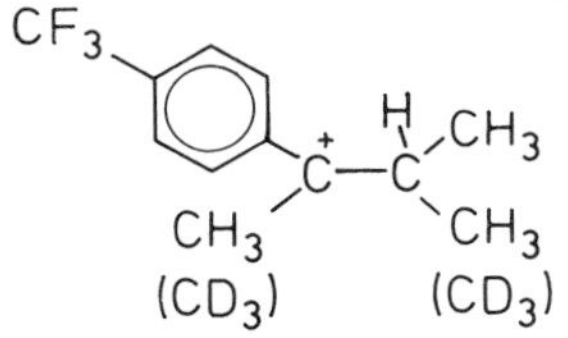

[142]

[143a] ⇌ [143b] (97)

The 2,2,3-trimethyl-2-butyl cation is an example where a rapid methyl shift as in (97) averages all methyl hydrogens leading to one averaged peak in the proton spectrum. When one methyl group is deuteriated in [143] the equilibrium is shifted towards [143b]. The remaining hydrogens continue to give a single averaged peak but this is found shifted downfield from the unlabelled position because the deuteriums disturb the statistical probability of protons residing in the chemically different environment. A value of

$\Delta H^{\circ} = -52 \text{ cal mol}^{-1}$ per D was determined from the temperature dependence of the equilibrium constants ($K = 1.55$–1.404 between $-92.7°C$ and $-53°C$). Equilibrium constants were calculated from the equation $K = (3\Delta + 2\delta)/(3\Delta - 2\delta)$ using an estimated Δ of 2.1 ppm as shift difference for the two types of methyl groups (Saunders and Vogel, 1971a).

2,3-Dimethyl-2-cyclopentyl cation. The equilibrium isotope effects in CD_3- and β-CD_2-dimethylcyclopentyl cations [116] and [104] have been measured by ^{13}C-nmr spectroscopy (Saunders *et al.*, 1977a). In [116] the averaged C^+/C—H carbons give a pair of peaks split by 81.8–45.4 ppm between $-142°C$ and $-45°C$. With $K = (\Delta + \delta)/(\Delta - \delta)$ and $\Delta = 261$ ppm, $K = 1.241$–1.124 (per D) and the temperature dependence of K gives $\Delta H^{\circ} = -60 \text{ cal mol}^{-1}$ and $\Delta S^{\circ} = -0.012 \text{ cal mol}^{-1} \text{ K}^{-1}$ per D. Cation [104] showed significantly larger isotope effects. The splitting for the C^+-/C—H carbons is 105.3–81.6 ppm giving $K = 1.534$–1.382 (per D) between $-130°C$ and $-81.6°C$ and $\Delta H^{\circ} = -137 \text{ cal mol}^{-1}$ and $\Delta S^{\circ} = -0.05 \text{ cal mol}^{-1} \text{ K}^{-1}$ per D.

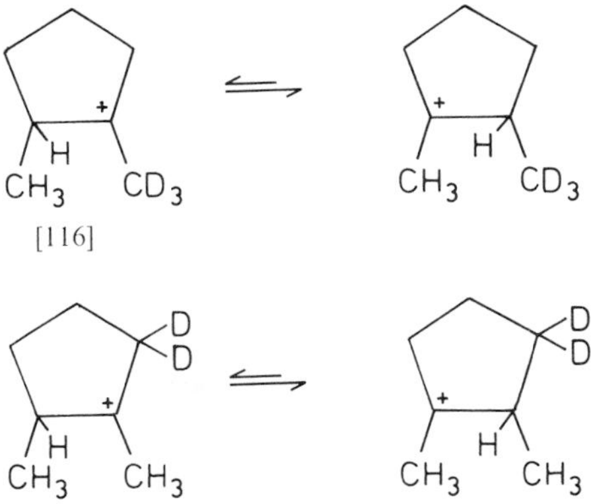

[116] (98)

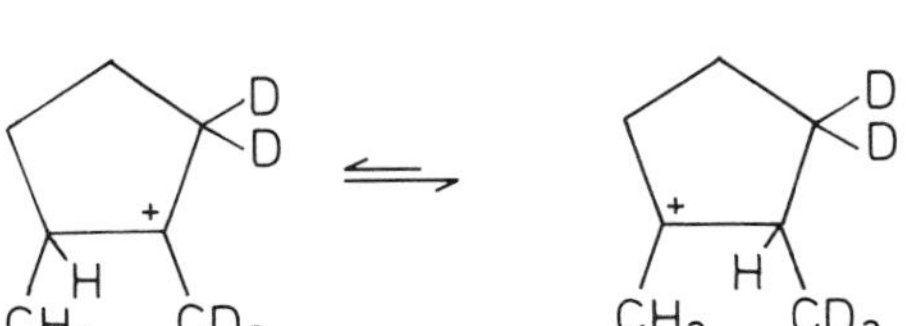

[104] (99)

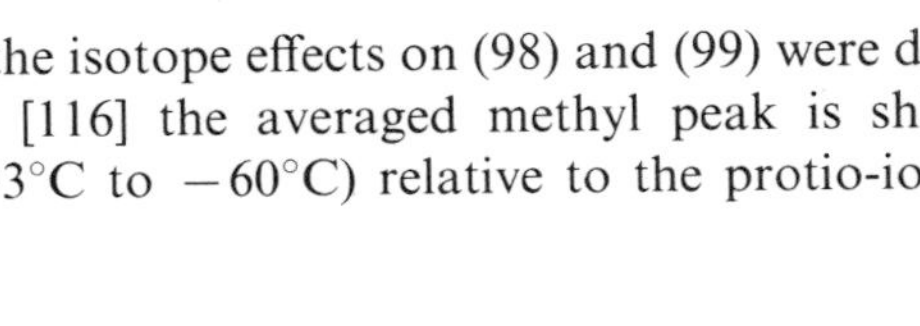

[144] (100)

The direction of the isotope effects on (98) and (99) were determined from proton spectra. In [116] the averaged methyl peak is shifted downfield 0.44–0.34 ppm ($-93°C$ to $-60°C$) relative to the protio-ion. Cation [104] showed two averaged methyl peaks with the downfield methyl peak significantly broader due to coupling to methylene protons. The downfield shift for the methyl protons in dimethylcyclopentyl cation [144] which was deuteriated at one methyl group and at the neighbouring methylene group is comparable to the sum of the shifts in [116] and [104] indicating a cumulative effect. It also confirms that the isotope effects in [116] and [104] have the same relative sign, protons preferring to be at the positions next to the positive charge (Telkowski, 1975).

A γ-isotope effect in 3,4-dimethyl-3-hexyl cation. The isotope effects observed in cations [116], [104] and [140]–[144] are differential β- *vs* γ-effects, i.e. $K_{observed} = K_{\beta} \times K_{\gamma}$. A differential γ- *vs* δ-equilibrium isotope effect favouring [145b] was measured by Kates (1978) in 1-D_3-3,4-dimethyl-3-hexyl cation [145] ($K = 1.032$ at $-126°C$; $\Delta H° = -12\ \text{cal mol}^{-1}$ per D). If one ignores a very small δ-effect these data can be used to extract the pure β-deuterium equilibrium isotope effects in the cations mentioned above.

$$CH_3\text{-}CH_2(CH_3)\overset{+}{C}\text{—}CH(CH_3)\text{-}CH_2\text{-}CD_3 \rightleftharpoons CH_3\text{-}CH_2(CH_3)CH\text{—}\overset{+}{C}(CH_3)\text{-}CH_2\text{-}CD_3 \quad (101)$$

[145a] [145b]

Equilibrium C^{13} isotope effects in 2,3-dimethyl-2-butyl and 1,2-dimethylcyclopentyl cations

A primary ^{13}C isotope effect on equilibrium (102) was measured by Saunders *et al.* (1981) in the ^{13}C nmr spectrum of 2,3-dimethyl-2-[2-^{13}C]-butyl cation [6]. The averaged C^+/C—H peak was found shifted downfield between 1.35 ppm ($-135°C$) and 0.79 ppm ($-62°C$) with respect to the corresponding peak in the dilabelled cation [146] which served as a reference since the equilibrium constant for (103) is exactly unity due to symmetry.

$$(CH_3)_2\overset{+}{C}\text{—}\overset{*}{C}H(CH_3)_2 \rightleftharpoons (CH_3)_2CH\text{—}\overset{+}{\overset{*}{C}}(CH_3)_2 \quad (102)$$

[6]

$$(CH_3)_2\overset{+}{\overset{*}{C}}\text{—}\overset{*}{C}H(CH_3)_2 \rightleftharpoons (CH_3)_2\overset{*}{C}H\text{—}\overset{+}{\overset{*}{C}}(CH_3)_2 \quad (103)$$

[146]

The equilibrium constant for [6], $K = 1.0197$ ($-135°C$) to 1.0114 ($-62°C$) was calculated from the equation $K = (\Delta + 2\delta)/(\Delta - 2\delta)$ using $\Delta = 277$ ppm. From the van't Hoff equation $\Delta H° = -6.0 \text{ cal mol}^{-1}$ and $\Delta S° = -0.056 \text{ cal mol}^{-1} \text{ K}^{-1}$. Any intrinsic isotope shifts were estimated from the precursor alcohol to be smaller than 0.06 ppm.

The observed downfield shift shows that the equilibrium favours the positive charge on ^{13}C. The vibrational frequencies associated with the charged carbon must therefore be higher than those of the methine carbon. The more confined bonding situation for the charged carbon despite the smaller number of attached atoms could indicate stiffer C—C bonding caused by C—H-hyperconjugation with the methyl groups.

*CH3 H + CH3 ⇌ *CH3 + H CH3 (104)

[147]

A secondary ^{13}C equilibrium isotope effect was measured in the 1,2-dimethylcyclopentyl cation [147] which was ^{13}C-labelled in one methyl group (Saunders *et al.*, 1977a). In the ^{13}C nmr spectrum of cation [147] the carbon next to the ^{13}C appeared as a doublet ($J_{CC} = 35$ Hz) offset downfield between 0.25 ppm ($-125°C$) and 0.10 ppm ($-65°C$) from the singlet of the other averaged C^+/C—H carbon. This shows that the labelled methyl group is preferred next to the charged carbon. Although the C—H frequencies are lower for this methyl group as evident from hydrogen isotope effects in [116] and [144], the ^{13}C isotope effect on (104) indicates a small net increase of all vibrational frequencies of the carbon of the CH_3-group at the C^+-position as compared to the other CH_3-group.

Both ^{13}C-equilibrium isotope effects in [6] and [147] have the same direction and may be regarded as another piece of evidence for C—H hyperconjugation. In valence bond terminology the bond between the charged carbon and the attached methyl group is stiffer because of partial double bond character. The heavier isotope (^{13}C) prefers this more confined bonding, whereas the lighter isotope (^{12}C) prefers to be at the remote position where the bonding is less stiff. Another secondary ^{13}C-equilibrium isotope effect has been observed in [5] using natural abundance ^{13}C nmr spectroscopy (Olah *et al.*, 1985a).

The magnitude of β-deuterium equilibrium isotope effects

The magnitude of the β-secondary deuterium kinetic isotope effects in solvolyses which proceed *via* a cationic transition state have been shown to be conformationally dependent (Sunko and Hehre, 1983). Shiner furnished two important representative examples relating the dihedral angle between an adjacent C—H or C—D bond and the vacant p-orbital of the cationic transition state with the magnitude of the kinetic isotope effect. In the

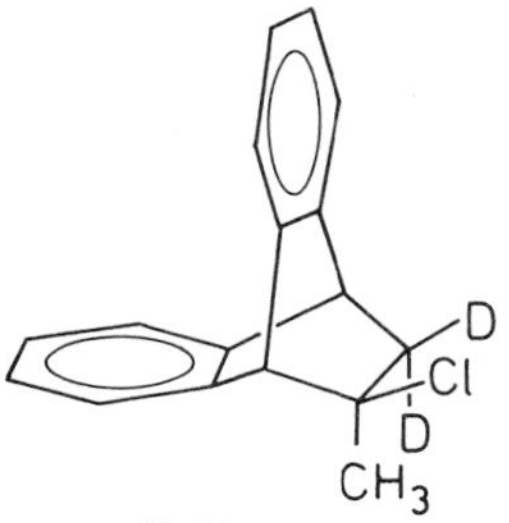

[148] 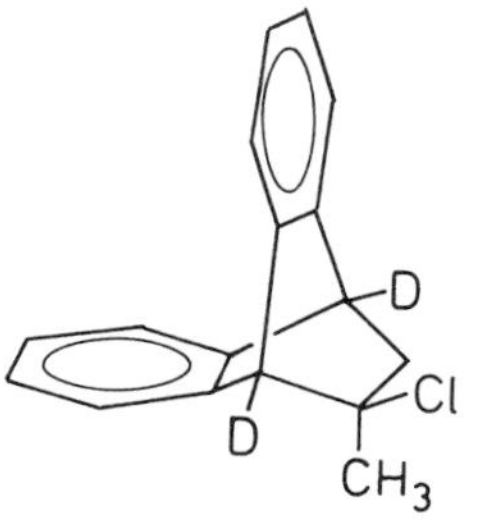 [149]

methylene-deuteriated, conformationally unambiguous, rigid dibenzobicyclo[2.2.2]octane derivative [148], the angle between the C—D bonds and the developing p-orbital of the cationic transition state is approximately 30° and a normal kinetic isotope effect of $k_H/k_D = 1.14$ at 45°C was observed. In [149] the β—C—D bond is essentially orthogonal to the developing p-orbital in the cationic transition state. Shiner and Humphrey (1963) reported an isotope effect of $k_H/k_D = 0.986 \pm 0.01$ obtained from two measurements giving considerably different rate constants, so that the error limit could include $k_H/k_D = 1$. This result was often interpreted as a small inverse β-effect but is really a combined β- and γ-effect. Further evidence for the conformational dependence of isotope effects comes from the low value of $k_H/k_D = 1.08$ for β-CD_2-2-chloro-2,4,4-trimethylpentane [151] compared to $k_H/k_D = 1.34$ for β-CD_2-2-chloro-2-methylpentane [150] (Shiner, 1961).

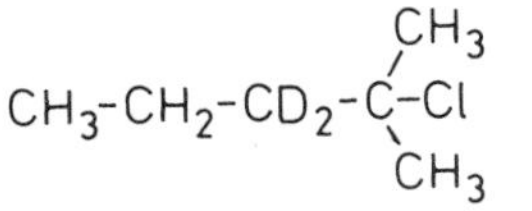

CH_3-CH_2-CD_2-C(CH_3)$_2$-Cl

[150]

CH_3-C(CH_3)$_2$-CD_2-C(CH_3)$_2$-Cl

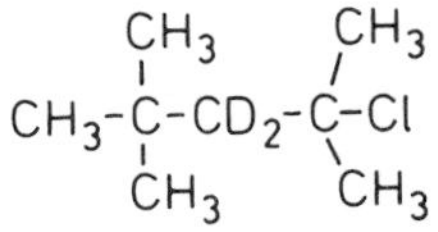

[151]

The magnitude of kinetic isotope effects in solvolysis may be distorted by possible multiple reaction pathways such as competing elimination, participation, ion pairing, rearrangements and other factors which might complicate unequivocal interpretation. Equilibrium isotope effect measurements in degenerate rearrangements of persistent carbocations involve only the properties of well-defined species and can be evaluated very accurately by nmr spectroscopy.

The 2,3-dimethylbicyclo[2.2.2]octyl cation [152] and the methyl-substituted 2-heptyl cations [95] and [97] are suitable models for different orientations of a β—C—D bond with respect to the vacant p-orbital at the C^+-carbon. The cations [152], [95] and [97] undergo fast 1,2- and 1,5-hydride shifts (105)–(107), which give averaged nmr-signals and establish an isotope exchange situation for β—H *vs* β—D. Both prerequisites for evaluation of the conformational dependence by the isotopic perturbation method are fulfilled.

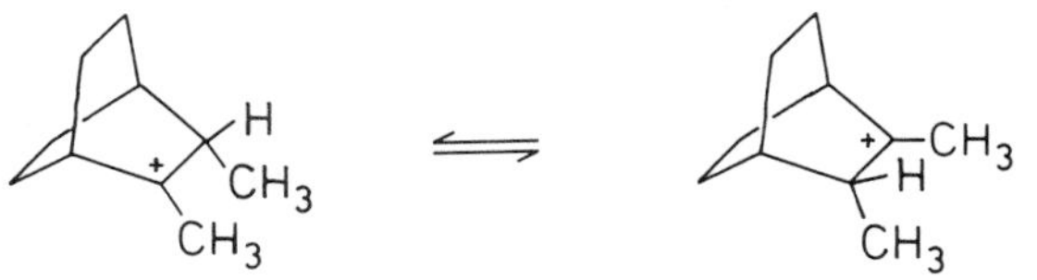

(105)

[152]

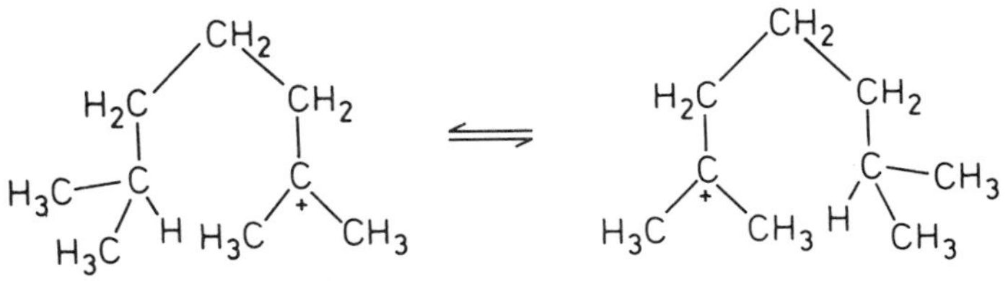

(106)

[95]

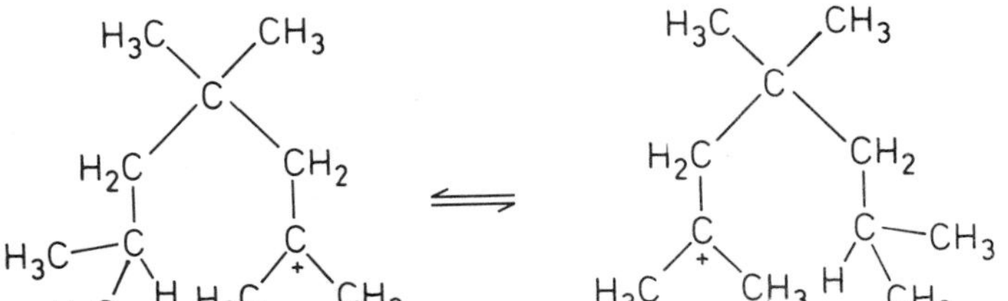

(107)

[97]

2,3-Dimethylbicyclo[2.2.2]octyl cation. The ^{13}C nmr spectrum of cation [153] shows typical splittings of the averaged peaks C-1/C-4, C-5/C-6 and C-7/C-8 due to the lifting of the degeneracy of the equilibrium (108) (Siehl and Walter, 1984). The signal for the averaged C-2/C-3 carbons is merged into the baseline due to the very large kinetic line broadening ($\Delta G^{\ddagger} = 4.7$ kcal mol^{-1} at -122°C). The isotope splitting for the CH_3/CD_3 groups is 6.4 ppm (-122°C) and not symmetrical with respect to the protio-cation [152]. The CD_3-group was found shifted upfield and has an additional intrinsic upfield shift of 0.9 ppm. This shows that cation [153b]

[153a] ⇌ [153b] (108)

with the CD_3 group further removed from the positive charge is favoured in equilibrium (108). The size of the equilibrium constant is $K_{CD_3} = 1.73$ at -120°C. The isotope splittings and the equilibrium constant are strongly temperature-dependent as expected. The thermodynamic parameters $\Delta H^{o} = -65$ cal mol^{-1} per D and $\Delta S^{o} = -0.06$ cal mol^{-1} K^{-1} per D were determined from the temperature dependence of K between -137°C and -92°C. These values are in good agreement with those reported for other β-methyl-deuteriated carbocations.

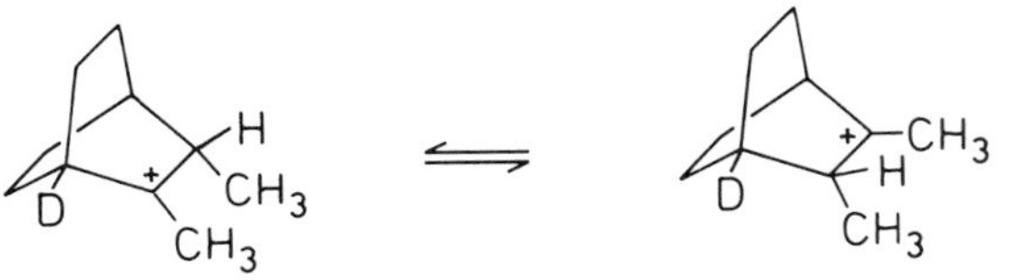

[154] (109)

In marked contrast, the bridgehead deuteriated cation [154] did not show any sizeable isotope splittings, except small temperature-independent intrinsic isotope shifts. The broadest peak, i.e. the averaged C_1/C_4 peak, has a line width of 42 Hz at -122°C, which gives a calculated maximum for K of 1.03 per D at -122°C. The absence of equilibrium isotope splittings in [154] was confirmed by the ^{1}H nmr spectrum, which is virtually identical with that of the unlabelled cation [152] except that the peak for the bridgehead protons has only half the intensity.

These results have been interpreted as a direct proof of the hyperconjugational origin of the dependence of the β-deuterium equilibrium isotope effect on dihedral angle in a persistent carbocation. In [153] maximum overlap between the vacant p-orbital and one of the methyl C—D or C—H bonds is always possible leading to large equilibrium isotope splittings. In [154] the dihedral angle of 90° permits no overlap of the vacant p-orbital with the bridgehead C—H and C—D bonds. Hyperconjugation with the bridgehead protons or deuteriums is sterically suppressed. In the parlance of valence bond theory, hyperconjugation involving the bridgehead position is inhibited because the "no bond" resonance structure would be a bridgehead olefin and thus too unstable to contribute significantly to the hyperconjugational stabilisation of the cation (Siehl and Walter, 1984).

2,6-Dimethyl- and 2,4,4,6-tetramethyl-heptyl cations

The total isotope effect of a CD_3-group is independent of conformation (Sunko *et al.*, 1977), and the CD_3-substituted cations [155] and [156] can therefore serve as a model for a dihedral angle of 0° whereas the cations [157] and [158] deuteriated in the methylene group provide information on the conformational dependence of the equilibrium isotope effect.

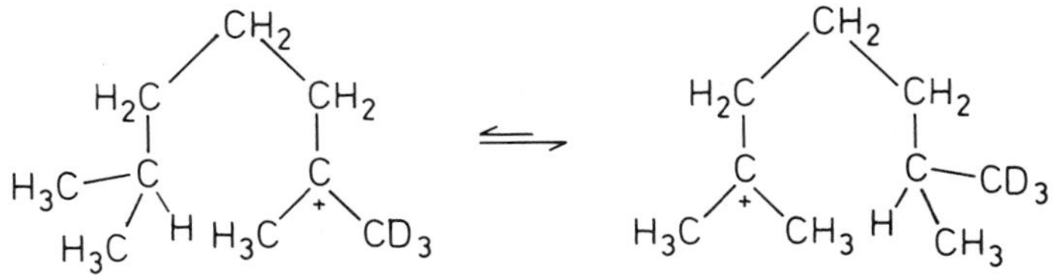

[155] (110)

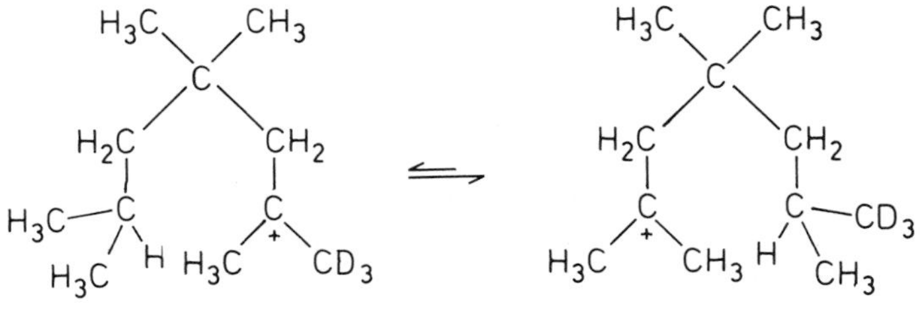

[156] (111)

The methyl groups on C-2/C-6 and the C-3/C-5 methylene groups give averaged peaks in the ^{1}H and ^{13}C nmr spectra of cations [95] and [97] (p. 153). The averaged peak for the C^{+}/C—H carbons is not visible due to large kinetic line-broadening caused by the fast 2,6-hydride shift (Saunders and Siehl, 1985; Siehl and Walter, 1985). The β-methyl deuteriated cations [155] and [156] experience large temperature-dependent isotope splittings for the averaged peaks. As expected the deuteriated methyl groups prefer the position remote from the positive charge.

Equilibrium constants for [155], $K(CD_3) = 1.69$ at $-120°C$, and for [156], $K(CD_3) = 1.91$ at $-123°C$, were obtained from the isotope splittings of the averaged CH_2- and CH_3-groups in ^1H and ^{13}C nmr spectra. Thermodynamic parameters for [155] are $\Delta H° = -62.4$ cal mol^{-1}, $\Delta S° = -0.06$ cal mol^{-1} K^{-1}, and for [156] $\Delta H° = -65.7$ cal mol^{-1}, $\Delta S° = -0.01$ cal mol^{-1} K^{-1} per deuterium, determined from the ^{13}C spectra.

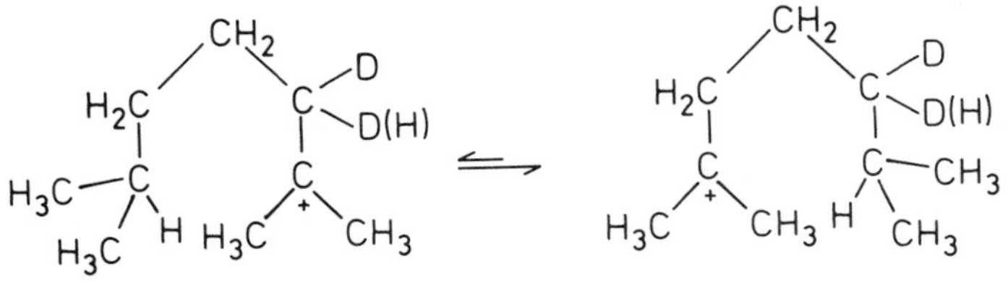

(112)

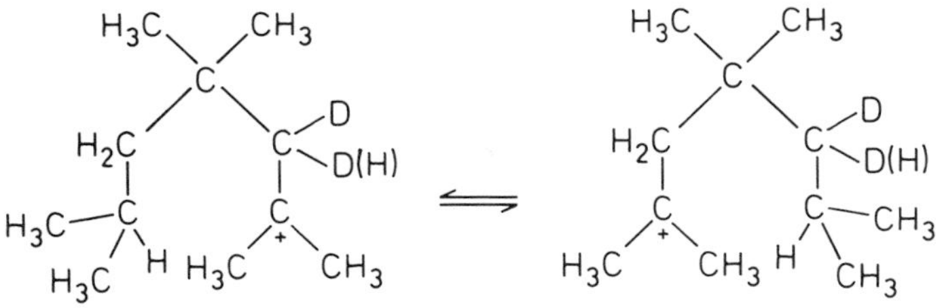

[157]

(113)

[158]

The methylene-deuteriated cations [157] also showed large temperature-dependent isotope splittings resulting from pertubation by one ($K = 1.31$ at $-110°C$) and two deuteriums ($K = 1.72$ at $-110°C$). The cations with deuterium remote from the positive charge are favoured in (112) equilibrium by $\Delta H° = -125$ cal mol^{-1} (d_1-[157]) and $\Delta H° = -129$ cal mol^{-1} per D (D_2-[157]) respectively.

No splittings appear in the spectra of the methylene-deuteriated cations [158] and only small temperature-independent intrinsic isotope shifts are observed in the ^{13}C nmr spectrum. The kinetically broadened signal of the averaged methyl groups in D_2-[158] has a line width of 118 Hz at $-130°C$ which gives a calculated maximum for K of 1.06 per D. The proton spectra are identical with the spectra of the unlabelled cation [97] except for the reduced intensity of the partially deuteriated methylene groups.

Different conformational preferences in [157] and [158] result in suppression of the equilibrium isotope effect in [158]. Two limiting conformations [159] and [160] for the cations [157] and [158] can be envisaged.

Conformation [160] has a dihedral angle of 30° between the empty p-orbital and the β-methylene C—D/C—H bonds and should show a large isotope effect, whereas conformation [159] with an angle of 60° should show a smaller effect (Sunko *et al.*, 1977; Hehre *et al.*, 1979). The conformational equilibrium between [159] and [160] is dependent on the size of R. In cation [157] (R = $CH_2CH_2CHMe_2$) the steric hindrance between R and the vicinal Me groups is less than in [158] where R is more sterically demanding (R = $CMe_2CH_2CHMe_2$). The large equilibrium isotope effect observed in [157] shows that the preferred conformation for this cation is [160] in which the β—C—H/C—D bonds are better aligned for hyperconjugation with the vacant p-orbital at the C^+ carbon than in [159]. The severe crowding prevents cation [158] from adopting conformation [160] and conformation [159] is preferred instead. The β—C—H/C—D methylene bonds are at unfavourable angles for hyperconjugation in [159] and thus the equilibrium isotope effect is suppressed.

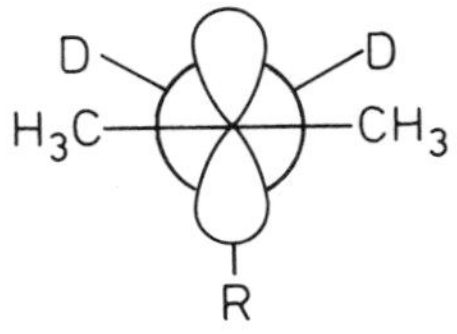

D D
H_3C CH_3
R

[159]

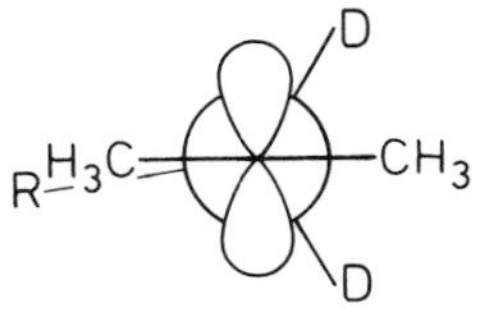

D
RH_3C CH_3
D

[160]

The cationic transition states in the solvolysis of [150] and [151] are closely related static analogues of cations [157] and [158]. They prefer conformation [160] (R = CH_2CH_3) and [159] (R = $C(CH_3)_3$) respectively. This shows that the same type of geometric requirements for C—H hyperconjugation apply to persistent carbocations as was concluded by Shiner for the transition state in solvolytic substitution reactions.

The similarity between the equilibrium isotope effects in [157] and the 1,2-dimethylcyclopentyl cation [104] points to a similar conformation in the vicinity of the C^+-carbon. The magnitude of the equilibrium isotope effect can thus be used as a new means of aiding in the assignment of conformation and structure in solution.

The other major factor which influences the extent of hyperconjugation besides conformational preferences is the degree of charge at the cationic carbon. The demand for C—H hyperconjugation varies according to what other modes of stabilisation are available, and this depends on the structure of the cation especially the substitution pattern at the cationic carbon. Variable demand for C—H hyperconjugation will in turn influence the magnitude of the β-deuterium equilibrium isotope effect. This is clearly evident comparing the β-CD_3-isotope effect in 1,2-dimethylcyclopentyl cation [115] with the decreasing effect with increasing σ-bridging in dimethylbicyclo[2.1.1]hexyl cation [116] and dimethylnorbornyl cation [117] (p. 132).

C—C-Hyperconjugation in carbocations with suitable structure and appropriate aligned conformations reduces the demand for C—H hyperconjugation and thus changes the magnitude of the β-deuterium isotope effect. The influence on vibrational frequencies of β-CH_3 and β-CH_2-groups could be different. Depending on these influences and on the preferred pathway (interior-CH_2 *vs* terminal-CH_3) for C—H hyperconjugation, discussed by Saunders *et al.* (1977a), the equilibrium isotope effect for β-methylene deuteriation can be larger, for example in [104] compared to [116] (p. 149), or smaller, for example in [158] compared to [156] (p. 155) (Siehl, 1986b), than predicted from the corresponding CD_3-effects using a $\cos^2$ function (Sunko and Hehre, 1983).

References

Ahlberg, P., Engdahl, C. and Jonsäll, G (1981). *J. Am. Chem. Soc.* **103**, 1583
Ahlberg, P., Jonsäll, G. and Engdahl, C. (1983a). *Adv. Phys. Org. Chem.* **19**, 223
Ahlberg, P., Engdahl, C. and Jonsäll, G. (1983b). *J. Am. Chem. Soc.* **105**, 891
Ahlberg, P., Jonsäll, G., Huang, M. B. and Goscinski, O. (1983c). *J. Chem. Soc. Perkin Trans 2* 305
Anet, F. A. L. and Dekmezian, A. H. (1979). *J. Am. Chem. Soc.* **101**, 5449
Anet, F. A. L. and Kopelevich, M. (1986a). *J. Am. Chem. Soc.* **108**, 1355
Anet, F. A. L. and Kopelevich, M. (1986b). *J. Am. Chem. Soc.* **108**, 2109
Anet, F. A. L. and Leyendecker, F. (1973). *J. Am. Chem. Soc.* **95**, 156
Anet, F. A. L. and Rawdah, T. N. (1978). *J. Am. Chem. Soc.* **100**, 7166
Anet, F. A. L., Cheng, A. K. and Krane, J. (1973). *J. Am. Chem. Soc.* **95**, 7877
Anet, F. A. L. Cheng, A. K., Mioduski, J. and Meinwald, J. (1974). *J. Am. Chem. Soc.* **96**, 2887
Askani, R., Kalinowski, H.-O. and Weuste, B. (1982). *Org. Mag. Res.* **18**, 176
Askani, R., Kalinowski, H.-O., Pelech, B. and Weuste, B. (1984). *Tetrahedron Lett.* **25**, 2321
Aydin, R. and Günther, H. (1981). *J. Am. Chem. Soc.* **103**, 1301
Batiz-Hernandez, H. and Bernheim, R. A. (1967). *Prog. NMR Spectrosc.* **3**, 63
Berger, S. and Diehl, B. W. K. (1986). *Mag. Res. Chem.* **24**, 1073
Berger, S. and Künzer, H. (1985). *J. Am. Chem. Soc.* **107**, 2804
Bigeleisen, J. and Mayer, M. G. (1947). *J. Chem. Phys.* **15**, 261
Booth, H. and Everett, J. R. (1980a). *Can. J. Chem.* **58**, 2709
Booth, H. and Everett, J. R. (1980b). *Can. J. Chem.* **58**, 2714
Brookhart, M., Lamanna, W. and Humphrey, M. B. (1982). *J. Am. Chem. Soc.* **104**, 2117
Brown, H. C. (1983). *Acc. Chem. Res.* **16**, 432
Brown, H. C. and Schleyer, P. v. R. (1977). *In* "The nonclassical ion problem", Chap. 5. Plenum Press, New York
Bywater, S., Brownstein, S. and Worsfold, D. J. (1980) *J. Organomet. Chem.* **199**, 1
Casey, C. P., Fagan, P. J. and Miles, W. H. (1982). *J. Am. Chem. Soc.* **104**, 1134

Chan, S. I., Lin, L., Clutter, D. and Dea, P. (1970). *Proc. Natl. Acad. Sci. U.S.A.* **65**, 816
Christl, M., Leininger, H. and Brückner, D. (1983). *J. Am. Chem. Soc.* **105**, 4843
Coates, R. M. and Fretz, E. R. (1975). *J. Am. Chem. Soc.* **97**, 2538
Collins, C. J. and Bowman, N. S. (eds.) (1970). "Isotope Effects in Chemical Reactions", Van Nostrand Reinhold, New York
DePuy, C. H., Fünfschilling, P. C. and Olson, J. M. (1976). *J. Am. Chem. Soc.* **98**, 276
DePuy, C. H., Fünfschilling, P. C., Andrist, A. H. and Olson, J. M. (1977). *J. Am. Chem. Soc.* **99**, 6297
Dewar, M. J. S. and Merz, K. M. Jr, (1986). *J. Am. Chem. Soc.* **108**, 5634
Dewar, M. J. S. and Reynolds, C. H. (1984). *J. Am. Chem. Soc.* **106**, 6388
Ernst, L., Hopf, H. and Wullbrandt, D. (1983). *J. Am. Chem. Soc.* **105**, 4469
Fong, F. K. (1974). *J. Am. Chem. Soc.* **96**, 7638
Forsén, S., Gunnarsson, G., Wennerström, H., Altman, L. J. and Laungani, D. (1978). *J. Am. Chem. Soc.* **100**, 8264
Forsyth, D. A. (1984). *In* "Isotopes in Organic Chemistry" (eds. E. Buncel and C. C. Lee), Vol. 6, Chap. 1. Elsevier, Amsterdam
Forsyth, D. A. and MacConnell, M. M. (1983). *J. Am. Chem. Soc.* **105**, 5920
Forsyth, D. A. and Pan, Y. (1985). *Tetrahedron Lett.* **26**, 4997
Forsyth, D. A. and Yang, J.-R. (1986). *J. Am. Chem. Soc.* **108**, 2157
Forsyth, D. A., Botkin, J. H. and Osterman, V. M. (1984). *J. Am. Chem. Soc.* **106**, 7663
Forsyth, D. A. Botkin, J. H. and Sardella, D. J. (1986). *J. Am. Chem. Soc.* **108**, 2797
Gold, V. (1968). *Trans. Faraday Soc.* **64**, 2770
Grob, C. A. (1983). *Acc. Chem. Res.* **16**, 426
Grubbs, R. H., Lee, B. J., Gajda, G. J., Schaefer, W. P., Howard, T. R., Ikariya, T. and Straus, D. A. (1981). *J. Am. Chem. Soc.* **103**, 7358
Günther, H. and Aydin, R. (1981). *Angew. Chem.* **93**, 1000
Günther, H. and Wesener, J. R. (1982). *Tetrahedron Lett.* **23**, 2845
Günther, H. and Wesener, J. R. (1985). *J. Am. Chem. Soc.* **107**, 7307
Günther, H., Pawliczek, J.-B., Ulsen, J. and Grimme, W. (1975). *Chem. Ber.* **108**, 3141
Günther, H., Joseph-Natan, P., Aydin, R., Wesener, J. R., Santillan, R. L. and Garibay, M. E. (1984). *J. Org. Chem.* **49**, 3847
Günther, H., Wesener, J. R. and Moskau, D. (1985). *Tetrahedron Lett.* 1491
Halevi, E. A. (1963). *Prog. Phys. Org. Chem.* **1**, 109
Halevi, E. A., Bary, Y. and Gilboa, H. (1979). *J. Chem. Soc. Perkin 2* 938
Hansen, H.-J. and Lang, R. W. (1980). *Helv. Chim. Acta* **63**, Fasc. 5, 1215
Hansen, P. E. (1983). *Ann. Rep. on NMR Spectra.* **15**, 105
Hansen, P. E. and Duus, F. (1984). *Org. Mag. Res.* **22**, 16
Hansen, P. E. and Lyčka, A. (1984). *Org. Mag. Res.* **22**, 569
Hansen, P. E., Duus, F. and Schmitt, P. (1982). *Org. Mag. Res.* **18**, 58
Hehre, W. J., DeFrees, D. J. and Sunko, D. E. (1979). *J. Am. Chem. Soc.* **101**, 2323
Hehre, W. J., Hout, Jr, R. F. and Levi, B. A. (1983). *J. Computational Chem.* **4**, 449
Hogeveen, H. and van Kruchten, E. M. G. A. (1981). *J. Org. Chem.* **46**, 1350
Ittel, S. D., Van-Catledge, F. A. and Jesson, J. P. (1979). *J. Am. Chem. Soc.* **101**, 6905
Jackman, L. M. and Cotton, F. A. (eds) (1975). "Dynamic Nuclear Magnetic Resonance Spectroscopy". Academic Press, New York
Jameson, C. J. (1977). *J. Chem. Phys.* **66**, 4977

Jameson, C. J. and Osten, H.-J. (1984a). *J. Chem. Phys.* **81**, 4293
Jameson, C. J. and Osten, H.-J. (1984b). *J. Chem. Phys.* **81**, 4300
Jensen, F. R. and Neese, R. A. (1971). *J. Am. Chem. Soc.* **93**, 6329
Jensen, F. R. and Smith, L. A. (1964). *J. Am. Chem. Soc.* **86**, 956
Jeremić, D., Milosavljević, S. and Mihailović, M. L. (1982). *Tetrahedron* **38**, 3328
Kalinowski, H.-O. (1984). *Nachr. Chem. Tech. Lab.* **32**, 874
Kalinowski, H.-O., Berger, S. and Braun, S. (1984). "13 C-NMR Spektroskopie". Thieme Verlag, Stuttgart
Karle, J., Lowrey, A. H., George, C. and D'Antonio, P. (1971). *J. Am. Chem. Soc.* **93**, 6399
Kates, M. R. (1978). Ph.D. Thesis. Yale University, New Haven, Connecticut
Koch, E.-W. (1985). Dissertation. University of Tübingen, Tübingen, Germany
Kresge, A. J., Arrowsmith, C. H., Baltzer, L., Powell, M. F. and Tang, Y. S. (1986). *J. Am. Chem. Soc.* **108**, 1356
Leipert, T. K. (1977). *Org. Mag. Res.* **9**, 157
Lippmaa, E., Pehk, T. and Laht, A. (1982). *Org. Mag. Res.* **19**, 21
Lloyd, J. R. (1978). Ph.D. Thesis. Yale University, New Haven, Connecticut
Maier, G., Kalinowski, H.-O. and Euler, K. (1982). *Angew. Chem.* **94**, 706 (Int. Ed. Engl., **21**, 693)
Majerski, Z., Zuanic, M. and Metelco, B. (1985). *J. Am. Chem. Soc.* **107**, 1727
McMurry, J. E. and Hodge, C. N. (1984). *J. Am. Chem. Soc.* **106**, 6450
Melander, L. (1960). "Isotope Effects on Reaction Rates". Ronald Press Company, New York
Melander, L. and Saunders, Jr. W. H. (1980). "Reaction Rates of Isotopic Molecules". John Wiley, New York
Moore, P., Howarth, O. W., McAteer, C. H. and Morris, G. E. (1981). *J. Chem. Soc. Chem. Commun.* 506
Myhre, P. C., Yannoni, C. S., Kruger, J. D., Hammond, B. L., Lok, S. M., Macho, V., Limbach, H. H. and Vieth, H. M. (1984). *J. Am. Chem. Soc.* **106**, 6079
Oki, M. (1985). "Applications of Dynamic NMR Spectroscopy to Organic Chemistry". VCH Publishers, Weinheim, Germany
Olah, G. A., Kelly, D. P., Jeuell, C. L. and Porter, R. D. (1970). *J. Am. Chem. Soc.* **92**, 2544
Olah, G. A., Jeuell, C. L., Kelly, D. P. and Porter, R. D. (1972). *J. Am. Chem. Soc.* **94**, 146
Olah, G. A., Prakash, G. K. S., Donovan, D. J. and Yavari, I. (1978a). *J. Am. Chem. Soc.* **100**, 7085
Olah, G. A., Roberts, J. D., Staral, J. S., Yavari, I., Prakash, G. K. S. and Donovan, D. J. (1978b). *J. Am. Chem. Soc.* **100**, 8016
Olah, G. A., Prakash, G. K. S. and Nakajima, T. (1982). *J. Am. Chem. Soc.* **104**, 1031
Olah, G. A., Saunders, M. and Prakash, G. K. S. (1983). *Acc. Chem. Res.* **16**, 440
Olah, G. A., Krishnamurthy, V. V., Prakash, G. K. S. and Iyer, P. S. (1985a). *J. Am. Chem. Soc.* **107**, 5015
Olah, G. A., Prakash, G. K. S. and Arvanaghi, M. (1985b). *J. Am. Chem. Soc.* **107**, 6017
Olah, G. A., Sommer, J. and Prakash, G. K. S. (1985c). *In* "Superacids", p. 144. John Wiley, New York
Petersen, S. B. and Led, J. J. (1979). *J. Mag. Res.* **3**, 603
Rappoport, Z., Biali, S. E. and Hull, W. E. (1985). *J. Am. Chem. Soc.* **107**, 5450
Reuben, J. (1984). *J. Am. Chem. Soc.* **106**, 6180

Reuben, J. (1985a). *J. Am. Chem. Soc.* **107**, 1756
Reuben, J. (1985b). *J. Am. Chem. Soc.* **107**, 5867
Reuben, J. (1985c). *J. Am. Chem. Soc.* **107**, 1433
Reuben, J. (1986a). *J. Am. Chem. Soc.* **108**, 1082
Reuben, J. (1986b). *J. Am. Chem. Soc.* **108**, 1735
Roberts, J. D., Brittain, W. J. and Squillacote, M. E. (1984). *J. Am. Chem. Soc.* **106**, 7280
Robinson, M. J. T. (1971). *Pure Appl. Chem.* **25**, 635
Robinson, M. J. T. and Baldry, K. W. (1977a). *J. Chem. Res (S)* 86; *J. Chem. Res (M)* 1001
Robinson, M. J. T. and Baldry, K. W. (1977b) *Tetrahedron* **33**, 1663
Robinson, M. J. T., Rosen, K. M. and Workman, J. D. B. (1977). *Tetrahedron* **3**, 1655
Sandström, J. (1982). "Dynamic NMR Spectroscopy". Academic Press, New York
Saunders, M. (1979). *In* "Stereodynamics of Molecular Systems" (ed. R. Sarma), p. 171. Pergamon Press, Oxford
Saunders, M. and Evilia, R. F. (1985). *Spectrosc. Lett.* **18**, 105
Saunders, M. and Handler, A. (1985). Unpublished results; *cf.* Saunders *et al.* Abstr. "Symposium on Advances in Carbocation Chemistry", American Chemical Society. Seattle meeting, March 1983
Saunders, M. and Jarret, R. M. (1985). *J. Am. Chem. Soc.* **107**, 2648
Saunders, M. and Jarret, R. M. (1986). *J. Am. Chem. Soc.* **108**, 7549
Saunders, M. and Kates, M. R. (1977). *J. Am. Chem. Soc.* **99**, 8071
Saunders, M. and Kates, M. R. (1978). *J. Am. Chem. Soc.* **100**, 7082
Saunders, M. and Kates, M. R. (1980). *J. Am. Chem. Soc.* **102**, 6867
Saunders, M. and Kates, M. R. (1983). *J. Am. Chem. Soc.* **105**, 3571
Saunders, M. and Siehl, H.-U. (1980). *J. Am. Chem. Soc.* **102**, 6868
Saunders, M. and Siehl, H.-U. (1985). Unpublished results; *cf.* Siehl and Walter (1985)
Saunders, M. and Vogel, P. (1971a). *J. Am. Chem. Soc.* **93**, 2559
Saunders, M. and Vogel, P. (1971b). *J. Am. Chem. Soc.* **93**, 2561
Saunders, M., Vogel, P. and Jaffe, M. H. (1971). *J. Am. Chem. Soc.* **93**, 2558
Saunders, M., Wiberg, K. B., Seybold, G. and Vogel, P. (1973). *J. Am. Chem. Soc.* **95**, 2045
Saunders, M., Telkowski, L. and Kates, M. R. (1977a). *J. Am. Chem. Soc.* **99**, 8070
Saunders, M., Wiberg, K. B., Kates, M. R. and Pratt, W. (1977b). *J. Am. Chem. Soc.* **99**, 8072
Saunders, M., Anet, F. A. L., Hewett, A. P. W. and Basus, V. J. (1980a). *J. Am. Chem. Soc.* **102**, 3945
Saunders, M., Faller, J. W. and Murray, H. H. (1980b). *J. Am. Chem. Soc.* **102**, 2306
Saunders, M., Schleyer, P. v. R. and Chandrasekhar, J. (1980c). Rearrangements of carbocations. *In* "Rearrangements in Ground and Excited States" (ed. P. de Mayo), Chap. 1. Academic Press, New York
Saunders, M., Walker, G. E. and Kates, M. R. (1981). *J. Am. Chem. Soc.* **103**, 4623
Saunders, M., Saunders, S. and Johnson, C. A. (1984). *J. Am. Chem. Soc.* **106**, 3098
Schleyer, P. v. R. and Neugebauer, W. (1980). *J. Organometal. Chem.* **198**, C1
Schleyer, P. v. R., Baborack, J. C. and Chari, S. (1971). *J. Am. Chem. Soc.* **93**, 5275
Schlosser, M. and Strähle, M. (1980). *Angew. Chem. 92*, 497 (Int. Ed. Engl. **19**, 487)
Schlosser, M. and Strähle, M. (1981) *J. Organometal. Chem.* **220**, 277
Schneider, J. (1985). Diplomarbeit. University of Tübingen, Tübingen, Germany

Schrock, R. R. and Holmes, S. J. (1981). *J. Am. Chem. Soc.* **103**, 4599
Servis, K. L. and Domenick, R. L. (1985). *J. Am. Chem. Soc.* **107**, 7186
Servis, K. L. and Domenick, R. L. (1986). *J. Am. Chem. Soc.* **108**, 2211
Servis, K. L. and Shue, F.-F. (1980). *J. Am. Chem. Soc.* **102**, 7233
Shapet'ko, N. N., Berestova, S. S., Lukovkin, G. M. and Bogachev, Y. S. (1975). *Org. Mag. Res.* **7**, 237
Shapiro, M. J., Kahle, A. D., Damon, R. and Coppola, G. M. (1981). *J. Org. Chem.* **46**, 1221
Shapley, J. R. and Calvert, R. B. (1977). *J. Am. Chem. Soc.* **99**, 5225
Shapley, J. R. and Calvert, R. B. (1978). *J. Am. Chem. Soc.* **100**, 7726
Shapley, J. R., Williams, J. M., Stucky, G. D., Calvert, R. B., Schultz, A. J. and Suib, S. L. (1978). *J. Am. Chem. Soc.* **100**, 6240
Shapley, J. R., Williams, J. M., Stucky, G. D., Calvert, R. B., and Schultz, A. J. (1979). *Inorg. Chem.* **18**, 319
Shapley, J. R., Cree-Uchiyama, M. and George, G. M. St. (1986). *J. Am. Chem. Soc.* **108**, 1316
Shiner, Jr., V. J. (1961). *J. Am. Chem. Soc.* **83**, 240
Shiner, Jr., V. J. and Hartshorn, S. R. (1972). *J. Am. Chem. Soc.* **94**, 9002
Shiner, Jr., V. J. and Humphrey, Jr., J. S. (1963). *J. Am. Chem. Soc.* **85**, 2416
Siehl, H.-U. (1985). *J. Am. Chem. Soc.* **107**, 3390
Siehl, H.-U. (1986a). Proc. 8th IUPAC Conference on Physical Organic Chemistry, Tokyo, Japan, Aug. 1986, Elsevier Science Publishers, Amsterdam, in press
Siehl, H.-U. (1986b). Habilitationsschrift. University of Tübingen, Tübingen, Germany
Siehl, H.-U. and Koch, E.-W. (1985). *J. Chem. Soc. Chem. Commun.* 496
Siehl, H.-U. and Walter, H. (1984). *J. Am. Chem. Soc.* **106**, 5355
Siehl, H.-U. and Walter, H. (1985). *J. Chem. Soc. Chem. Commun.* 76
Sorensen, T. S. and Kirchen, R. P. (1977). *J. Am. Chem. Soc.* **99**, 6687
Sorensen, T. S. and Kirchen, R. P. (1978). *J. Chem. Soc. Chem. Commun.* 769
Sorensen, T. S. and Kirchen, R. P. (1979). *J. Am. Chem. Soc.* **101**, 3240
Sorensen, T. S. and Ranganayakulu, K. (1984). *J. Org. Chem.* **49**, 4310
Sorensen, T. S. and Schmitz, L. R. (1980). *J. Am. Chem. Soc.* **102**, 1645
Sorensen, T. S., Ranganayakulu, K. and Huang, E. (1972). *J. Am. Chem. Soc.* **94**, 1780
Sorensen, T. S., Haseltine, R. and Wong, N. (1975). *Can. J. Chem.* **53**, 1891
Sorensen, T. S., Kirchen, R. P. and Wagstaff, K. (1978). *J. Am. Chem. Soc.* **100**, 6761
Sorensen, T. S., Kirchen, R. P., Ranganayakulu, K., Rauk, A. and Singh, B. P. (1981a). *J. Am. Chem. Soc.* **103**, 588
Sorensen, T. S., Kirchen, R. P., Okazawa, N., Ranganayakulu, K. and Rauk, A. (1981b). *J. Am. Chem. Soc.* **103**, 597
Sorensen, T. S., Kirchen, R. P., Ranganayakulu, K. and Singh, B. P. (1981c). *Can. J. Chem.* **59**, 2173
Stern, M., Spindel, W. and Monse, E. U. (1968). *J. Chem. Phys.* **48**, 2908
Stone, G. A., Dawkins, G. M., Green, M. and Orpen, A. G. (1982). *J. Chem. Soc. Chem. Commun.* 41
Sunko, D. E. and Hehre, W. J. (1983). *Prog. Phys. Org. Chem.* **14**, 205
Sunko, D. E., Boričić, S. and Nikolertić, M. (1962). *J. Am. Chem. Soc.* **84**, 1615
Sunko, D. E., Humski, K., Malojčić, R. and Borčić, S. (1970). *J. Am. Chem. Soc.* **92**, 6534
Sunko, D. E., Szele, I. and Hehre, W. J. (1977). *J. Am. Chem. Soc.* **99**, 5000

Swain, C. G., Stivers, E. C., Reuver, J. F. and Schaad, L. J. (1958). *J. Am. Chem. Soc.* **80**, 5885

Telkowski, L. (1975). Ph.D Thesis. Yale University, New Haven, Connecticut

Vogel, P. (1979). Carbokationen. *In* Wentrup, C. "Reaktive Zwischenstufen" (ed. C. Wentrup), Vol. 2, Chap. 7. Thieme Verlag, Stuttgart

Vogel, P. (1985). "Carbocation Chemistry", Elsevier Science Publishers, Amsterdam

Walling, C. (1983). *Acc. Chem. Res.* **16**, 448

Walter, H. (1985). Dissertation. University of Tübingen, Tübingen, Germany

Wehrli, F. W. and Wirthlin, T. (1976). "Interpretation of Carbon-13 NMR Spectra", p. 108. Heyden, London

Wehrli, F. W., Jeremic, D., Milhailović, M. L. and Milosavljević, S. (1978). *J. Chem. Soc. Chem. Commun.* 302

Whipple, E. B., Andrews, G. C. and Chmurny, G. N. (1981). *Org. Mag. Res.* **15**, 324

Willi, A. V. (1983). "Isotopeneffekte bei chemischen Reaktionen". Thieme Verlag, Stuttgart

Williams, I. H. (1986). *J. Chem. Soc. Chem. Commun.* 629

Williams, J. M., Ittel, S. D., Brown, R. K., Schultz, A. J., Stucky, G. D. and Harlow, R. L. (1980). *J. Am. Chem. Soc.* **102**, 981

Wolfsberg, M. (1969). *Ann. Rev. Phys. Chem.* **20**, 449

Wolfsberg, M. (1972). *Acc. Chem. Res.***5**, 225

Wolfsberg, M. and Kleinman, L. I. (1973). *J. Chem. Phys.* **59**, 2043

Wolfsberg, M. and Kleinman, L. I. (1974a). *J. Chem. Phys.* **60**, 4740

Wolfsberg, M. and Kleinman, L. I. (1974b). *J. Chem. Phys.* **60**, 4749

Yannoni, C. S., Myhre, P. C. and Macho, V. (1982). *J. Am. Chem. Soc.* **104**, 7380

Yannoni, C. S., Myhre, P. C. and McLaren, K. L. (1985). *J. Am. Chem. Soc.* **107**, 5294

Yashiro, M., Yano, S. and Yoshikawa, S. (1986). *J. Am. Chem. Soc.* **108**, 1096

The Mechanisms of Reactions of β-Lactam Antibiotics

MICHAEL I. PAGE

Department of Chemical and Physical Sciences, The Polytechnic, Queensgate, Huddersfield HD1 3DH, U.K

ADVANCES IN PHYSICAL ORGANIC CHEMISTRY
VOLUME 23 ISBN 0 12 033523 9

1 Introduction

Nearly sixty years have now passed since Sir Alexander Fleming observed antibiosis between a *Penicillium* mould and bacterial cultures and gave the name penicillin to the active principle (Fleming, 1929). Although it was proposed in 1943 that penicillin contained a β-lactam ring [1] (Abraham *et al.*, 1943), this was not generally accepted until an X-ray crystallographic determination of the structure had been completed (Crowfoot *et al.*, 1949).

Penicillin was the first naturally occurring antibiotic to be characterized and used in clinical medicine. It can now be seen as the progenitor of the β-lactam family of antibiotics which are characterized by the possession of the four-membered β-lactam ring. Penicillins were originally detected in fungi but subsequently were found in streptomycetes (Miller *et al.*, 1962). The cephalosporins [2], the second member of the β-lactam antibiotic family, were also originally discovered in fungi but later detected in streptomycetes (Newton and Abraham, 1955; Higgins *et al.*, 1974).

Until 1970 penicillins and cephalosporins were the only examples of naturally occurring β-lactam antibiotics. The discovery of 7-α-methoxy-cephalosporins [3] from *Streptomyces* in 1971 (Nagarajan *et al.*, 1971) stimulated the search for novel β-lactam antibiotics of microbial origin using sensitive new screening procedures. At present, β-lactam antibiotics can be classified into several groups according to their structure:

1	Penicillins (penams)	[1]	β-lactams fused to a thiazolidine.
2	Cephalosporins (cephems)	[2]	β-lactams fused to a dihydrothiazine.
3	Cephamycins	[3]	7-α-methoxycephalosporins.
4	Oxacephems	[4]	replacement of cephalosporin ring S by O.
5	Penems	[5]	double bond in 5-membered thiazolidine ring.
6	Clavulanic acid	[6]	oxapenam derivatives, β-lactam fused to an oxazolidine.
7	Thienamycin (carbapenems)	[7]	no heteroatom in the ring fused to the β-lactam.
8	Nocardicins	[8]	monocyclic β-lactams.
9	Monobactams	[9]	monocyclic β-lactams of sulphamic acid.

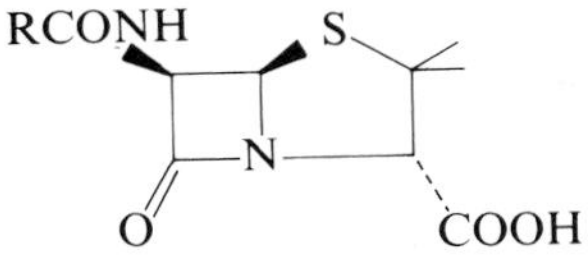

RCONH S N O COOH

[1]

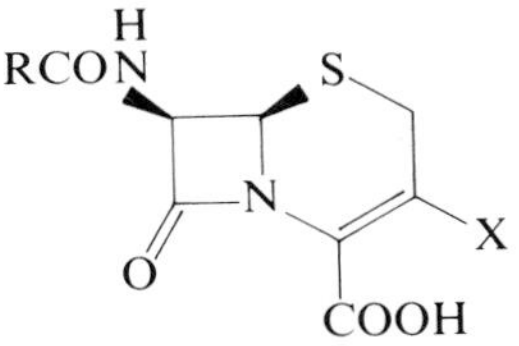

H RCON S N X O COOH

[2]

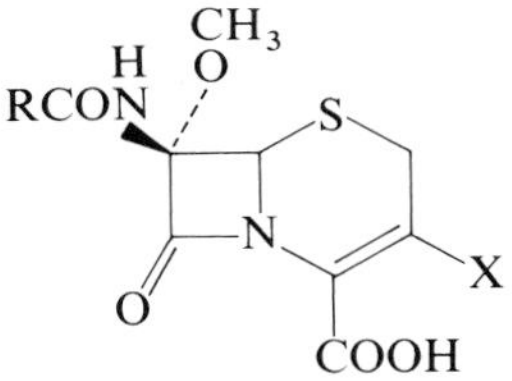

CH_3 H O RCON S N X O COOH

[3]

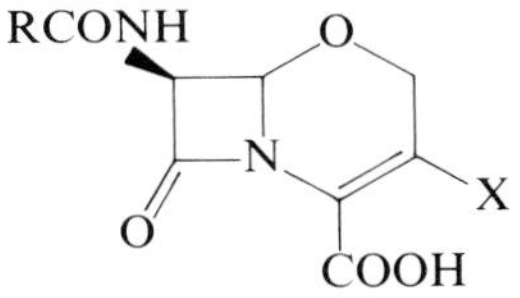

RCONH O N X O COOH

[4]

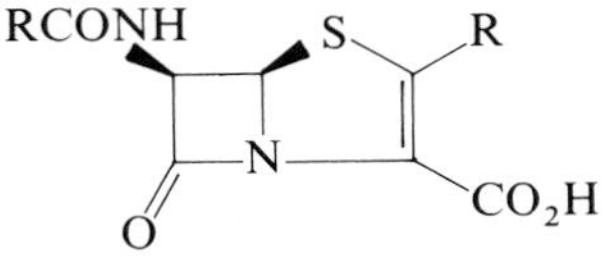

RCONH S R N O CO_2H

[5]

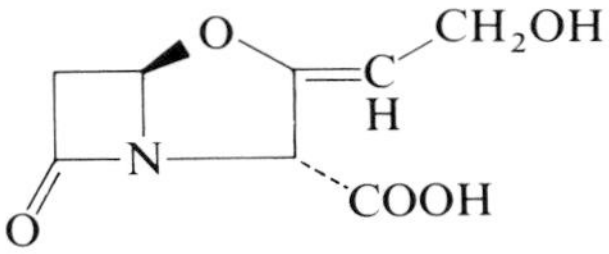

O CH_2OH C H N O COOH

[6]

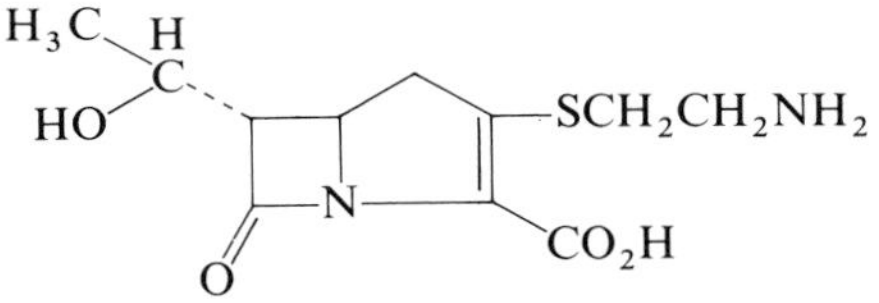

H_3C H C HO $SCH_2CH_2NH_2$ N O CO_2H

[7]

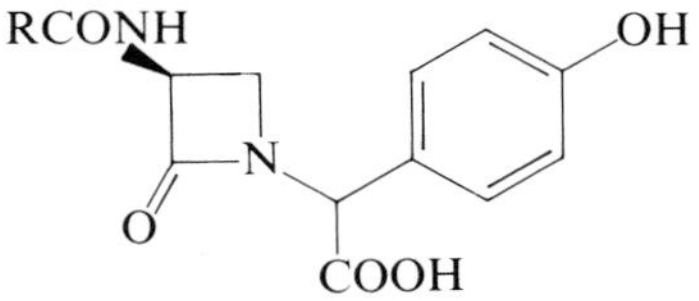

RCONH OH N O COOH

[8]

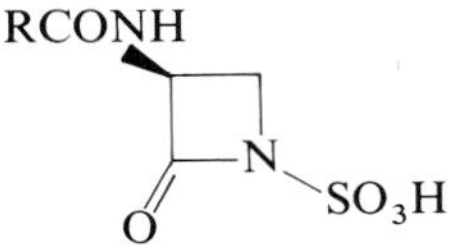

RCONH N O SO_3H

[9]

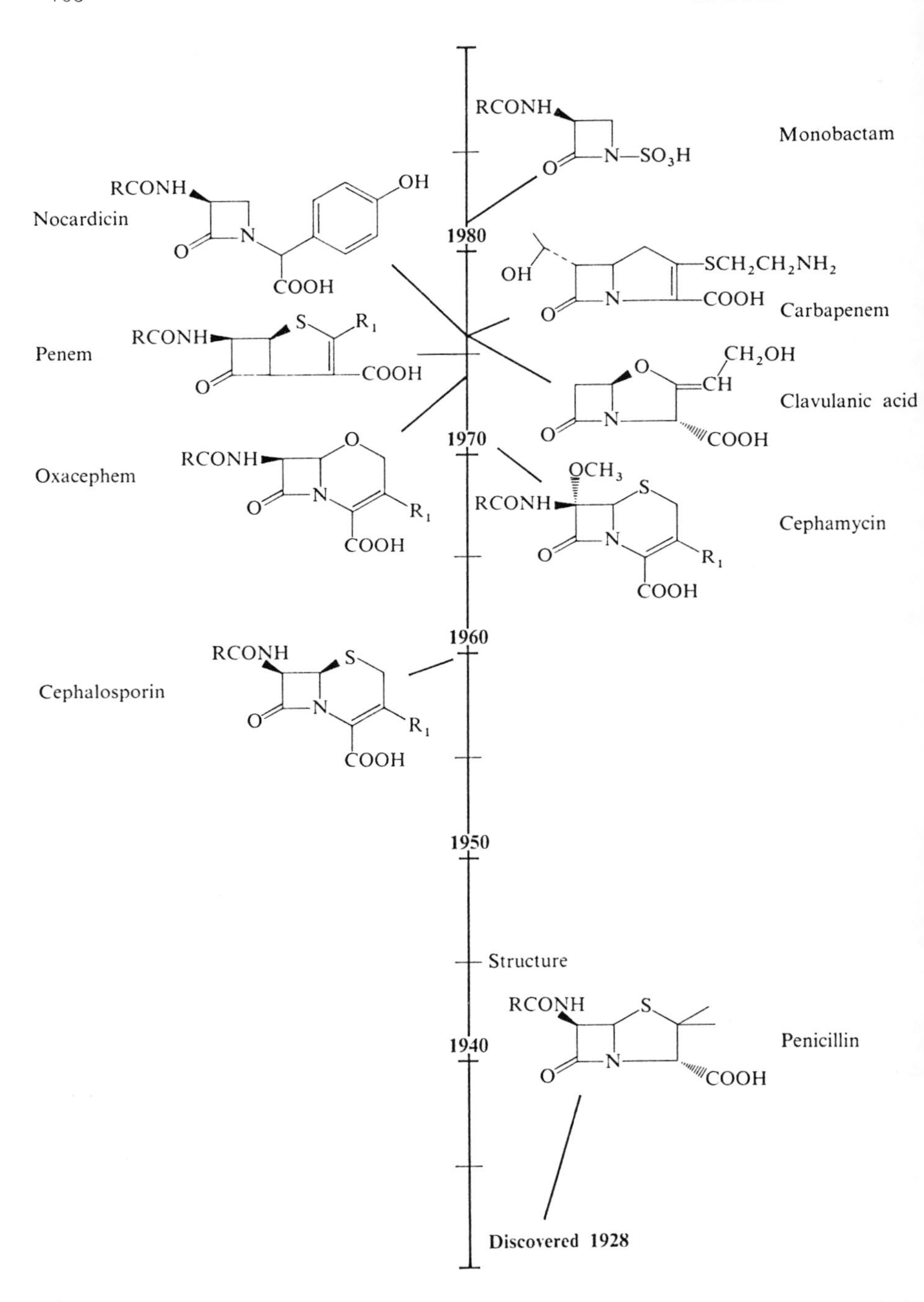

FIG. 1 The discovery of β-lactam antibiotics displayed against Anno Domini

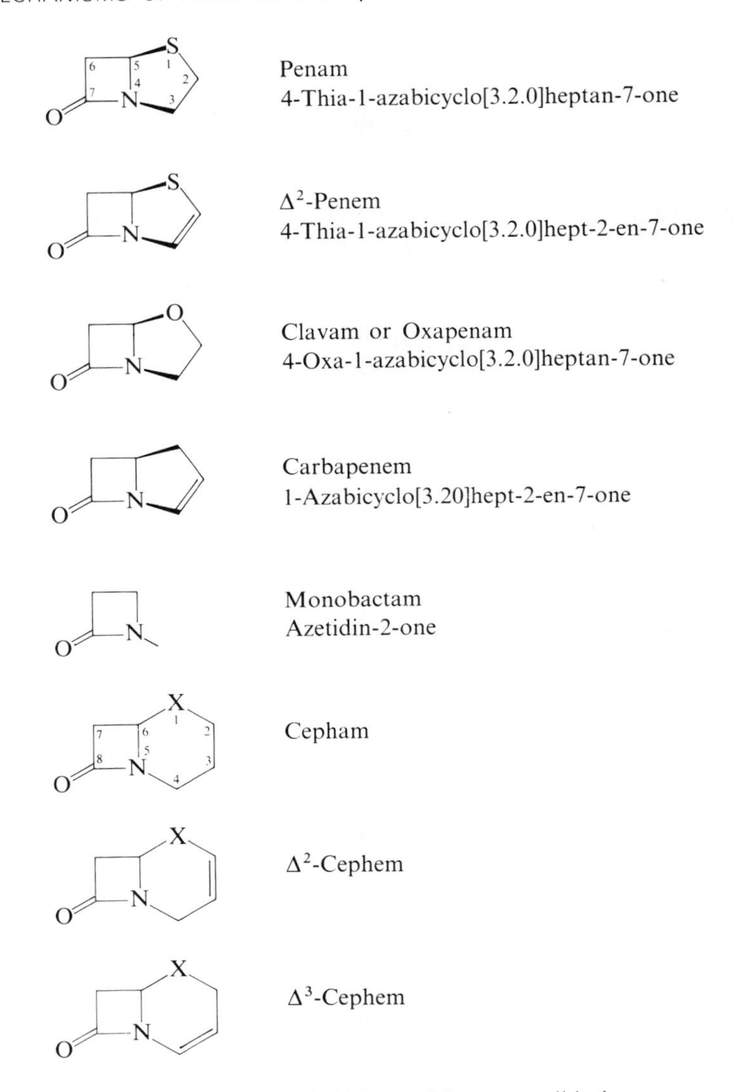

FIG. 2 Basic structural units of common β-lactam antibiotics

The chronology of these antibiotics is illustrated in Fig. 1, and some trivial names used to describe the ring structures are shown in Fig. 2. β-Lactams have now been found in eukaryotic fungi, actinomycetes (mostly in the genus *Streptomyces*) and even in bacteria, but their *in vivo* role is not yet understood. Recently β-lactones have also been isolated from various micro-organisms (Sykes *et al.*, 1985).

A large number of nuclear analogues of the β-lactam antibiotics have been prepared by complete chemical synthesis or by partial synthesis starting from a naturally occurring β-lactam. Of these only the oxacephems [4] and the penems [5] appear to have been investigated clinically. Most semisynthetic penicillins are made by the acylation of 6-aminopenicillanic acid [10] whilst cephalosporins can be prepared to give a variety of side chain substituents at C(7) and C(3) [11]. The latter illustrates a problem with nomenclature. Classically β-lactam antibiotics have been described using the ring sulphur as position-1. Hence for penicillins and cephalosporins the numbering is as illustrated in [12] and [11], respectively. Similarly substituents in these bicyclic systems are usually described by the prefix α- or β- rather than the equivalent *exo* or *endo* description.

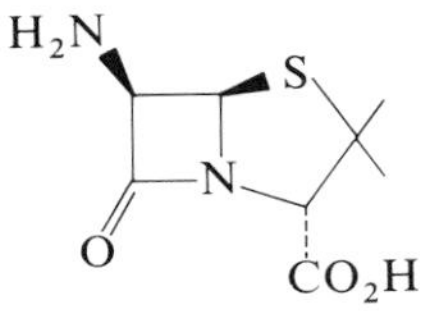

[10]

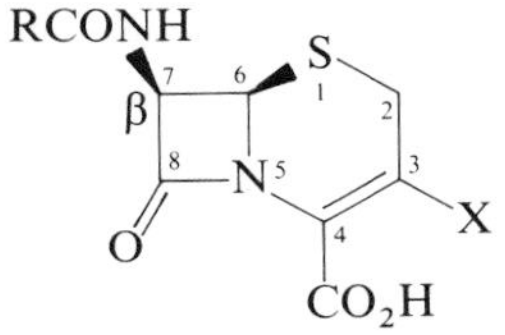

[11]

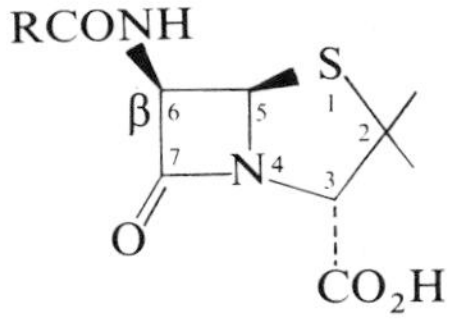

[12]

6-β-Side chains of penicillins may confer stability to acids and have allowed the development of orally administered penicillins, [12]. These derivatives are unaffected by gastric juice and may be absorbed from the gut. In contrast, although most cephalosporins are intrinsically stable at the low pH of gastric juice, the number of 7-β-side chains allowing oral absorption is limited. Some examples of oral cephalosporins are given in Table 1. The so-called first and second generation injectable cephalosporins are 3,7-modified cephalosporins (Table 1). Although 7-β-side chains may improve potency and breadth of antibacterial activity their potential is only expressed when the molecule also contains an appropriate 3-side chain. Third generation cephalosporins differ in their increased spectrum of activity, potency and high cost; "And they did worship strange Lactams, even unto the third generation" (Berger, 1984).

TABLE 1 Names and structures of common penicillins and cephalosporins

(a) Penicillins

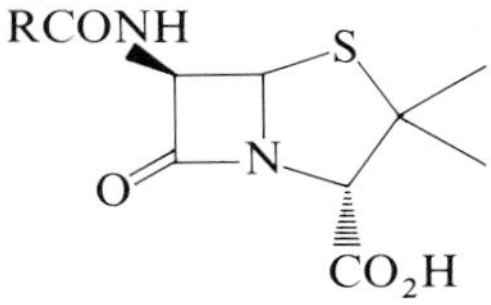

RCONH S N O CO_2H

Penicillin	R	Penicillin	R
Benzylpenicillin (Pen G)	$PhCH_2$—	Carbenicillin	PhCH— with CO_2H
Pen F	$C_2H_5CH{=}CHCH_2$—	Oxacillin	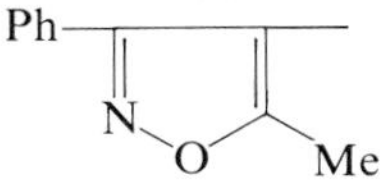 Ph— N O Me
Pen X	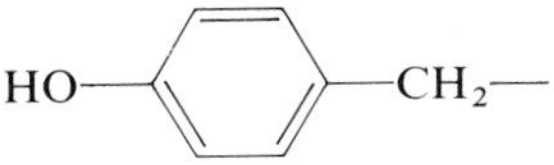 HO—(benzene ring)—CH_2—		
Pen K	$CH_3(CH_2)_6$—	Cloxacillin	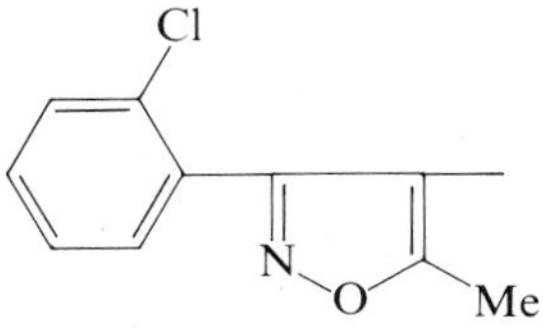 Cl N O Me
Pen V	$PhOCH_2$—	Dicloxacillin	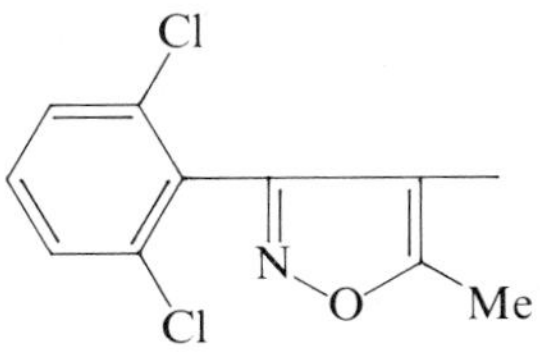 Cl Cl N O Me
Pen N	$H_3\overset{+}{N}-\underset{CO_2^-}{CH}-(CH_2)_3-$		
Propicillin	PhOCH(Et)—	Nufcillin	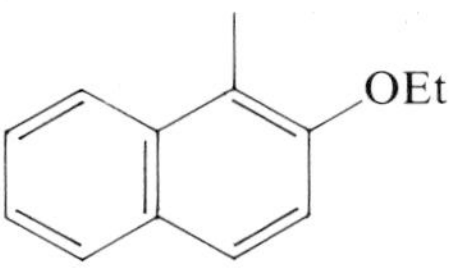 OEt
Phenbenicillin	PhOCH(Ph)—	Diphenicillin	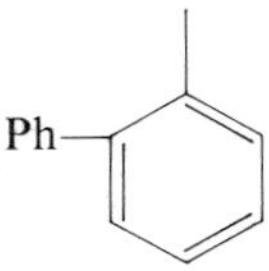 Ph—

TABLE 1 (*continued*)

Penicillin	R	Penicillin	R
Methicillin	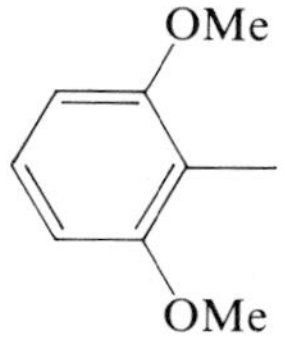 OMe OMe	Quinacillin	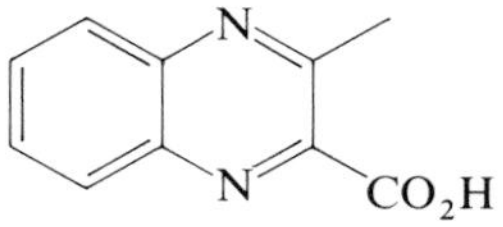 N N CO_2H
Ampicillin	PhCH— NH_2	Ticarcillin	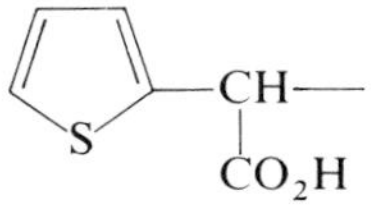 S —CH— CO_2H

(b) Cephalosporins

RCONH S N O CH_2L CO_2H

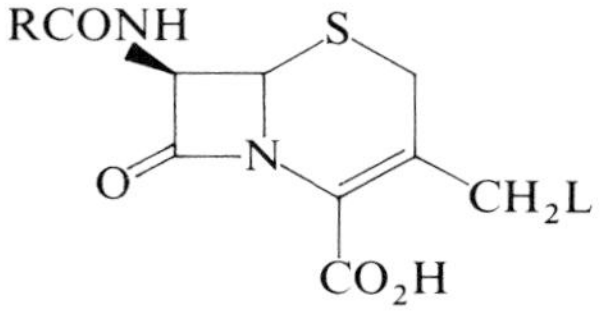

Cephalosporin	R	L
Oral cephalosporins		
cephaloglycin	Ph-CH(NH_2)—	-OAc
cephalexin	Ph-CH(NH_2)—	-H
cefaclor	Ph-CH(NH_2)—	-Cl
cefatrizine	HOC_6H_4-CH(NH_2)—	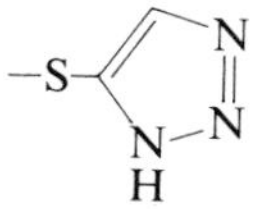 —S— N N N N H
First generation		
cephalothin	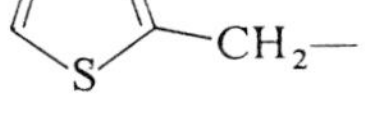 S CH_2—	-OAc
cephaloridine	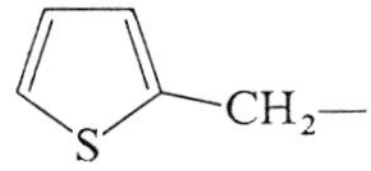 S CH_2—	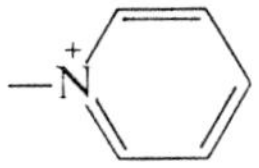 —$\overset{+}{N}$
cefazolin	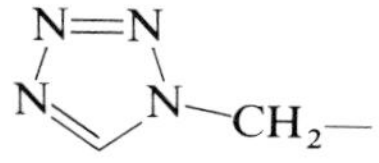 N═N N N—CH_2—	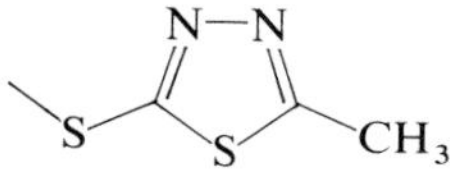 S N—N S CH_3

TABLE 1 (*continued*)

Cephalosporin	R	L
Second generation		
cefamandole	PhCH(OH)$^-$	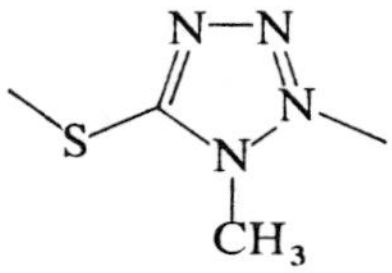 N—N, N, S, N, CH_3
cefuroxime	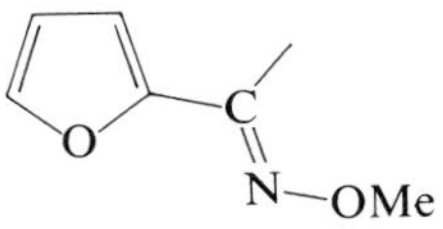 O, C, N—OMe	—O—$CONH_2$
cephamycin	$HO_2C—CH(CH_2)_3—$ (with NH_2 on CH)	$—OCONH_2$
Third generation		
cefotaxime	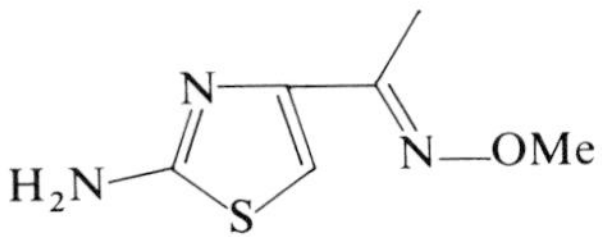 N, H_2N, S, N—OMe	—OAc
cefoperazone	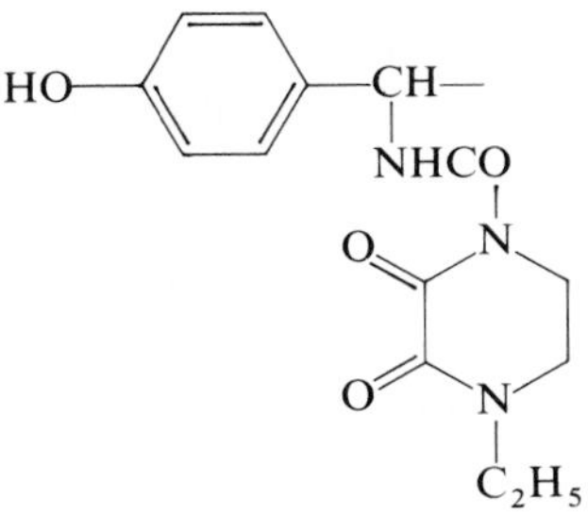 HO, CH—, NHCO, N, O, O, N, C_2H_5	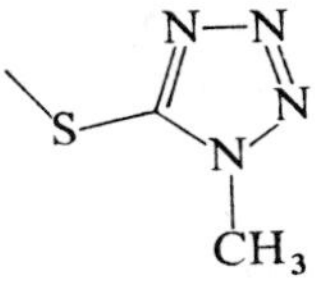 N—N, N, S, N, CH_3

2 Mode of action of β-lactam antibiotics

To understand how β-lactams kill bacteria requires a little knowledge of how a bacterial cell wall is formed and why bacteria need cell walls. Bacteria, unlike eukaryotic cells, possess a cell wall which is a complex structure in both Gram-positive and Gram-negative types. Cell walls help to maintain the shape of bacteria whilst the cell membranes are the osmotic barriers that allow the retention of nutrients and the exclusion of other compounds. Gram-positive and Gram-negative bacteria have internal osmotic pressures which are 10 to 30 times and 3 to 5 times, respectively, the external osmotic pressure. The rigid structure of the bacterial wall of both Gram-positive and Gram-negative species is due to the cross linking of linear polysaccharide chains by short segments of peptides called peptidoglycan.

It is accepted generally, but not universally (Reinicke *et al.*, 1985), that the biochemical targets for the β-lactams are some of the enzymes concerned with cell-wall synthesis. What *is* undoubtedly universally accepted is the increasing complexity of drug action as one moves from target enzyme to target cell and from target cell in a culture tube to target pathogen in the complex environment of an infected host. An understanding of the mode of action of any drug requires a knowledge of:

1 the pathway by which the drug reaches its biochemical target,
2 the identity and structure of the biochemical target and the chemistry of how the target is inhibited,
3 the cell's physiological response to the inhibition of the drug-sensitive reaction,
4 how and why inhibition of this reaction leads to interference of the life cycle of the cell.

β-Lactams exert their lethal action only on growing bacterial cells. Within a short time of adding penicillin to a bacterial culture, small bulges form in the bacteria often near mid cell where cell division is expected to occur. With time, the bulge enlarges and ultimately the cell membrane ruptures, resulting in death of the cell. These morphological and lytic effects of penicillin are accompanied by the covalent binding of the antibiotic to proteins, called penicillin binding proteins, PBP (Spratt, 1980). The PBPs are located on the inner membrane and, to reach these, the antibiotic has to traverse a variety of chemical and physical barriers (Fig. 3). In Gram-negative bacteria the diffusion of the β-lactam through the outer (plasma) membrane and across the periplasm may be a rate-limiting factor in determining antibacterial effectiveness (Nikaido, 1981).

Before detailing the nature of PBPs and their relationship to the mode of action of β-lactam antibiotics, it is worth noting the high toxicity of β-lactam antibiotics against bacteria compared with their low toxicity against eukaryotic cells. The lowest concentration of penicillin capable of inhibiting the growth of a pneumococcal culture is about 6×10^{-9}g per 10^8 bacteria per cm^3 (Tomasz, 1983). In contrast the recommended therapeutic dose for carbenicillin is 40 g per day which illustrates the insensitivity of eukaryotic

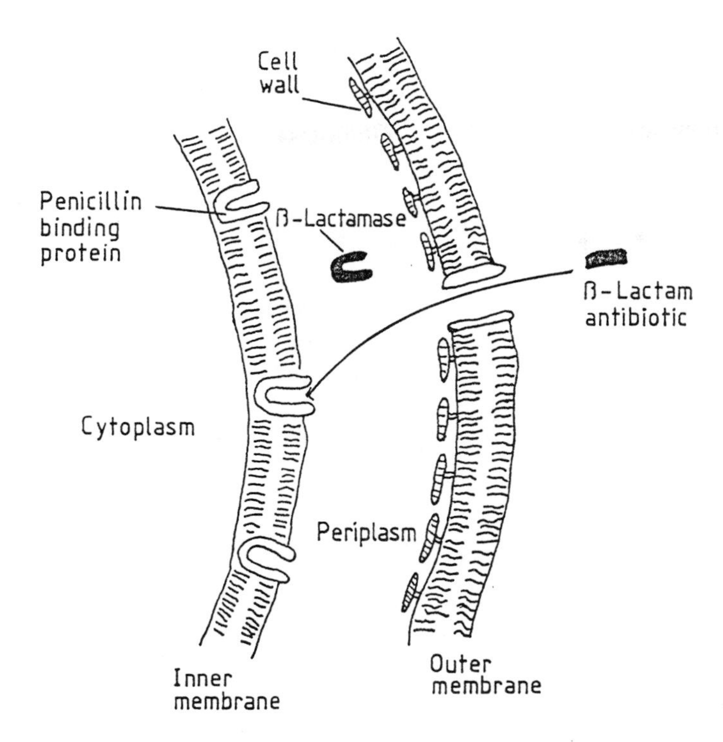

FIG. 3 The passage of a β-lactam antibiotic to its killing target in Gram-negative bacteria

cells to β-lactam antibiotics. The selective action of these drugs is also compatible with the notion that their action is related to a unique bacterial component, e.g. the cell wall found in prokaryotes but not in the more complex eukaryotes.

The bacterial cell wall envelops the cytoplasm and consists of peptidoglycan which is made up of polysaccharide or "glycan" strands that are cross-linked by branched peptide chains. The glycan structure is universal in murein and consists of alternating β-1,4-linked residues of N-acetyl-D-glucosamine (NAG) and N-acetylmuramic acid (NAM). The glycan is a modified form of chitin and is drawn in the flat ribbon conformation in which chitin is constrained by hydrogen bonds (Tipper, 1970). The short peptide crosslinks are attached to the carboxyl groups on the NAM residues (Fig. 4). The length and nature of the peptide crosslinks vary with bacteria. During biosynthesis of the bacterial cell wall all peptidoglycans carry a pentapeptide which commonly has the sequence: L-ala-D-glu-γ-L-X-D-ala-D-ala, in which X is usually an amino carrying residue such as L-lysine or *meso*-diamino-

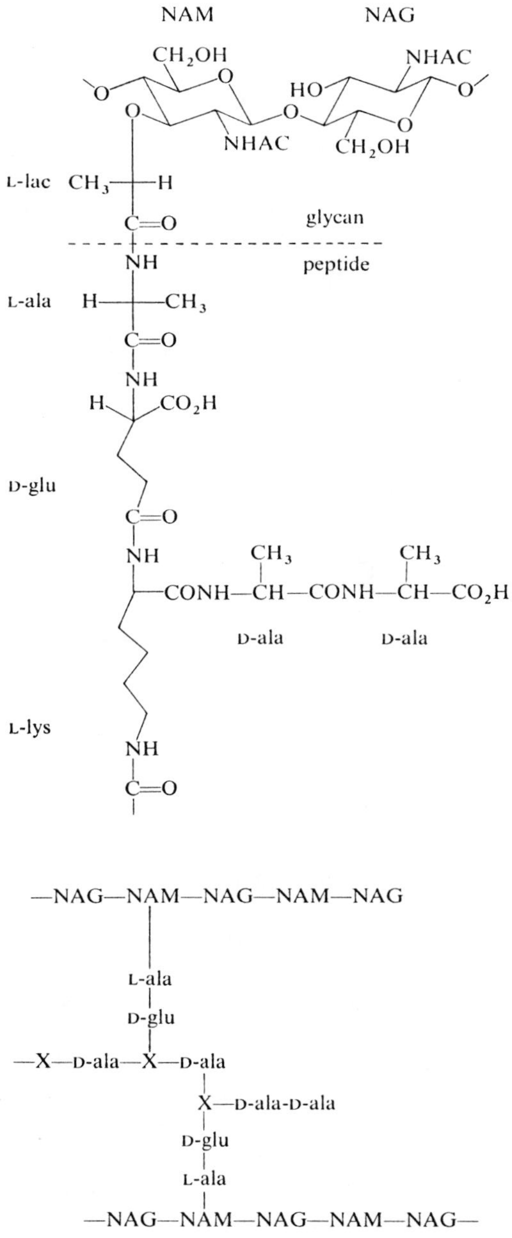

NAM NAG

CH_2OH NHAC HO O O NHAC CH_2OH

L-lac CH_3—H

C=O glycan

NH peptide

L-ala H—CH_3

C=O

NH

H CO_2H

D-glu

C=O

NH CH_3 CH_3

—CONH—CH—CONH—CH—CO_2H

D-ala D-ala

L-lys

NH

C=O

—NAG—NAM—NAG—NAM—NAG

L-ala

D-glu

—X—D-ala—X—D-ala

X—D-ala-D-ala

D-glu

L-ala

—NAG—NAM—NAG—NAM—NAG—

FIG. 4 General structure of peptidoglycans

pimelic acid. Crosslinking occurs by displacing the terminal D-alanine residue with the free amino group of X on an adjacent pentapeptide, i.e. by a transpeptidation reaction. The actual extent of crosslinking varies with the bacterial species and can be as low as 25% to greater than 80%, but no peptidoglycan is completely crosslinked. In the uncrosslinked sections, the D-ala-D-ala terminal residues may be removed by carboxypeptidase action.

The sequence of events involved in the maturing of peptidoglycan during extension of the bacterial cell wall is only partially understood and remains a matter of controversy. The early stages in the biosynthesis of peptidoglycan in which the glycan polymer is uncrosslinked are not penicillin-sensitive. However, the transpeptidation process, which makes the peptidoglycan insoluble, is highly penicillin-sensitive. These reactions are catalysed by a range of peptidases (Ghuysen *et al.*, 1985):

(*i*) DD-transpeptidases – catalyse the cross linking of adjacent NAM pentapeptides with loss of the terminal D-alanine residue (Scheme 1).
(*ii*) DD-carboxypeptidases – catalyse the cleavage of the terminal D-alanine from pentapeptide side chains by a hydrolysis reaction. (Scheme 1).
(*iii*) endopeptidases – catalyse the hydrolysis of interpeptide links formed by transpeptidation.

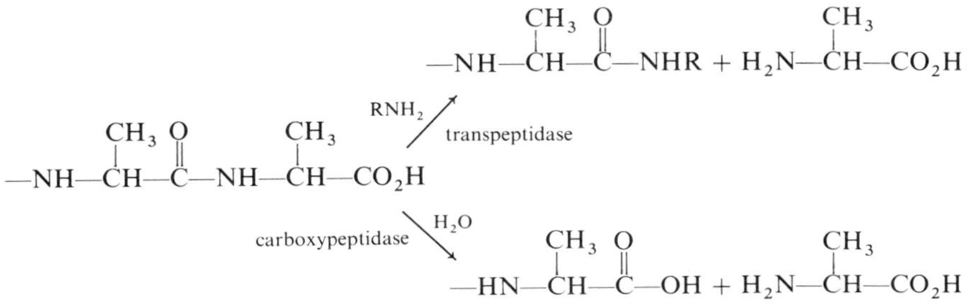

$$—NH—CH(CH_3)—C(=O)—NH—CH(CH_3)—CO_2H \xrightarrow[\text{transpeptidase}]{RNH_2} —NH—CH(CH_3)—C(=O)—NHR + H_2N—CH(CH_3)—CO_2H$$

$$—NH—CH(CH_3)—C(=O)—NH—CH(CH_3)—CO_2H \xrightarrow[\text{carboxypeptidase}]{H_2O} —HN—CH(CH_3)—C(=O)—OH + H_2N—CH(CH_3)—CO_2H$$

Scheme 1

There is a basic similarity in the carboxypeptidase and transpeptidase reactions in that in both cases the carbonyl of the penultimate D-alanine is transferred to an exogeneous nucleophile. If the latter is water then hydrolysis results, whereas if it is an amino group of another peptide then aminolysis occurs and the product is a crosslinked dimer of the two peptides.

Historically, a plausible model for the molecular basis of the selectivity of β-lactam action was proposed 20 years ago (Tipper and Strominger, 1965). It was suggested that penicillin may be a structural analogue of a conformational isomer of D-alanyl-D-alanine, i.e. the carboxyterminal portion of the cell wall precursor disaccharide-pentapeptide (Fig. 5). It could be noted in

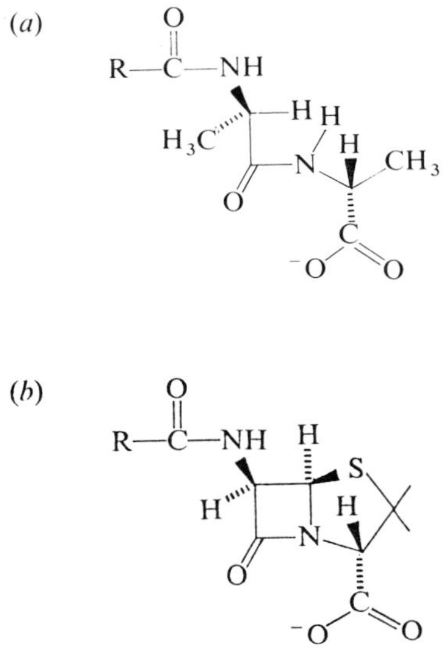

(a)

(b)

FIG. 5 Comparison of the configuration of N-acylated D-alanyl-D-alanine (*a*) with a 3*S*, 5*R*, 6*R*-penicillin (*b*)

passing that the configuration at C(6) in penicillin is the opposite of that predicted by a complete structural analogy with the D-alanyl residue. Tipper and Strominger also suggested that transpeptidation was a two step reaction involving an acyl-enzyme intermediate formed by displacement of the terminal D-alanine using a serine hydroxyl group on the enzyme (Fig. 6). The acyl-D-alanyl residue was then transferred from this ester intermediate to the free amino group of an acceptor substrate. It was proposed that penicillin may be an active site-directed inhibitor capable of acylating bacterial transpeptidases because of the relatively high reactivity of its β-lactam bond (Fig. 6). At that time the relationship of this primary event to cell death and lysis was predicted to be a direct consequence of weakening of the peptidoglycan. These effects are now known to be indirect because disruption of cell-wall biosynthesis is thought to activate "autolysins" which are peptidoglycan hydrolases that hydrolyse the cell wall causing lysis (Tomasz, 1979).

Following Tipper's and Strominger's suggestions, there followed an intensive period of activity to purify and identify the targets for penicillin action.

Exposure of bacterial plasma membrane preparations to radioactively labelled penicillin gives protein complexes covalently linked to the antibiotic. All bacterial membranes give rise to these complexes which can be detected by sodium dodecyl sulphate (SDS) – gel electrophoretic separation and

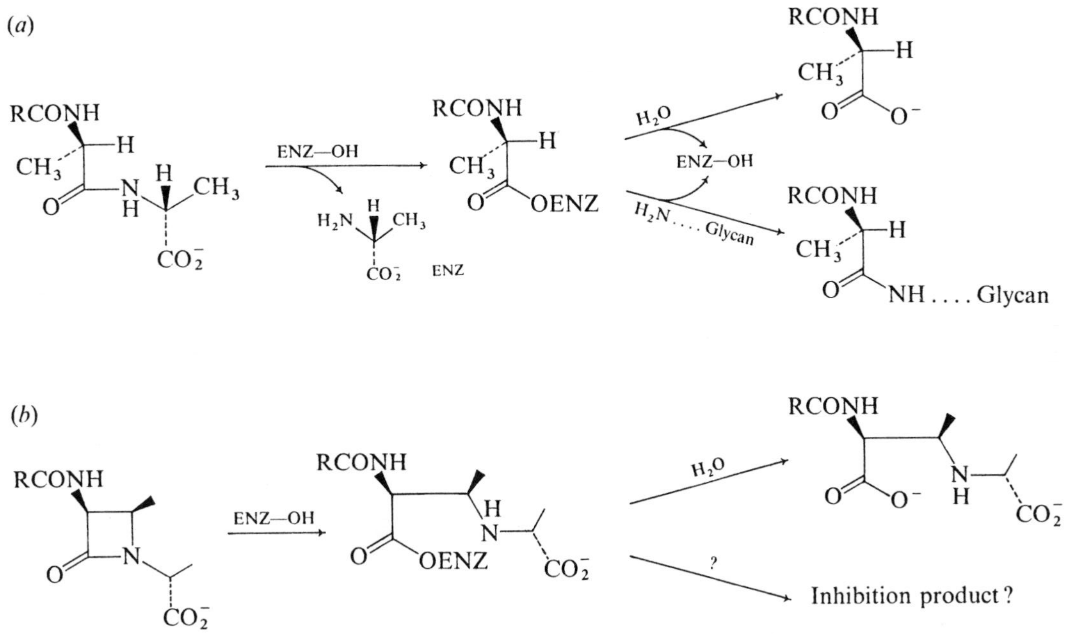

(*a*)

(*b*)

Figure 6

FIG. 6 The transpeptidation and hydrolysis reactions of N-acylated D-alanyl-D-alamine catalysed by serine enzymes (*a*) compared with a β-lactam antibiotic's reaction with serine enzymes (*b*)

detection of the radioactively labelled proteins by autoradiography. This procedure was developed by Spratt (1975) and subsequently resulted in the detection of a number of bacterial proteins capable of binding penicillin which vary in number, molecular size and binding ability from one bacterium to another. For example, seven penicillin binding proteins (PBPs) have been isolated from *E. coli* (Spratt, 1977) as shown in Table 2 (Waxman and Strominger, 1983). The numerical connotation of PBPs refers to their relative molecular size within the group of PBPs detected in a microorganism (PBP 1 being the slowest moving on the gel and having the highest molecular size). PBP 1 of *E. coli* is not necessarily similar to PBP 1 of gonococci. PBPs constitute only 1% of the membrane-bound protein and number 10^3 to 10^4 per cell. They vary in type from three to ten per bacterium and in molecular weight from 30 000 to 100 000 (Spratt, 1980). PBPs fall into two main categories: the first comprises abundant, relatively low molecular weight proteins with *in vitro* D,D-carboxypeptidase activity, and the second, much less abundant, high molecular weight components which are difficult to purify and which frequently lack demonstrable *in vitro* activities. β-Lactam antibiotics appear to be lethal to growing *E. coli* cells by binding to and inactivating the "penicillin-sensitive enzymes" which correspond to the high molecular weight proteins PBP 1A, IV, 2 and 3. All of these PBPs appear to be transpeptidases *in vitro* and all, but PBP 2, have also been shown to be transglycosylases *in vitro* (Tipper, 1985).

The situation in Gram-positive bacteria is more complex because of the wide variation in peptidoglycan structure (Ghuysen, 1977). The major PBP is a low molecular weight D,D-carboxypeptidase but in none has it been demonstrated unequivocally to be the lethal target.

The selective inactivation of individual PBPs leads to strikingly unique morphological changes (Table 2) which could imply that these proteins perform a range of distinct physiological functions.

The distinction between penicillin-sensitive enzymes (PSEs) and PBPs appears arbitrary and is probably a reflection of the interests of the investigators. Three non-membrane bound water-soluble D-alanyl-D-alanine peptidases have been studied in detail (Frère and Joris, 1985): R39, R61 and *albus G* enzymes. The R61 and *albus G* enzymes have been crystallised and the crystal structure for R61 solved to 2.8 Å resolution (Kelly *et al.*, 1985). Although for both R61 and R39 it is very unlikely that these exocellular enzymes are the killing targets of β-lactam antibiotics (Dusart *et al.*, 1973), their study has provided useful models for the mechanism of inhibition of the enzymes anchored in the bacterial plasma membrane. The *albus G* enzyme contains zinc(II) and catalyses the hydrolysis of the C-terminal D-alanine of D-alanyl-D-alanine terminated peptides but it is only very weakly inactivated by penicillins and cephalosporins (Frère and Joris, 1985).

TABLE 2

Penicillin Binding Proteins (PBPs) of *E. coli*: properties and roles in cell-wall metabolism[a]

PBP	Apparent M.W. (daltons)	Abundance (% total PBPs)	Morphological effects of inactivation (where known)	Activities (where known)	Cell-survival/ viability	Proposed *in vivo* function
1A	90 000	6	–	–	Non-essential	Minor transpeptidase, can compensate for PBP 1Bs
1B	87 000	2	Rapid lysis	Transpeptidase and possible trans-glycosylase	Essential	Major transpeptidase of cell elongation
2	66 000	0.7	Ovoid cell formation	–	Essential	Cell shape determination
3	60 000	2	Filamentation	–	Essential	Implicated in cell division and specifically in cross-wall formation
4	49 000	4	–	Carboxypeptidase Transpeptidase Endopeptidase	Non-essential	Secondary transpep-tidase to increase cross-linking
5	42 000	65	–	Carboxypeptidase (transpeptidase)	Non-essential	Regulation of cross-linking
6	40 000	21	–	Carboxypeptidase (transpeptidase)	Non-essential	

[a] Waxman and Strominger, 1983.

The overall mechanism of interaction of the R39 and R61 enzymes (E) with β-lactams (L) can be represented by (1). The rate constant for the

$$E + L \underset{}{\overset{K}{\rightleftharpoons}} EL \xrightarrow{k_2} EI \xrightarrow{k_3} E + \text{products} \qquad (1)$$

irreversible formation of the enzyme-inhibitor complex (EI) is k_2 and the association constant for the reversible binding of the β-lactam to the enzyme is K. The enzyme may be slowly regenerated with a rate constant k_3. To be an efficient inactivator of the transpeptidase enzymes the β-lactam should show a high value of k_2K and a low value of k_3. There have been several reports on the values of these kinetic constants with different β-lactam antibiotics (Frère *et al.*, 1975a; 1982, 1984; Kelly *et al.*, 1981; Laurent *et al.*, 1984). Although caution has been advised when examining the individual values of K and k_2 (Jencks and Page, 1972; Jencks 1975; Page, 1980a, 1981, 1984b), it has been tempting to interpret K as a measure of the strength of binding between the enzyme and substrate. Consequently the observed low association constants have been taken as evidence that recognition of the β-lactam inhibitor by the enzyme is very inefficient. Similar and low values of K do not necessarily imply that discrimination between a group of substrates is poor. It is misleading to separate the values of K and k_2 because some of the intrinsic binding energy between the enzyme and substrate may be used to compensate for unfavourable energy changes which are necessary for efficient reaction within the enzyme-substrate complex (Jencks 1975; Page 1984b). Hence, a high intrinsic binding energy between enzyme and substrate can actually appear not as a high observed or apparent association constant but as an increased rate of decomposition of the enzyme-substrate complex. This could explain the variation in k_2 of up to 10^5 with β-lactam structure.

Furthermore, *in vivo* specificity results from a *competition* between substrates/inhibitors for the active site of the enzyme. The important parameter then becomes the free energy of activation as measured by k_2K. Specificity between competing substrates is given by the relative values of k_2K and not by the individual values of k_2 or K. In fact k_2K varies by up to 10^8 over a range of β-lactams with the R39 and R61 transpeptidase enzymes (Frère and Joris, 1985). This variation emphasises one of the difficulties of finding relationships between structure, chemical reactivity and activity towards enzymes.

The R61 and R39 enzymes have been shown to have active site serine residues which are acylated by penicillin and other β-lactams (Georgopapadakou *et al.*, 1981), to give penicilloyl enzymes [13] as shown in Scheme 2 (Frère *et al.*, 1976b; Degelaen *et al.*, 1979b). Interestingly, ester [13] does not hydrolyse to give penicilloic acid [14], although this does occur if the

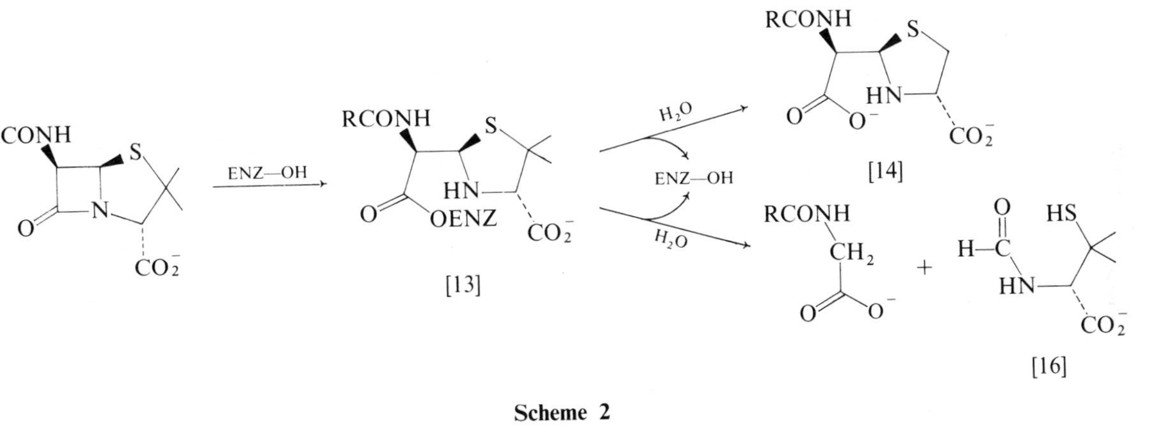
RCONH
S
O
N
CO2
ENZ—OH
RCONH
S
HN
O
OENZ
CO2
[13]
H2O
ENZ—OH
H2O
RCONH
S
HN
O
O
CO2
[14]
RCONH
CH2
O
O
+
O
HS
H—C
HN
CO2
[16]

Scheme 2

enzyme-complex is first denatured. At neutral pH the ester intermediate undergoes an enzyme catalysed cleavage of the C(5)—C(6) bond and the identified products are an acylated glycine [15] and N-formylpenicillamine [16] (Frère *et al.*, 1974, 1975b, 1976a, 1978; Adriaens *et al.*, 1978). The precursor of the isolated [16] is not known although a thiazoline is a possibility. All R61 and R39 enzyme-β-lactam complexes do not degrade by this C(5)—C(6) fragmentation pathway. The reaction of the R61 enzyme with cephalosporins appears to give simple hydrolysis products (Kelly *et al.*, 1981).

In summary, the main mechanistic studies of β-lactam antibiotics with penicillin-sensitive enzymes have been performed with soluble exocellular enzymes which act as hydrolytic catalysts. Unfortunately, mechanistic studies with membrane-bound PBPs which act as penicillin-sensitive enzymes are far less advanced and so the molecular mechanism for the lethal action of β-lactam antibiotics remains speculative.

3 Is the antibiotic β-lactam unusual?

Because of the relatively rare occurrence of β-lactams in nature, it is not surprising that the biological activity of these compounds should be attributed to the chemical reactivity of the β-lactam ring. Shortly after the introduction of penicillin to the medical world it was suggested that the antibiotic's activity was due to the inherent strain of the four-membered ring (Strominger, 1967) or to reduced amide resonance (Woodward, 1949). The latter is conceivable because the butterfly shape of the penicillin molecule [17] prevents the normal planar arrangement of the O, C and N atoms assumed to be necessary for the effective delocalisation of the nitrogen lone pair. Both of these ideas are, of course, intuitively appealing and they have remained unchallenged for several decades. Indeed, these two proposals have dominated the thoughts of synthetic chemists who were, and to some extent still are, convinced that more effective antibiotics may be made by making the β-lactam system more strained or non-planar. However, the evidence to support an unusually strained or an amide resonance inhibited β-lactam in penicillin is ambiguous.

The assessment of the nature of the β-lactam ring in penicillins and in cephalosporins and of its contribution to the reactivity of these molecules has been based on structural and kinetic studies. Before reviewing these, it should be noted that the potential contributions of strain energy release and inhibition of amide resonance are not trivial.

It is estimated that resonance stabilises amides by about 18 kcal mol^{-1} (Fersht and Requena, 1971). The reason for the greater stability of amides compared with other carbonyl containing functionalities is attributed to the

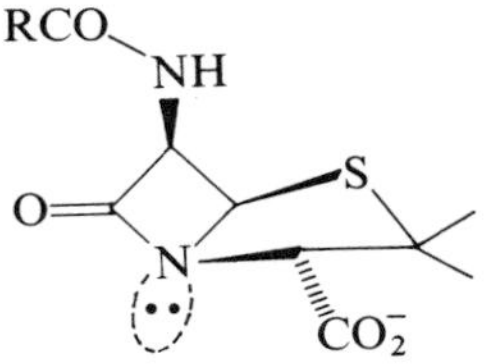

[17]

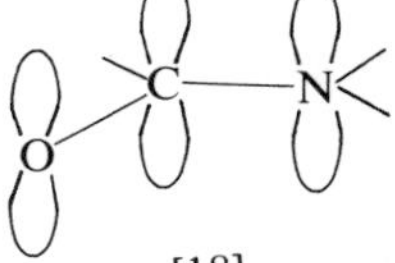

[18]

unfavourable loss of resonance when nucleophiles attack the amide carbonyl carbon to form a tetrahedral intermediate (Scheme 3). If delocalisation is completely inhibited in an amide then the rate of a reaction, which would normally involve loss of resonance stabilisation in the transition state, could occur up to 10^{13}-fold (antilog 18/2.303 $\boldsymbol{RT}$) faster than the analogous resonance stabilised system. The strain energy of a four membered ring is 26–29 kcal mol^{-1} (Page, 1973) and therefore a reaction involving opening of the β-lactam ring could take place faster by a factor of up to 10^{20} (antilog 27.5/2.303 $\boldsymbol{RT}$) than the analogous bond fission process in a strain-free amide. If strain or resonance inhibition is even slightly significant in penicillins and cephalosporins their effects should therefore be easily observable.

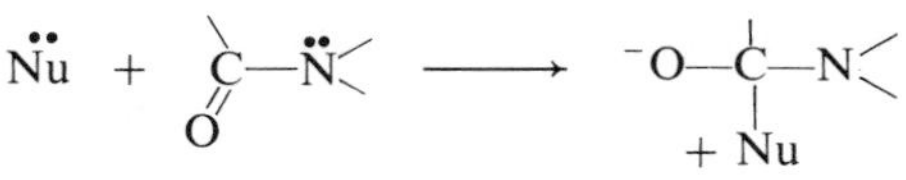

Scheme 3

The treatment of amide resonance as a result of delocalisation of the nitrogen lone pair by overlap of the 2p-orbitals on the participating atoms [18] has been used to predict that a pyramidal amide nitrogen will cause loss of resonance energy (Woodward, 1949). Pyramidalisation of the nitrogen will, however, not necessarily produce the same effect as that caused by rotation about the C—N bond, i.e. a change in the dihedral angle between the p-orbitals with the O, C and N atoms remaining coplanar. These contrasting effects are illustrated, respectively, by the Newman projection formulae [19] and [20] obtained by looking along the C—N bond. Despite these differences, the assumption that amide resonance in penicillins and cephalosporins is inhibited has been generally accepted and several experimental observations have been used to support this suggestion.

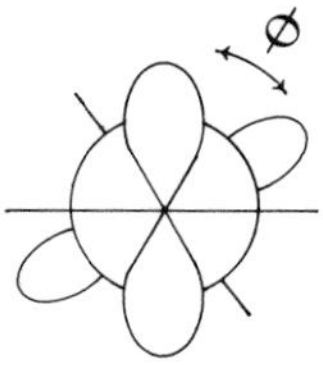

[19] 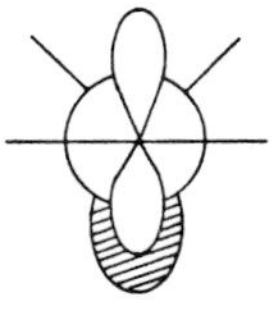 [20]

STRUCTURAL AND GROUND-STATE EFFECTS

Amide resonance is usually depicted by the canonical forms [21] and [22].

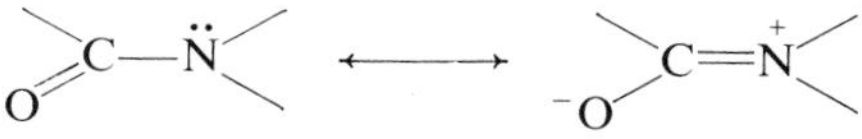

[21] [22]

Inhibition of amide resonance should make the amide resemble [21] at the expense of [22]. The reasonable conclusion would be that, compared with a normal amide, resonance inhibition would:

(*i*) increase the C—N bond length and decrease the C—N bond strength.
(*ii*) decrease the C—O bond length and increase the C—O bond strength.
(*iii*) decrease the positive charge density on nitrogen.
(*iv*) decrease the negative charge density on oxygen.

Planarity of the nitrogen and bond lengths

X-ray crystallography has been invaluable in providing detailed three-dimensional structures of the β-lactam antibiotics. Despite the obvious criticism that solid state conformations may not be relevant to solution conformations, the geometrical picture presented from X-ray data has been the basis for many discussions relating chemical structure and biological activity.

Although acetamide has C_s symmetry in the gas phase and solution, the carbonyl carbon is pyramidalised in the crystalline state (Jeffrey *et al.*, 1980). The degree of coplanarity of the β-lactam nitrogen with its three substituents can be expressed either by the perpendicular distance, *h*, of the nitrogen from the plane of its substituents or by the sum of the bond angles about nitrogen. The former is easier to visualise and the nitrogen ranges from being essentially in the plane of its three substituents in monocyclic β-lactams to being 0.5 Å out of the plane in bicyclic systems. Examples of *h*-values are given in Table 3 and there have been several reviews on

TABLE 3

Structural parameters of some β-lactams

Compound	C=O stretch/ cm^{-1}	Distance of N from plane h/Å	β-lactam C=O bond length/Å	β-lactam C—N bond length/Å
Penicillins[a]	1770–1790			
ampicillin[b]		0.38	1.20	1.36
benzylpenicillin[c]		0.40	1.17	1.34
phenoxymethylpenicillin[d]		0.40	1.21	1.46
Δ^3-*Cephalosporins*[e]	1760–1790			
cephaloridine[f]		0.24	1.21	1.38
Δ^2-*Cephalosporins*[e]	1750–1780			
phenoxymethyl Δ^2-cephalosporin[f]		0.06	1.22	1.34
Anhydropenicillins[g]	1810			
phenoxymethylanhydro-penicillin[h]		0.41	1.18	1.42
Monocyclic β-lactams	1730–1760	0	1.21	1.35
Amides	1600–1680	0	1.24	1.33

[a] Morris and Jackson, 1970
[b] James *et al.*, 1968
[c] Pitt, 1952
[d] Abrahamson *et al.*, 1963
[e] Green *et al.*, 1965
[f] Sweet and Dahl, 1970
[g] Wolfe *et al.*, 1968
[h] Simon *et al.*, 1972

structural data determined by X-rays (Sweet and Dahl, 1970; Simon *et al.*, 1972; Sweet, 1973; Boyd, 1982). Until recently it had been generally assumed that a more pyramidal nitrogen decreased amide resonance and increased biological activity. Great effort was therefore discharged in making non-planar β-lactams. A 1-carba-1-penem shows the highest h-value, 0.54 Å (Woodward, 1980), and yet this compound is biologically inactive. Furthermore, there is no direct correlation between h-values and chemical reactivity (see Section 4). It has been claimed that bond length data for penicillins and cephalosporins show evidence for the inhibition of amide resonance (Sweet, 1973). The C—N bond length of planar monocyclic β-lactams (1.35 Å) is generally longer than that of amides (1.33 Å). The converse is true for C=O bond lengths, 1.24 Å for amides compared with 1.21 Å for monocyclic β-lactams. In non-planar penicillins and cephalosporins there is a general trend for the C—N bond length to increase as the C=O bond length decreases (Sweet, 1973; Simon *et al.*, 1972). However, this trend is by no means linear. Bond lengths for C=O vary from 1.17 to 1.24 Å and for C—N from 1.33 to 1.46 Å. There is also a tendency for the C—N bond length to increase with h.

It is difficult to discern reasons and consequences of these bond length differences. Penicillin V [1; R = $PhOCH_2$] shows the longest C—N bond length of 1.46 Å (Abrahamsson *et al.*, 1963) and yet the C=O bond length is identical to that commonly found in planar monocyclic β-lactams (1.21 Å). In monocyclic β-lactams the nitrogen is coplanar with its three substituents and yet the bond length differences are also in the direction predicted by inhibition of amide resonance. The degrees of non-planarity in penicillin V [1; R = $PhOCH_2$] and ampicillin [1; R = $PhCH(NH_2)$] are similar (h = 0.40 and 0.38 Å, respectively) and yet the C—N bond length in the former is 0.10 Å longer than in the latter (Abrahamsson *et al.*, 1963; James *et al.*, 1968).

[23]

Structural data have also been used to support the suggestion that enamine resonance [23] is important in cephalosporins and that this also reduces amide resonance (Sweet, 1973). However, there is no significant difference in the C—O and C—N bond lengths of cephalosporins from that general trend exhibited by penicillins. Although it is claimed (Sweet, 1973) that the C(4)—N(5) of Δ^3-cephalosporins is significantly shorter than the "expected" values this is not generally true. For example, C(4)—N(5) of cephaloglycine is 1.51 Å (Sweet and Dahl, 1970) which is *longer* than the quoted expected value of 1.47 Å for C—N. Furthermore, the C(4)—N(5) bond length in the Δ^3-cephalosporin, cephaloglycin, is longer than that of 1.45 Å in Δ^2-cephems (Sweet and Dahl, 1970) and that of 1.46 Å in cephams (Vijayan *et al.*, 1973) where enamine resonance cannot occur.

It would seem logical to conclude that variations in bond lengths within penicillin and cephalosporin derivatives are caused by the nature of substituents and the minimisation of unfavourable strain energies caused by the geometry of the molecule. To attribute these differences to the inhibition of amide resonance seems speculative and is only supported by the selection of examples.

Nmr chemical shifts

The conformation of the substituent on nitrogen relative to the carbonyl group has a significant effect on the carbonyl ^{13}C chemical shift in amides. For example, a difference of 4 ppm is observed in the carbonyl ^{13}C

resonances of the *E*- and *Z*-isomers of N-methylformamide (Levy and Nelson, 1972a). This is a reflection of both steric and anisotropic differences in the environment. There does not appear to be a simple relationship between ^{13}C chemical shifts and the local electron density distribution. However, there have been several attempts to correlate antibacterial activity with chemical shifts.

The β-lactam carbonyl carbon usually resonates between 160–167 ppm in a ^{13}C nmr spectrum (Bose and Srinivasan, 1979). This is in the same region in which the carbonyl resonances of formamide and its N-methylated derivatives also appear (Levy and Nelson, 1972b). It is interesting to note that the carbonyl resonances of 5- or larger-membered lactams appear between 170 and 180 ppm (Williamson and Roberts, 1976). Replacement of an alkyl substituent on the β-lactam nitrogen by an aryl substituent causes shielding of the lactam carbonyl resonance by *ca* 4 ppm (Bose and Srinivasan, 1979). Dipole moment (Malihowski *et al.*, 1974) and uv spectral studies (Manhas *et al.*, 1968) indicate that the lone pair on the β-lactam nitrogen is conjugated with the aromatic ring [24]. This resonance interaction presumably alters the electron density at the β-lactam carbonyl and could account for the shielding of the resonance due to that carbon.

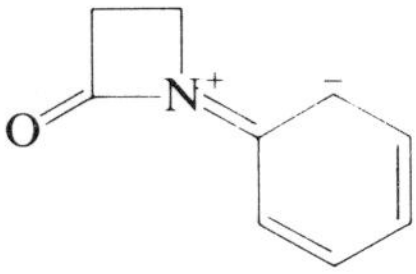

[24]

There is little variation in the chemical shifts of the β-lactam carbonyl carbon of penicillins and cephalosporins (Mondelli and Ventura, 1977; Paschal *et al.*, 1978; Dereppe *et al* 1978; Schanck *et al.*, 1979). The carbonyl carbon of the β-lactam in penicillins resonates about 10 ppm to lower field than that in cephalosporins. Surprisingly, the shifts in the biologically active Δ^3- and the inactive Δ^2- cephalosporins are similar (Mondelli and Ventura, 1977). Inhibition of amide resonance may be expected to make the carbonyl carbon more electron deficient (Dhami and Stothers, 1964). Although the difference in chemical shifts between penicillins and Δ^3-cephalosporins support this proposal if amide resonance is inhibited by a pyramidal β-lactam nitrogen, it is not apparent from the Δ^2-/Δ^3- cephalosporin comparison. In penicillins the nitrogen is 0.4 Å from the plane of its substituents compared with 0.2 Å in Δ^3-cephalosporins, whereas the ceph-2-em systems are planar (Flynn, 1972). The similarity of the values of the ^{13}C shifts found for the β-lactam carbonyl carbons in ceph-3-ems and ceph-2-ems indicates that the charge density and bond order at the carbonyl carbons in both systems is approximately the same. This argues against enamine type resonance [23] in ceph-3-ems.

^{15}N-Chemical shifts of the β-lactam nitrogen in ceph-3-ems are almost invariant ($< \pm 1$ ppm) with the nature of the substituent at C(3) (Paschal *et al.*, 1978) and therefore also do not indicate significant enamine type resonance in these systems. Not surprisingly there is a large difference of 15 ppm in the ^{15}N-chemical shifts of the β-lactam nitrogen in ceph-2-ems and ceph-3-ems. Interestingly, there is an *upfield* shift of 30 ppm in the β-lactam nitrogen on going from non-planar penicillins to planar ceph-2-ems (Lichter and Dorman, 1976). Increased amide conjugation in the planar system would be expected to have induced a downfield shift. No doubt, since the factors determining ^{15}N-chemical shifts are not very well understood, it could be argued that the shifts are insensitive to minor changes in the degree of amide delocalisation.

Infrared carbonyl stretching frequency

The β-lactam infrared stretching frequency ($\nu_{c=o}$) has been regarded as an important index for both inhibition of amide resonance and for investigating structure-activity relationships of the β-lactam antibiotics (Morin *et al.*, 1969; Sweet and Dahl, 1970; Demarco and Nagarajan, 1973; Indelicato *et al.*, 1974; Murakami *et al.*, 1980; Takasuka *et al.*, 1982; Nishikawa *et al.*, 1982).

In normal penams the β-lactam carbonyl stretching frequency occurs in the 1770–1790 cm^{-1} range compared with 1730–1760 cm^{-1} for monocyclic unfused β-lactams and about 1600–1680 cm^{-1} for amides. In general, the non-planar 3-cephems show higher stretching frequencies, 1786–1790 cm^{-1}, than the planar 2-cephems which absorb at 1750–1780 cm^{-1}. The frequency in cephalosporins increases by *ca* 5 cm^{-1} when the ring sulphur is replaced by oxygen but decreases by a similar amount when the 7-α-hydrogen is substituted by a methoxy group (Takasuka *et al.*, 1982). It is difficult to make generalisations about the observed β-lactam frequency since different conditions of measurement (KBr, film, solution, etc.) may cause variations comparable with those produced by structural changes. There is a tendency for a high carbonyl stretching frequency to be associated with a shorter β-lactam C=O bond length and a more pyramidal nitrogen. Furthermore it has been tempting to associate a high carbonyl stretching frequency with increased strain, increased double bond character and reduced amide resonance (Morin *et al.*, 1969; Simon *et al.*, 1972; Pfaendler *et al.*, 1981). However, the evidence again is ambiguous. Although selected examples may show some of these interrelationships there are many exceptions; for example, the carbonyl stretching frequency for some penems *decreases*

$20\,cm^{-1}$ whilst the β-lactam nitrogen becomes *more* pyramidal by 0.12 Å (Pfaendler *et al.*, 1981).

The direct interpretation of carbonyl stretching frequencies in terms of bond strengths or electron density distributions is not straightforward. Many subtle effects can alter the frequency even if the force constant for C═O stretching, which is presumably the best indicator of bond strength, remains constant. For example, in the system X—C═O the carbonyl stretching frequency can be increased by decreasing the C—X bond length, by increasing the C—X stretching force constant or by increasing the XCO bending force constant (Collings *et al.*, 1970).

Ultraviolet absorption and circular dichroism spectra

The 3-cephem systems show two characteristic absorptions in their uv absorption spectra, a strong band near 260 nm and a weaker one at about 230 nm. The chromophore near 260 nm is not associated with the presence of sulphur (Wolfe *et al.*, 1974) and it appears that the amide and enamine systems contribute to this absorption. Semi-empirical MO calculations (Boyd, 1972) suggest that the 260 nm absorption is associated with the excitation from the ground state with a doubly occupied enamine π MO to an excited state described mainly by a configuration with the excited electron in an MO with both π*(C═O) and π*(C═C) character.

Circular dichroism (CD) spectra of penams show a strong positive Cotton effect at about 230 nm and a weaker one of variable sign at about 207 nm (Busson *et al.*, 1976a; Richardson *et al.*, 1977). A CD study of a 1-carbapenam, (5*S*)-1-azabicyclo[3.2.0]heptan-7-one, in which the β-lactam chromophore is free of sulphur, carboxyl and other amido groups, also shows the Cotton effect at 231 nm (Busson and Vanderhaeghe, 1978). It is unlikely therefore that the band in penicillins is due to charge transfer from sulphur to amide as suggested by MO calculations (Boyd *et al.*, 1976). Furthermore, monocyclic β-lactams show similar bands (Busson *et al.*, 1978) and the simplest interpretation of the observations is that the Cotton effects are associated with n → π* transitions characteristic of all amides.

Theoretical calculations

Theoretical geometry optimisation of β-lactams at a semi-empirical level and a limited *ab initio* study using minimal basis STO-3G calculations at fixed geometries have been reported (Glidewell and Mollison, 1981). The STO-3G basis set underestimates valence angles at heteroatoms and is expected to exaggerate the degree of non-planarity in amides whilst the split valence 4-31G basis set characteristically overestimates these valence angles and consequently overestimates the tendency of amides to be planar. Given these reservations the calculated STO-3G energy of formamide in a penicillin-like geometry is only 2.8 kcal mol^{-1} higher than the planar geometry (Vishveshwara and Rao, 1983). Furthermore, in general, the geometrical parameters associated with the β-lactam ring vary only slightly with changes in the hybridisation at nitrogen. An exception is the C—N bond length which becomes longer as the nitrogen becomes pyramidal.

The barrier to inversion at nitrogen in ammonia is 5.8 kcal mol^{-1} which is much greater than that in molecules like formamide (Radom and Riggs, 1980; Carlsen *et al.*, 1979). Formamide lies in a potential well which is very flat with respect to inversion at nitrogen. The inversion barrier is lower for molecules favouring a large angle at nitrogen (amides) and higher for systems adopting a small angle at nitrogen (e.g. aziridine) (Stackhouse *et al.*, 1971). It appears that the nitrogen in amides can be made pyramidal without severe changes in energy.

Calculations of charge distributions of β-lactam rings, and in particular the Mulliken overlap populations for the CO and CN bonds, have been reported for several systems (Boyd, 1982). The calculated charge density on nitrogen becomes less negative, by 0.013 of atomic charge, as the nitrogen of a β-lactam is made more pyramidal. However, the calculated charge on the carbonyl carbon becomes less positive as the nitrogen becomes pyramidal (Glidewell and Mollison, 1981) which would presumably suggest that β-lactams with a pyramidal nitrogen are less susceptible to nucleophilic attack.

Basicity of β-lactam nitrogen

Inhibition of amide resonance in bicyclic β-lactams will make the amide resemble canonical form [21] at the expense of [22]. A consequence of increased localisation of the lone pair on nitrogen would be to increase the basicity of nitrogen.

It is well known that torsional strain in amides can increase the basicity of nitrogen in amides. For example, 6,6-dimethyl-1-azabicyclo[2.2.2]octan-2-one [25] presumably has the nitrogen lone pair almost orthogonal to the carbonyl π system and amide resonance is consequently inhibited. The carbonyl absorption for [25] is at 1762 cm^{-1} (Pracejus *et al.*, 1965) which lies

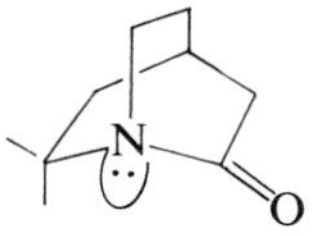

N O

[25]

between the normal amide carbonyl stretch and that for some N-acetyltrialkylammonium tetrafluoroborates at about 1816 cm^{-1} (Paukstelis and Kim, 1974). Amides are normally only very weakly basic and the pK_a-values of their conjugate acids are around zero. By contrast, [25] is half protonated at pH 5.3, consistent with the increased basicity of the amide nitrogen.

If amide resonance in penicillins is inhibited because of the pyramidal nature of the β-lactam nitrogen, penicillins should also show enhanced basicity compared with normal amides. There is no evidence to suggest that this is the case. In fact, penicillins appear to show reduced basicity and cannot be detectably protonated even in 12 M hydrochloric acid. For N-protonation of penicillins pK_a must be < -5 (Proctor *et al.*, 1982). This reduced basicity is discussed in Section 5.

Another indication of increased nitrogen basicity would be a large binding constant of penicillin to metal ions. However, the equilibrium constant for metal-ion coordination between the carboxyl group and β-lactam nitrogen [26] is only about 100–200 M^{-1} for copper(II), zinc(II), nickel(II) and cobalt(II) (Gensmantel *et al.*, 1978, 1980). This is the order of magnitude expected for coordination between a normal amide and a carboxyl group.

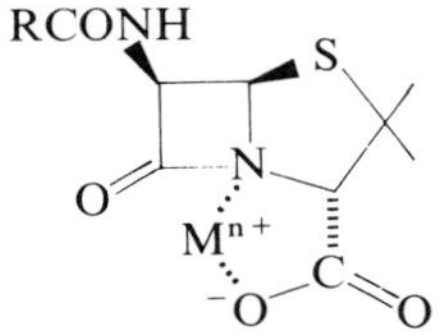

[26]

There appears, therefore, to be no evidence of substantial inhibited amide resonance in penicillins and the β-lactam nitrogen shows no enhanced electron pair donating ability to either a proton or to metal ions.

Resonance and non-planarity

The degree of amide resonance that is lost by distortion of the normal planar geometry is not immediately apparent. The angular dependence of resonance interactions, particularly steric inhibition of resonance, has been of long-term interest (Hammond, 1956; Wheland, 1955; Wittig and Steinhoff, 1964; Dewhirst and Cram, 1958; Ree and Martin, 1970; Buss *et al.*, 1971; Doering *et al.*, 1984). The importance of residual resonance stabilisation even when the interacting p-orbitals are orthogonal is highlighted by *ab initio* studies. For example, nearly 50% of the possible charge donation from oxygen to the ring carbons in phenol occurs even in the perpendicular conformation (Hehre *et al.*, 1972).

Rotation about bonds which increases the dihedral angle between p-orbitals on adjacent atoms does decrease resonance interaction. For example, the increased basicity of the quinuclidin-2-one [25] may be attributed to decreased amide resonance. Similarly, the energy barrier to rotation about the C—N bond in amides of 20–30 kcal mol^{-1} (Kessler, 1970; Stewart and Siddall, 1970) indicates a significant torsional dependence for resonance.

However, resonance interaction appears to remain significant if nitrogen adjacent to a π or p system is made pyramidal. For example, π-sp^3 conjugation in aromatic amines can be appreciable and the configuration about nitrogen in most simple anilines is nearly pyramidal rather than trigonal (Bottini and Nash, 1962; Aroney *et al.*, 1968). The geometry of the bicyclo[3.3.1]nonane system [27] does not completely inhibit resonance stabilisation of the developing bridgehead cation in solvolysis reactions (Krabbenhoft *et al.*, 1974). The α-amino substitutent in [27; Y = NCH_3] increases reactivity relative to the carbon analogue [27; Y = CH_2] by a factor of 10^7. The geometry of the 1-adamantyl-type cation [28] imposes a perpendicular twist on the 2-heteroatom's lone pair relative to the vacant carbon p-orbital and would be expected to minimise resonance stabilisation.

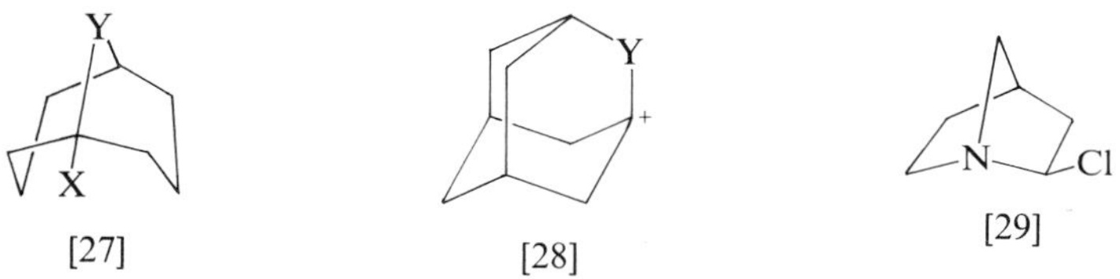

However, despite this unfavourable geometry there is still significant interaction (18% of σ_R^+ used to correlate rates) between the cation and non-bonding electrons of the nitrogen, and [28; Y = NCH_3] is 356-fold more reactive than [28; Y = CH_2] (Meyer and Martin, 1976). Only in the case of the much more rigid bicyclo[2.2.1]heptane system [29] has an α-amino substituent been found to decrease reactivity (20-fold) (Gassman *et al.*, 1972) but even here resonance stabilisation must still occur because an α-nitrogen is expected inductively to destabilise a carbocation leading to a decrease in reactivity by a factor of *ca* 10^5. So even though a carbocation, because of its lower stability, probably makes a greater and less stereospecific demand upon adjacent nitrogen lone pairs than a carbonyl group these observations suggest that pyramidal distortion of nitrogen in amides will not cause significant loss of resonance.

Even though the nitrogen in penicillins is pyramidal there is little evidence from ground-state effects to support the often quoted claim that this causes significant inhibition of amide resonance. The next section shows that kinetic observations are in agreement with this conclusion.

KINETIC EFFECTS

Nucleophilic substitution at the carbonyl group of an amide invariably occurs in a stepwise manner by initial formation of a tetrahedral intermediate (Scheme 3, p. 185). Conversion of the three-coordinate, sp^2-hybridised carbonyl carbon to a four-coordinate sp^3-hybridised carbon in the intermediate must be accompanied by the loss of amide resonance. This contribution to the activation energy will be reduced if amide resonance is inhibited and a rate enhancement is expected in such cases. Similarly, the release of strain energy will increase the rate if the four-membered ring is opened or has been opened in the transition state. The total strain energy of four-membered rings is probably not released until there is significant bond extension (Earl *et al.*, 1983).

A simple reaction to see if either of these effects is apparent is the hydrolysis of the β-lactam antibiotics. The alkaline hydrolysis of benzylpenicillin opens the β-lactam ring to give benzylpenicilloate [30]. With respect to hydroxide ion, benzylpenicillin shows a reactivity similar to that of ethyl acetate. The pK_a-value of the protonated amine in the thiazolidine derivative [30] is 5.2 and because of this weakly basic nitrogen the leaving group ability of the amine is expected to be improved. Therefore, in order to assess any special reactivity of the β-lactam antibiotics, the dependence of the rate of hydrolysis of simple amides and β-lactams upon substituents must be known.

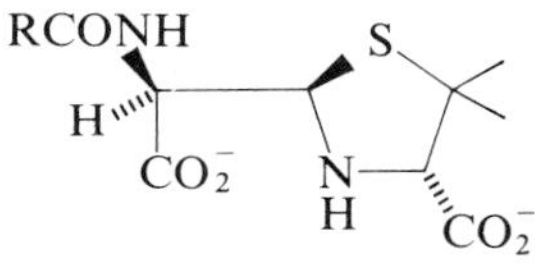

RCONH, H, CO_2^-, S, N, H, CO_2^-

[30]

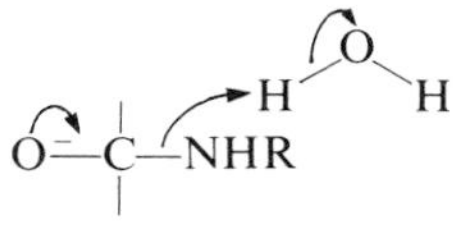

O, H, H, O—C—NHR

[31]

A Brønsted plot of the second order rate constants for the hydroxide ion catalysed hydrolysis of acyclic amides, monocyclic β-lactams and bicyclic β-lactams is shown in Fig. 7 (Proctor *et al.*, 1982). The Brønsted β_{lg} value for N-substituted acyclic amides and anilides is -0.07. Since the nitrogen of amides has an effective charge of $+0.6$ in the ground state, this is indicative of a transition state in which the nitrogen behaves as if it has *ca* 0.5 positive charge in the transition state (Morris and Page, 1980c). The alkaline hydrolysis of acyclic anilides is characterised by ^{18}O exchange with solvent occurring faster than hydrolysis (Bender and Thomas, 1961) and by a second-order dependence on hydroxide ion concentration (Schowen *et al.*, 1966). Both results indicate rate-limiting breakdown of the tetrahedral intermediate. The Brønsted β_{lg} is incompatible with rate-limiting expulsion of the amine anion or with a transition state in which the nitrogen has a unit positive charge, i.e. is fully protonated. The observations are consistent with water acting as a general acid catalyst in the breakdown of the tetrahedral intermediate [31] (Morris and Page, 1980c; Proctor *et al.*, 1982). The rates of alkaline hydrolysis of β-lactams exhibit a first-order dependence on hydroxide ion concentration and show a Brønsted β_{lg} value of -0.44, which is

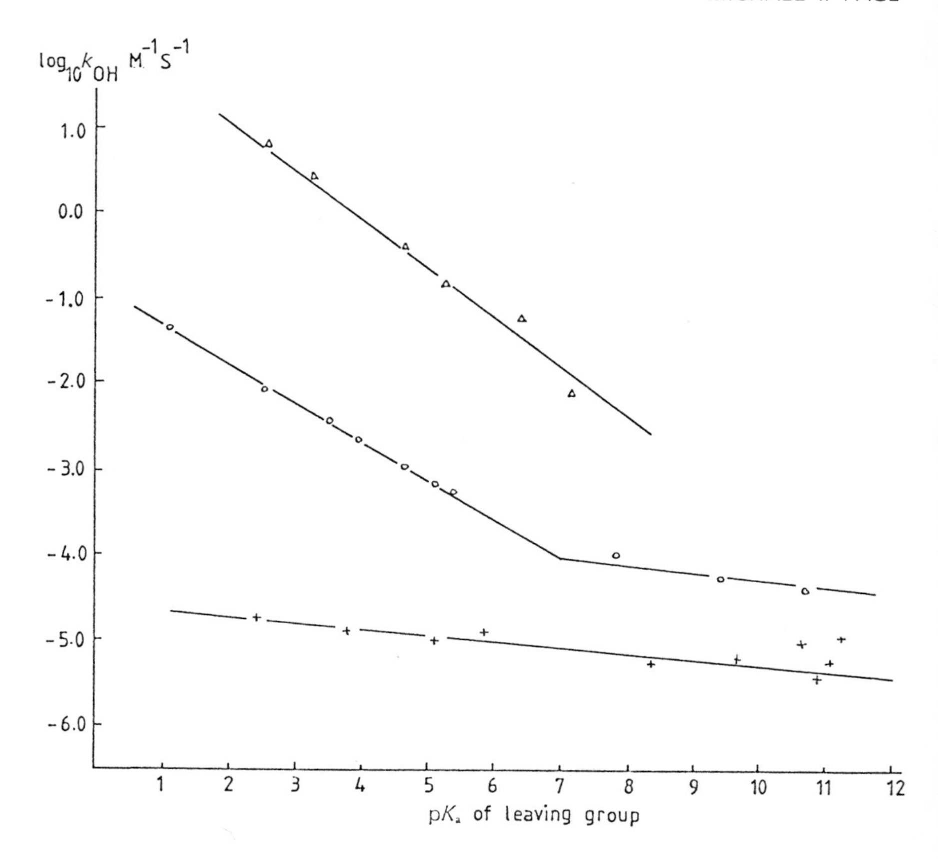

FIG. 7 Brønsted plot of the second-order rate constants for the hydroxide-ion catalysed hydrolysis of acyclic amides (+), monocyclic β-lactams (○), and bicyclic β-lactams (△) against pK_a for the leaving group amine. Data refer to 30°C and are taken from Proctor *et al.*, 1982

indicative of rate-limiting formation of the tetrahedral intermediate (Blackburn and Plackett, 1972). However, β-lactams of the more basic amines show a positive deviation from this line and exhibit a smaller dependence upon basicity of the leaving group which may signify a change in rate-limiting step to breakdown of the intermediate. (Proctor *et al.*, 1982).

A consequence of the different dependence upon leaving group basicity is that the rate enhancement of β-lactams compared with acyclic amides depends upon the basicity of the leaving group amine. β-Lactams of weakly basic amines are *ca* 500-fold more reactive than an acyclic amide of the same amine. However, β-lactams of basic amines are only slightly more reactive than an analogous acyclic amide. Similarly, β-lactones are only about 10-fold more reactive than analogous esters (Blackburn and Dodds, 1974).

Crystallographic (Luche *et al.*, 1968; Parthasarathy, 1970; Chambers and Doedens, 1980) and spectroscopic evidence (Manhas *et al.*, 1968) show that N-substituted β-lactams are planar and resonance-stabilised as in amides. The rate enhancement of 30–500-fold shown by β-lactams of amines of $pK_a < 6$ may be adequately rationalised by the change in coordination number/hybridisation of the carbonyl carbon as the tetrahedral intermediate is formed (Page, 1973; Gensmantel *et al.*, 1981). The magnitude is similar to the 500-fold faster rate of reduction of cyclobutanone by borohydride compared with acetone (Brown and Ichikawa, 1957).

The conversion of three-coordinate to four-coordinate carbon in four-membered rings is accompanied by the release of $11.4\,\text{kJ mol}^{-1}$ of strain energy (Page, 1973; Allinger *et al.*, 1972). A rate enhancement of 100-fold can therefore be expected in the conversion of the β-lactam carbonyl carbon to a tetrahedral intermediate. The change in strain energy will be even greater for a resonance stabilised β-lactam which has some endocyclic double bond character within the four-membered ring.

As the rate-limiting steps for the alkaline hydrolysis of amides and β-lactams are different, the relatively small rate enhancement shown by β-lactams indicates that the energy of the transition state for breakdown of the tetrahedral intermediate in amide hydrolysis is not significantly greater than that for formation of the intermediate.

There is nothing unusual about the chemical reactivity of the monocyclic β-lactam antibiotics nocardicin [8] and the monobactams [9]. The second order rate constants for their alkaline hydrolysis fit the Brønsted plot (Fig. 7) (Proctor *et al.*, 1982).

β-Lactams of basic amines (pK_a of $R\overset{+}{N}H_3 > 7$) are only *ca* 10-fold more reactive than analogous acyclic amides. The Brønsted β_{lg} values for both of these systems are compatible with rate-limiting breakdown of the tetrahedral intermediate. It follows that there can be little or no β-lactam C—N bond fission in the transition state. The release of strain energy accompanying the opening of the β-lactam ring could increase the rate by up to 10^{20} so this energy must still be present in the transition state for the hydrolysis of β-lactams.

Fusing the β-lactam ring to a five-membered ring to make 1-aza-bicyclo[3.2.0]heptan-2-ones increases the reactivity by *ca* 100-fold but does not significantly change the Brønsted β_{lg} value which is -0.55 for the bicyclic system (Proctor *et al.*, 1982). Although the rate enhancement is substantial, it is hardly of the magnitude expected from the release of strain energy in opening a four-membered ring or from a system in which amide resonance is significantly inhibited. Ring opening does not lower the activation energy because the rate-limiting step for the alkaline hydrolysis of penicillins is formation of the tetrahedral intermediate. The Brønsted β_{lg} of -0.55 indicates that the nitrogen behaves as if it has no charge in the transition state and has lost all of the expected 0.6 positive charge present in the resonance stabilised β-lactam. This is compatible with a transition state that very much resembles the tetrahedral intermediate. It cannot be argued that the transition state is 'early' and involves little bond formation between the carbonyl carbon and the attacking hydroxide ion and that therefore any effect of amide resonance inhibition is not manifest in the transition state. Furthermore, the β_{lg} of -0.55 indicates that the positive charge density on the β-lactam nitrogen in penicillins is similar to that in monocyclic β-lactams and amides where resonance is established.

It is worth noting that monocyclic β-lactams of weakly basic amines can be as chemically reactive as penicillins and cephalosporins. It is not necessary to make the β-lactam part of a bicyclic system to have a reactive amide.

Although it has been suggested that the reactivity of penicillins may be due to intramolecular nucleophilic attack of the 6-acylamido side chain on the β-lactam carbonyl (Doyle and Nayler, 1964; Moll, 1968), there is no significant dependence of k_{OH^-} on the nature of the side chain (Yamana *et al.*, 1974b; Bundgaard, 1972).

SUMMARY

Both kinetic and ground-state effects do not indicate a significant degree of inhibition of amide resonance in penicillins and cephalosporins. The bicyclic β-lactam antibiotics do not exhibit exceptional chemical reactivity. Monocyclic β-lactams with suitable electron withdrawing substituents may be as reactive as the bicyclic systems. A pyramidal geometry of the β-lactam nitrogen does not necessarily give a chemically more reactive β-lactam. Strained β-lactams are not necessarily better antibiotics and biological activity is not directly related to chemical reactivity.

4 Alkaline hydrolysis and structure: chemical reactivity relationships

It is well known that minor substituent changes in β-lactam antibiotics can have a dramatic effect upon antibacterial activity and susceptibility to β-lactamase catalysed hydrolysis. In order to identify the binding energy effects between the β-lactam and enzymes, the effect of substituents on chemical reactivity must be known.

The effects of structural changes on the rates of alkaline hydrolysis of penicillin and cephalosporin derivatives are summarised in Table 4.

TABLE 4

Second-order rate constants ($M^{-1}s^{-1}$) for the hydroxide-ion catalysed hydrolysis of penicillins, cephalosporins and other β-lactam antibiotics at 30°C; $I = 1.0$ M (KCl)[a]

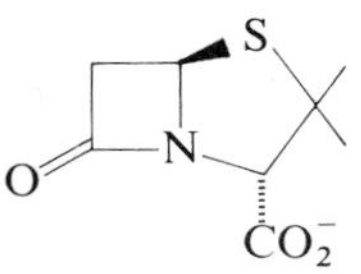

S, O, N, CO_2^-

7.40×10^{-3}

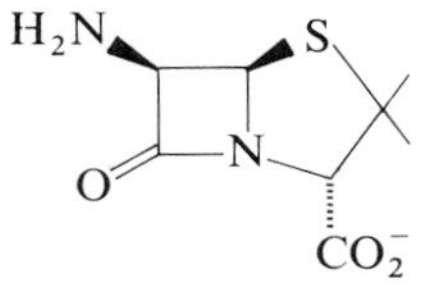

H_2N, S, O, N, CO_2^-

6.31×10^{-2}

$PhCH_2CONH$, S, O, N, CO_2^-

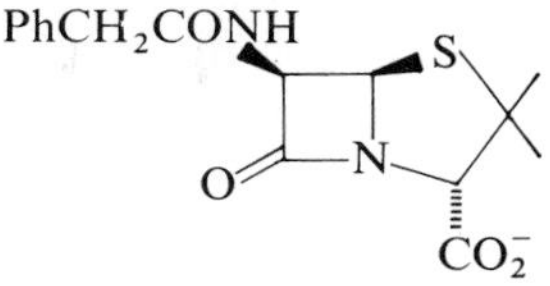

1.54×10^{-1}

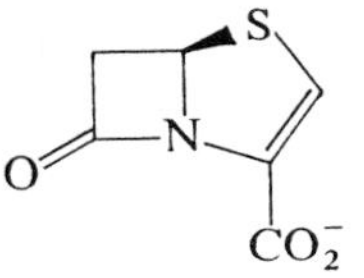

S, O, N, CO_2^-

5.10×10^{-1}

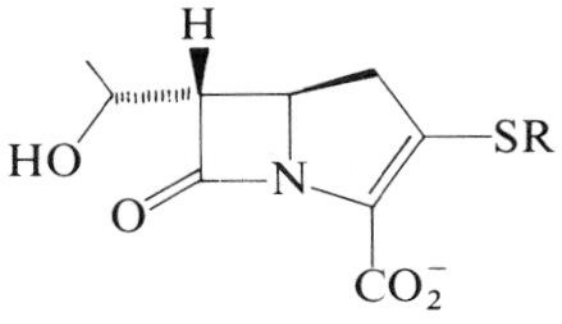

H, HO, SR, O, N, CO_2^-

1.59

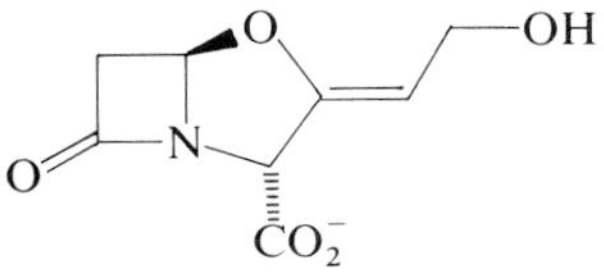

O, OH, O, N, CO_2^-

2.39

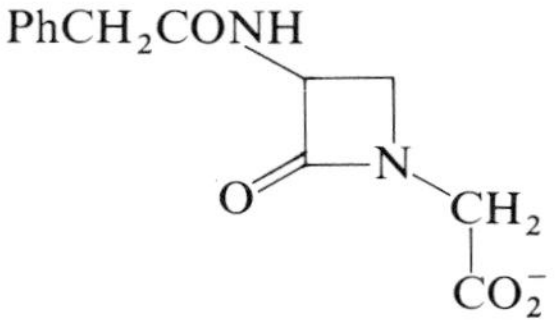

$PhCH_2CONH$, O, N, CH_2, CO_2^-

1.02×10^{-4}

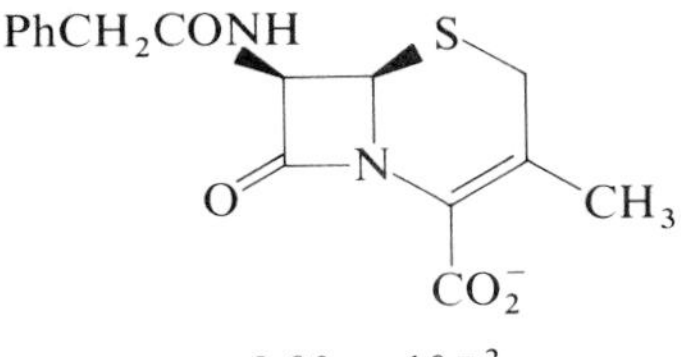

$PhCH_2CONH$, S, O, N, CH_3, CO_2^-

2.90×10^{-2}

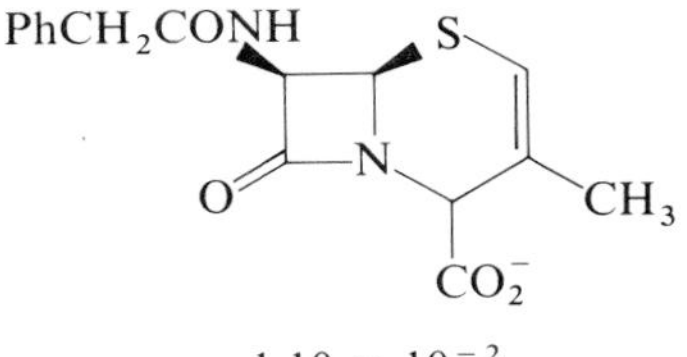

$PhCH_2CONH$, S, O, N, CH_3, CO_2^-

1.10×10^{-2}

$PhCH_2CONH$, O, O, N, CH_3, CO_2^-

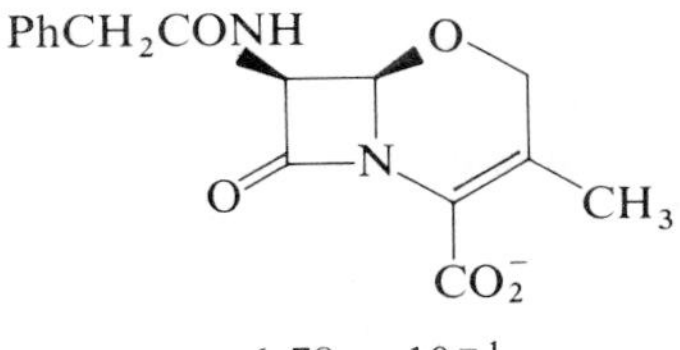

1.78×10^{-1}

---

[a] See also Table 5, p. 205

PENICILLINS

The mechanism of the alkaline hydrolysis was reviewed in the previous section and it has already been mentioned that decreasing the basicity of the leaving group amine increases the rate of alkaline hydrolysis of penicillins. The Brønsted β_{lg} of -0.55 (Proctor *et al.*, 1982) is indicative of rate-limiting formation of a tetrahedral intermediate [32].

Electron withdrawing substituents at C(6) also increase the rate of hydroxide ion hydrolysis and give a Hammett ρ_I-value of $+2.0$ (Proctor *et al.*, 1982). This is only slightly less than the value of 2.7 reported for acyclic amides (Bruylants and Kezdy, 1960). Substituents at C(6) affect the rate of nucleophilic substitution by their effect upon both the electrophilicity of the carbonyl carbon and the leaving group amine. Although the acylamido side chain at C(6) is important for biological activity and increases the rate of alkaline hydrolysis 20-fold relative to penicillanic acid, its effect on chemical reactivity is purely inductive. 6-α-Chloropenicillanic acid undergoes alkaline hydrolysis 2.7 times faster than the 6-β-acylamido derivative.

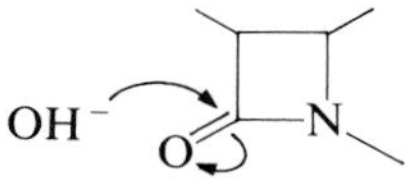

[32] 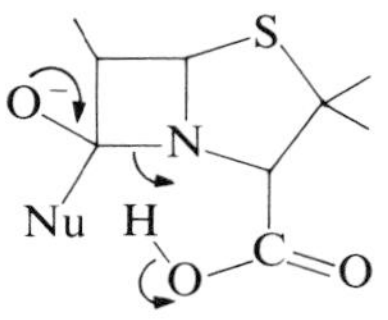 [33] 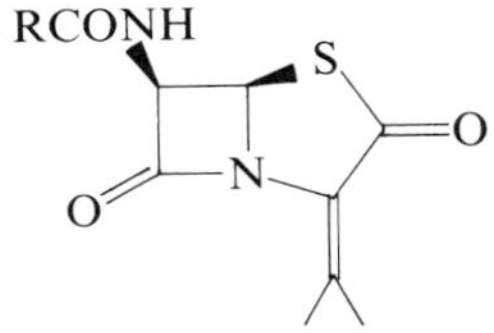 [34]

It is conceivable that the carboxyl group at C(3) which is also very important for biological activity, could act as a general acid catalyst in the reaction of nucleophiles with penicillins [33]. However, there is no evidence that the carboxyl group facilitates hydrolysis and, as expected on the basis of a purely inductive effect, the esterification of this group increases the rate of reaction 10- to 100-fold (Proctor *et al.*, 1982; Gensmantel *et al.*, 1978).

The replacement of the thiazolidine S by CH_2 to give a carbapenam increases the rate by a factor of 3, whereas substitution by O as in the oxapenams increases the rate *ca* 5-fold (Table 4).

The effect of the replacement by oxygen is that expected on the basis of an inductive effect. The CSC bond angle is relatively small and the CS bond length relatively long compared with the carbon analogue. Replacement of S by CH_2 will decrease the leaving group ability of the β-lactam amine [$\overset{+}{N}H_3CH_2CH_2SMe$ has pK_a 9.2 compared with 10.6 for $\overset{+}{N}H_3(CH_2)_3Me$]. Presumably these effects must cancel so that there is little difference in reactivity between the penams and carbapenams.

The incorporation of a double bond into the thiazolidine ring of a penam to give the corresponding penem system [7] increases the rate of hydrolysis by *ca* 25-fold. This is the order of magnitude expected from the decrease in basicity of the leaving group amine brought about by the introduction of a conjugated amine in the tetrahedral intermediate (Proctor *et al.*, 1982). Conversion of a Δ^2-carbapenem to a Δ^1-carbapenem similarly decreases the reactivity 25-fold (Pfaendler *et al.*, 1981).

Anhydropenicillins [34] caused some anxiety to early workers because their *apparent* chemical stability could not be reconciled with the assumed strain in the system (Wolfe *et al.*, 1963, 1968). The β-lactam nitrogen is pyramidal, $h = 0.41$ Å (Simon *et al.*, 1972), and the carbonyl stretching frequency is higher than that found in normal penicillins. The characterisation of anhydropenicillins as chemically stable compared with penicillins was based on the observation that they were recovered unchanged from refluxing solvents such as water and ethanol (Wolfe *et al.*, 1968). In fact, it was later shown (Bundgaard and Angelo, 1974) that anhydropenicillins are *ca* 100-fold more reactive towards alkaline hydrolysis than are normal penicillins. This is as expected because of the effect of the electron withdrawing sulphoxide group on the leaving group amine. Bundgaard and Angelo (1974) suggested that nucleophilic attack was on the thiolactone of [34], but it has since been demonstrated that this occurs on the β-lactam carbonyl (Pratt *et al.*, 1983).

The rate of the alkaline hydrolysis of penicillins is not greatly affected by ionic strength. The second-order rate constant increases by about 30% up to $I = 0.5$ M (KCl) but is then independent of I up to $I = 4.0$ M (Morris and Page, 1978).

The observed pseudo first-order rate coefficient for the hydrolysis of benzylpenicillin is first order in hydroxide ion up to 2 M sodium hydroxide. Above this concentration it begins to level off (Minhas and Page; 1982). This is probably attributable to ionisation of the benzylamido side chain. Presumably hydroxide ion attack on the penicillin with a 6-amido anion side chain is retarded. In support of this, the observed rate constant for the hydrolysis of phenoxymethylpenicillin shows a non-linear dependence upon hydroxide ion above 0.1 M sodium hydroxide (Minhas and Page, 1982; Pratt *et al.*, 1983). The more electron-withdrawing phenoxymethyl group decreases the pK_a-value of the amide side chain to 13.3. The observed first-order rate constants for the hydrolysis of 6-aminopenicillanic acid are, as expected, linear in hydroxide ion concentration.

Solvent isotope effects on the alkaline hydrolysis of penicillins are consistent with rate-limiting formation of the tetrahedral intermediate; $k_{OH^-}/k_{OD^-} = 0.65$ at 30°C and 0.59 at 35°C (Gensmantel *et al.*, 1978; Yamana *et al.*, 1977).

CEPHALOSPORINS

The major structural differences between cephalosporins [11] and penicillins [12] are that the five-membered thiazolidine ring of penicillins is replaced by a six-membered dihydrothiazine ring in cephalosporins and that the degree of pyramidalisation of the β-lactam nitrogen is generally smaller in cephalosporins. In addition, many of the cephalosporins have a leaving group, e.g. acetate, pyridine and thiol, at C(3′) and expulsion of these groups occurs during the hydrolysis of the β-lactam as shown in Scheme 4 (Hamilton-Miller *et al.*, 1970a; O'Callaghan *et al.*, 1972). There have been many

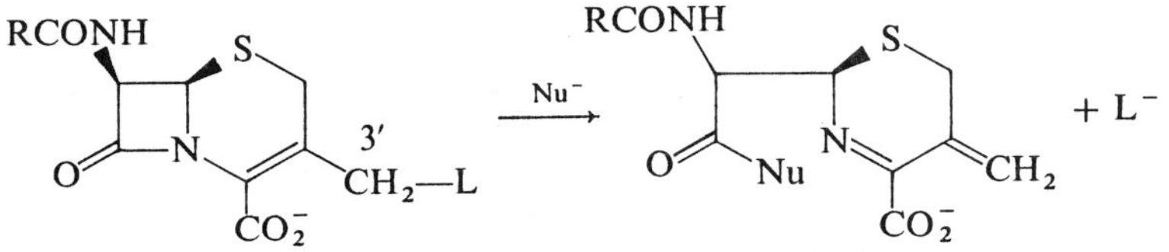

RCONH S 3′ CH_2—L CO_2^- $\xrightarrow{Nu^-}$ RCONH S Nu N CH_2 CO_2^- + L^-

[37]

Scheme 4

suggestions (Bundgaard, 1975), apparently supported by theoretical calculations (Boyd *et al.*, 1975; Boyd and Lunn, 1979), that a nucleophilic attack on the β-lactam carbonyl carbon is concerted with departure of the leaving group at C(3′) [35]. It is common to read that the presence of the leaving group at C(3′) enhances chemical reactivity (Wei *et al.*, 1983). Furthermore, it has been proposed that biological activity is related to the leaving group ability of the C(3′) substituent (Boyd *et al.*, 1980).

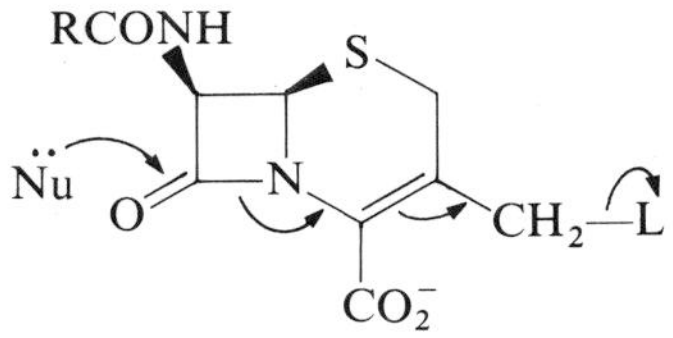

RCONH S Nu O N CH_2—L CO_2^-

[35]

In general, the second-order rate constants for the hydroxide-ion catalysed hydrolysis of cephalosporins are similar to those of penicillins (Proctor *et al.*, 1982; Yamana and Tsuji, 1976). This similarity indicates that the non-planarity of the β-lactam nitrogen does not significantly affect amide resonance since the nitrogen is 0.4 Å out of the plane defined by its substituents in penicillins (Sweet and Dahl, 1970) whereas in the cephalosporins it deviates by 0.2–0.3 Å (Sweet, 1973). The kinetic similarity also indicates that having a leaving group at C(3′) does not significantly affect the reactivity of cephalosporins. The rate-limiting step in the alkaline hydrolysis of cephalosporins appears to be formation of the tetrahedral intermediate [32]. Electron-withdrawing substituents attached to the β-lactam nitrogen increase the rate of hydrolysis and give a Brønsted β_{lg} of -0.6 (Proctor *et al.*, 1982). In the stepwise mechanism, breakdown of the tetrahedral intermediate generates the enamine [36] followed by expulsion of the leaving group at C(3′) to give the conjugated imine [37] (Agathocleous *et al.*, 1985). It has been observed that the rate of appearance of the leaving group L is identical to the rate of β-lactam ring opening (Coene *et al.*, 1984; Bundgaard, 1975).

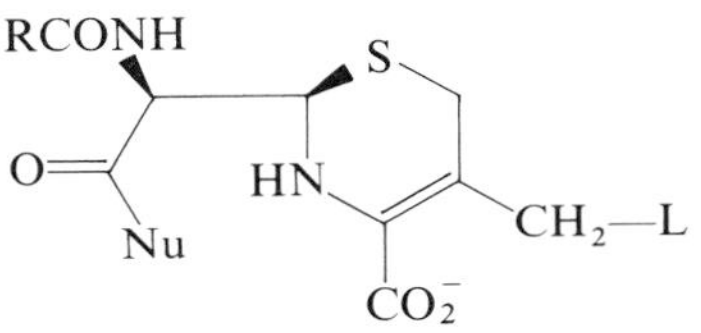

[36] 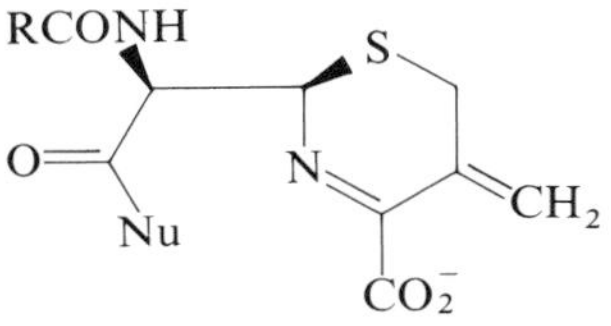 [37]

Although this has been interpreted as supporting the concerted mechanism, it is, of course, also consistent with a stepwise process involving rate-limiting formation of the tetrahedral intermediate if expulsion of the leaving group from [36] occurs faster than formation of [36]. Evidence to show the presence of the enamine [36] can be obtained by increasing its rate of formation and decreasing its rate of decomposition (Agathocleous *et al.*, 1985).

There are several experimental observations which indicate that the reaction is not concerted and that expulsion of the leaving group at C(3′) occurs *after* β-lactam ring opening.

The second-order rate constants for the hydroxide-ion catalysed hydrolysis of cephalosporins are correlated with σ_I for C(3) substituents and give a Hammett ρ_I of 2.5 for CH_2L and of 1.35 for L (Fig. 8; Table 5). Several substituents at C(3), e.g. CH_3, H, CH_2CO_2Et, are not expelled during hydrolysis or cannot be expelled directly by a concerted mechanism, e.g. Cl. Substituents which are and those which are not expelled are controlled by the same linear free energy relationship (Proctor *et al.*, 1982; Page 1984a; Indelicato *et al.*, 1974; Bundgaard, 1975). Leaving groups of different nucleofugalities (Stirling, 1979) influence the rate of reaction only by their inductive effect (Proctor *et al.*, 1982; Bundgaard, 1975). A series of cephalosporins with substituted pyridines and thiol leaving groups at C(3′) covering a range of 10 pK_a units show a Brønsted β_{lg} of 0.1 (Buckwell and Page, 1985). There is little or no change in the effective charge on the leaving group on going from the ground to the transition state.

Esterification of the C(4) carboxylate group will make the β-lactam carbonyl carbon more electrophilic and facilitate β-lactam C—N bond

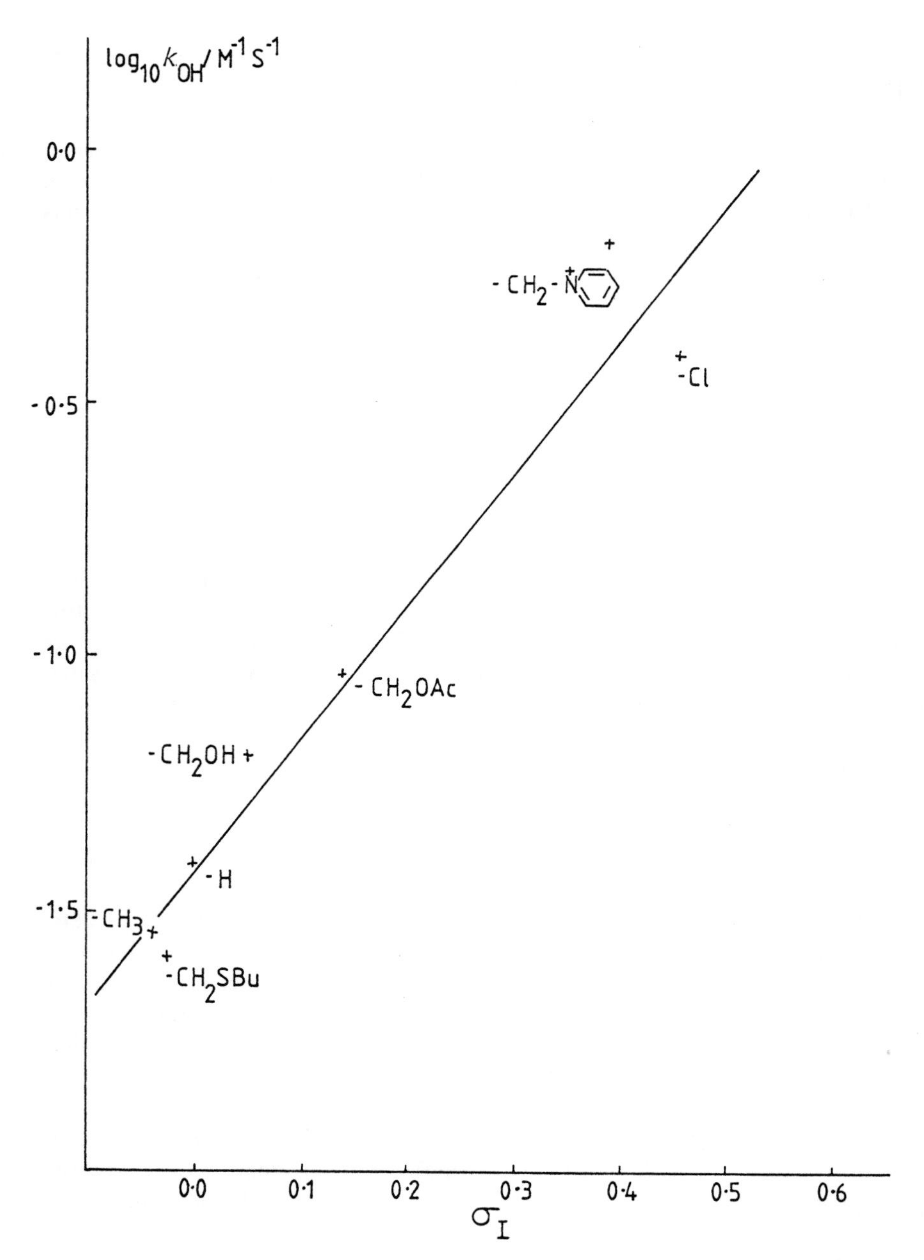

FIG. 8 Hammett plot of the second-order rate constants for the hydroxide-ion catalysed hydrolysis of cephalosporins [11] against Charton's σ_I constant for the C(3) substituent.

cleavage but decrease the rate of fission of the C(3′)—L bond. The carbomethoxy group at C(4) induces biphasic kinetic behaviour in the alkaline hydrolysis of cephalosporins (Agathocleous *et al.*, 1985). The two consecutive reactions observed spectroscopically are compatible; a stepwise mechanism for hydrolysis-fission of the β-lactam C—N bond to give the enamine [36] followed by expulsion of the leaving group to give [37].

TABLE 5

Second-order rate constants (k_{OH}) at 30°C for the hydroxide-ion catalysed hydrolysis of cephalosporins as a function of substituents at C(3); $I = 1.0$ M(KCl)

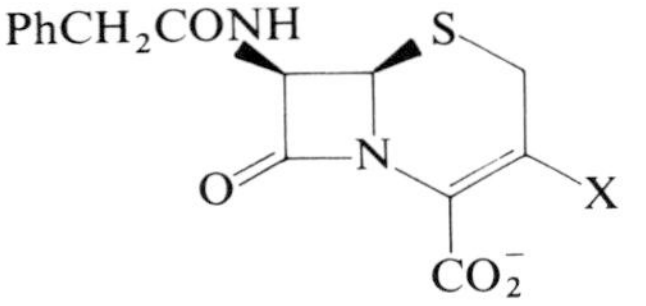

PhCH$_2$CONH, S, N, O, X, CO_2^-

X	$k_{OH}/M^{-1}s^{-1}$
H	3.86×10^{-2}
CH_3	2.90×10^{-2}
Cl	3.90×10^{-1}
$CH_2S(CH_2)_3CH_3$	2.65×10^{-2}
CH_2SCOCH_3	3.30×10^{-2}
CH_2OCOCH_3	9.36×10^{-2}
CH_2OH	6.51×10^{-2}
$CH_2\overset{+}{N}C_5H_5$	6.49×10^{-1}

It is conceivable that the conversion of the enamine [36] to the α,β -unsaturated imine [37] is reversible. The addition of thiolate anions to the hydrolysis products of cephalosporins generated using β-lactamase as a catalyst indicates that the enamine and α,β -unsaturated imine are in equilibrium (Buckwell and Page, 1985). (See Section 12.)

There is spectroscopic and kinetic evidence that the aminolysis of cephalosporins proceeds by a stepwise mechanism (see Section 12), and, in general, it appears that 3′-eliminations are not concerted with β-lactam C—N bond cleavage when cephalosporins react with nucleophiles.

It has been frequently suggested (Flynn, 1972; Nishikawa and Tori, 1984) that the rate of basic hydrolysis of the β-lactam is correlated with the antibacterial activity of the antibiotic. Consequently, there have been many attempts to investigate possible correlations between structural parameters and chemical and biological reactivity. Biological activity is, of course, also very dependent upon the ability of the compound to penetrate into bacteria.

There is a linear correlation between the logarithm of the rate constant for alkaline hydrolysis and the infrared carbonyl stretching frequency (Morin *et al.*, 1969; Indelicato *et al.*, 1974; Takasuka *et al.*, 1982). Although the correlation has been claimed to be poor (Boyd, 1982), an extensive and careful examination (Nishikawa and Tori, 1981) supports the earlier claims. The higher the β-lactam $\nu_{C=O}$ value the greater is the β-lactam reactivity.

The ^{13}C nmr chemical shifts of the C(8) β-lactam carbonyl carbon varies over only a very narrow range (Paschal *et al.*, 1978; Dereppe *et al.*, 1978; Schanck *et al.*, 1979). However, there does appear to be a good linear relationship between the logarithms of the rate constants, k_{OH}, for the base catalysed hydrolysis of cephalosporins and the differences between the ^{13}C chemical shifts at C(3) and at C(4), $\Delta\delta(4-3)$ (Nishikawa and Tori, 1981, 1984; Mondelli and Ventura 1977; Schanck *et al.*, 1983).

In addition to attempts to correlate hydrolysis rates with β-lactam C—N and C=O bond lengths and the degree of β-lactam nitrogen pyramidalisation (Sweet and Dahl, 1970), there have also been reports of the relationship between chemical reactivity and theoretically calculated parameters such as the net atomic charge on the β-lactam carbonyl oxygen, the overlap population of the β-lactam carbonyl and transition state energies (Indelicato *et al.*, 1974; Boyd, 1983; Boyd *et al.*, 1980; Petrongolo *et al.*, 1980).

In addition to the leaving group at C(3′), many other structural parameters within cephalosporins have been varied. Although the change from a Δ^3- to a Δ^2-cephem system causes the β-lactam nitrogen to become planar there is little difference, only 2- to 3-fold, in the chemical reactivity (Proctor *et al.*, 1982).

Other effects of structural changes are given in Table 4. Replacement of the dihydrothiazine S by O increases the rate of alkaline hydrolysis about 6-fold (Narisada *et al.*, 1983). 1-Oxacephems sometimes show greater antibacterial activity than the corresponding cephalosporins (Firestone *et al.*, 1977; Narisada *et al.*, 1977; Murakami *et al.*, 1981). Replacement of the ring S by CH_2 increases the rate of hydrolysis about 3-fold. It is interesting to note that this increase in reactivity is accompanied by a *decrease* in the β-lactam carbonyl stretching frequency which is contrary to the correlation described earlier (Nishikawa *et al.*, 1982). Increased antibacterial activity of 1-oxacephalosporins may result from a higher rate of penetration through the bacterial cell membrane because of increased hydrophilicity (Murakami and Yoshida, 1982).

The addition of a methyl, methoxy or thiomethyl group at the 6-α-position of penicillin results in a reduction in antibacterial activity (Ho *et al.*, 1973). In contrast, the addition of a 7-α-methoxy group to a cephalosporin results in compounds that are better transpeptidase enzyme inhibitors although they do not necessarily show better antibacterial properties (Indelicato and Wilham, 1974).

The introduction of a 7-α-methoxy group has an almost insignificant effect, less than 2-fold, upon the susceptibility of cephalosporins to alkaline hydrolysis (Indelicato and Wilham, 1974; Nishikawa *et al.*, 1982; Narisada *et al.*, 1983). Inductively a 7-α-methoxy group should increase the rate 3-fold (Proctor *et al.*, 1982) but unfavourable steric interactions in the tetrahedral intermediate must lower the rate. This steric effect is supported by the effect of a 7-α-methyl group which decreases the rate by a factor of 9 in cephalosporins (Narisada *et al.*, 1983) whereas a 6-α-methyl decreases k_{OH} for penicillins by a factor of 10 (Indelicato and Wilham, 1974). It is worth noting that 7-α-methoxycephalosporins have a less pyramidal β-lactam nitrogen (Applegate *et al.*, 1974) and so its minimal effect upon chemical reactivity again indicates that the degree of pyramidalisation of nitrogen is not a major influence.

Substituent changes in the 7-β-acylamido side chain have little effect upon chemical reactivity and yet can enormously change biological activity. For example, the incorporation of a *syn*-oxime function, as in cefuroxime, confers both high antibacterial activity and β-lactamase resistance (Bucourt *et al.*, 1978; Schrinner *et al.*, 1980). The oxime substituent, irrespective of its configuration, does not affect k_{OH} for alkaline hydrolysis of the β-lactam but is highly enzyme specific. For example, the *syn*-isomer is 35-fold less reactive, as measured by k_{cat}/K_m, towards β-lactamase, whereas, the *anti*-isomer is twice as reactive compared with an analogous cephalosporin lacking the oxime function (Laurent *et al.*, 1984).

As the β-lactam ring of cephalosporins generally has a reactivity comparable with that of ethyl acetate, it is not surprising that hydrolysis of an acetoxy ester side chain at C(3) is competitive with hydrolysis of the β-lactam. The second-order rate constant for the base-catalysed conversion of the ester to the 3-hydroxylmethylcephalosporin is usually similar to that for β-lactam hydrolysis (Yamana and Tsuji, 1976; Berge *et al.*, 1983; Bundgaard, 1975).

5 Acid hydrolysis

The pH-rate profile for the hydrolysis of benzylpenicillin is shown in Fig. 9 (Gensmantel *et al.*, 1978). There is no significant spontaneous hydrolysis but the β-lactam does undergo an acid catalysed degradation. Also shown in Fig. 9 is the pH-rate profile for the hydrolysis of cephaloridine. There are two immediate differences: cephalosporins exhibit a spontaneous pH-independent hydrolysis and are less reactive towards acid than penicillins by a factor of about 10^4 (Proctor *et al.*, 1982).

In addition to the expected hydrolysis product, benzylpenicilloic acid [30], the acid catalysed degradation of benzylpenicillin gives benzylpenicillenic

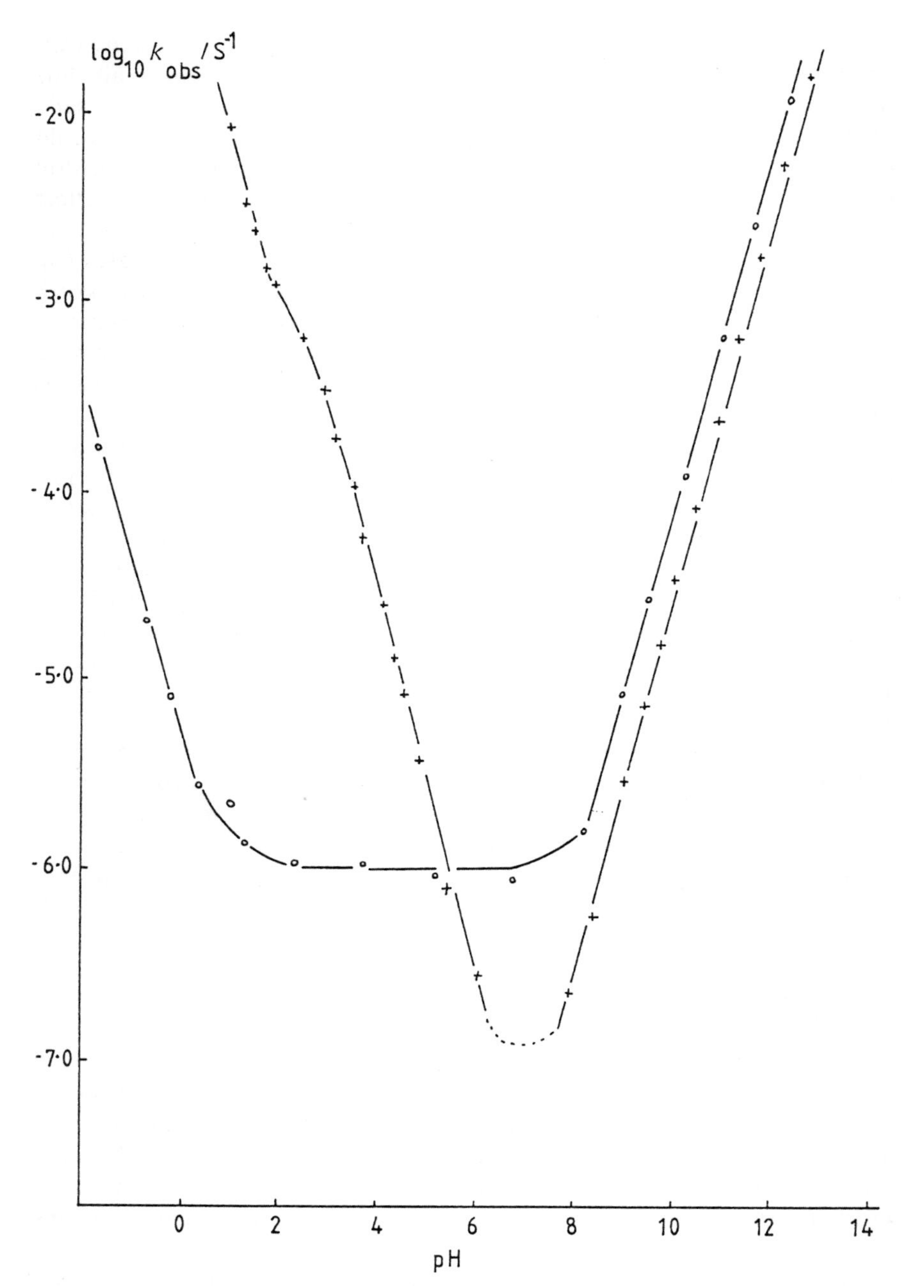

FIG. 9 pH-Rate profiles for the hydrolysis of benzylpenicillin (+) and cephaloridine (○) at 30°C; I = 1.0 M(KCl)

acid [38], benzylpenamaldic acid [39], benzylpenillic acid [40] and benzylpenilloic acid [41]. The proportion of each product formed depends upon the pH (Schwartz, 1965; Dennen and Davis, 1962). Although several kinetic studies have been reported on the degradation of penicillins in acidic media (Schwartz, 1965; Longridge and Timms, 1971; Degelaen *et al.*, 1979a; Blaha *et al.*, 1976; Bundgaard, 1980; Kessler *et al.*, 1983) there is still considerable uncertainty about the details of the reaction pathway.

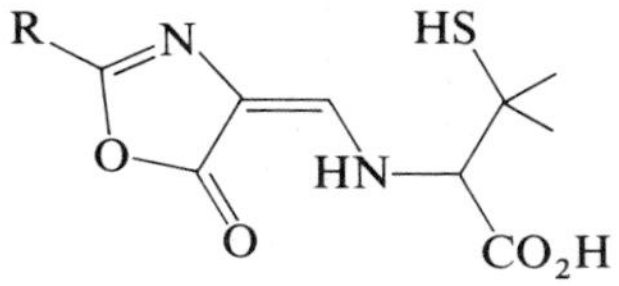

[38]

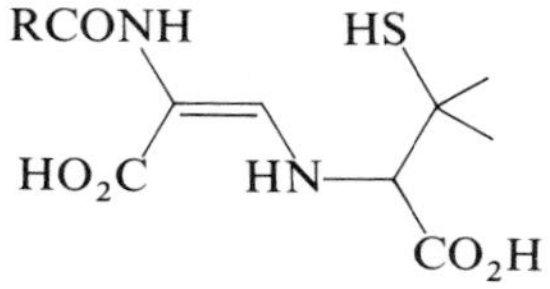

[39]

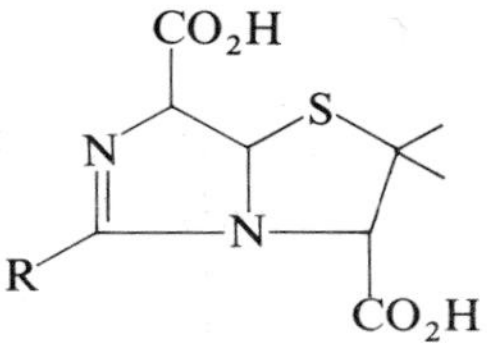

[40]

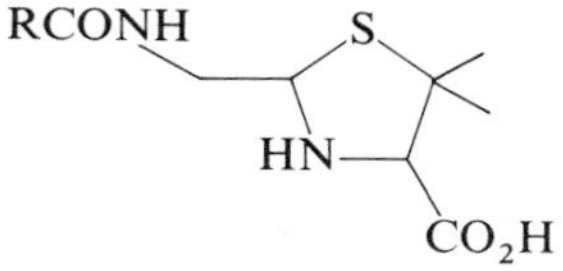

[41]

It has been suggested (Schwartz, 1965) that one reaction pathway involves specific acid catalysed hydrolysis of the unionised penicillin to give penicilloic acid [30] and, by decarboxylation of this product, penilloic acid [40]. In a second, parallel reaction penicillin rearranges to penicillenic acid [38] which subsequently degrades to penillic acid [40]. The formation of penicillenic acid was suggested to be the result of a specific acid catalysed reaction of ionised penicillin or the kinetically equivalent spontaneous rearrangement of the unionised acid. Another study (Blaha *et al.*, 1976) proposed that penicillenic acid is the first formed product and that all other degradation products are formed from this. Based on an nmr spectroscopic study and the extent of deuteriation at C(6), it was concluded that three parallel degradation reactions occur and penicillin initially gives directly penicilloic acid, penillic acid and penicillenic acid. In addition to these detectable intermediates it has also been proposed (Bundgaard, 1980; Proctor *et al.*, 1982) that an oxazolone-thiazolidine intermediate [42] is formed which is the precursor of the degradation products.

The proposal (Degelaen *et al.*, 1979a) that penillic acid is formed directly from penicillin has been criticised and the optimisation of fitting kinetic data to the nmr observations suggests that all the penicillenic acid is transformed

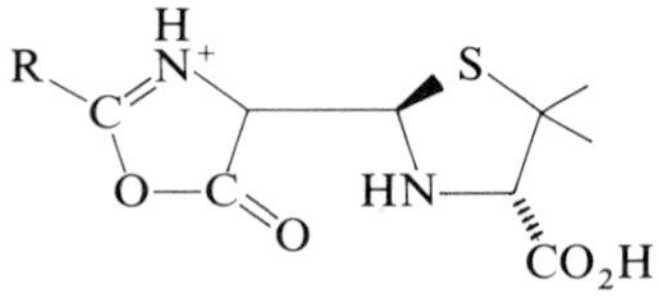

[42]

into penamaldic acid [39] (Kessler *et al.*, 1983). Furthermore, it was suggested that penillic acid is the major degradation product of penicilloic acid.

A study of substituent effects and the acidity dependence of the rate of degradation shows that most of these problems can be resolved into an acceptable reaction pathway (Proctor *et al.*, 1982). Thermodynamically the most basic site for the protonation of normal amides is oxygen and the pK_a of O-protonated amides is 0 to -3, (Liler, 1969; Yates and Stevens, 1965), although it is claimed that N-protonation occurs in dilute acid and O-protonation in strong acid (Liler 1975). Whether the mechanism of the acid catalysed hydrolysis of simple amides proceeds *via* O- or N- protonation seems to have been resolved in favour of the former (Williams 1976; Modro *et al.*, 1977; McLelland and Reynolds, 1974; Kresge *et al.*, 1974), although there remain some dissenting voices (Liler, 1975).

Remarkably, the logarithms of the pseudo first-order rate constants for the hydrolysis of some β-lactam antibiotics and derivatives increase linearly with decreasing H_0 values up to -5 (Proctor *et al.*, 1982). This is quite unlike the behaviour of other amides for which the rate of hydrolysis passes through a maximum, attributed to complete conversion of the amide into its O-conjugate acid and to decreasing water activity (O'Connor, 1970; Smith and Yates, 1972). This indicates that the β-lactams are far less basic than normal amides for O-protonation and that a different mechanism of hydrolysis is operating. Neither the nitrogen nor the oxygen of the bicyclic β-lactams is sufficiently basic for substantial conversion to the conjugate acid; pK_a for O- or N- protonation must be < -5. This behaviour is not peculiar to bicyclic β-lactams since *monocyclic* β-lactams show similar reactivities and behaviour (Proctor *et al.*, 1982; Wan *et al.*, 1980). The reduced basicity of cyclobutanones compared with other ketones has been previously noted (McLelland and Reynolds, 1976) and the very weak basicity of β-lactams may have a similar origin (Bouchoux and Houriet, 1984). The slopes of plots of the logarithms of the pseudo first-order rate constants against H_0 are -1 to -1.3 and, since water activity decreases with increasing acidity, it appears that water is not involved in the transition state.

This can be explained by a unimolecular A-1 type mechanism with N-protonation of the β-lactam (Scheme 5). That N-protonation takes place

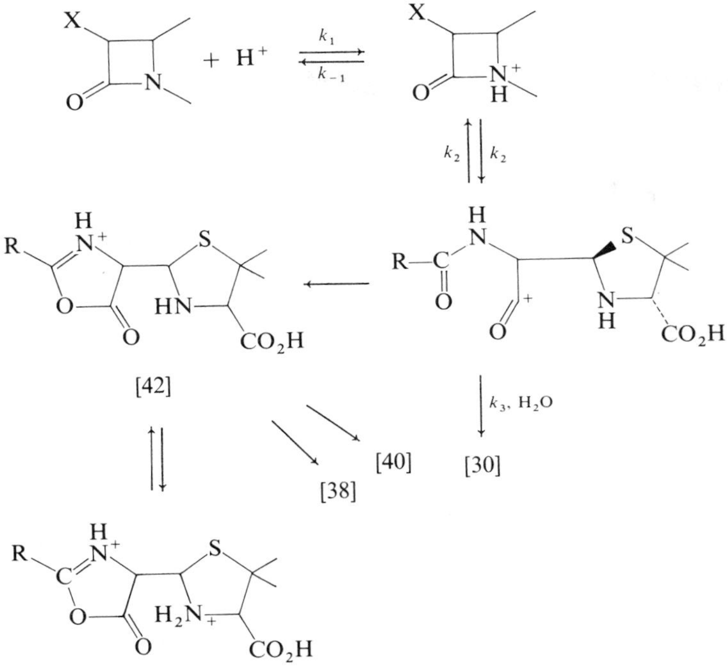

Scheme 5

is not the result of reduced amide resonance in penicillins and cephalosporins but must be an intrinsic property of β-lactams. The introduction of the A-1 mechanism could result because the normal A-2 mechanism is retarded or because the A-1 pathway is favoured. The most likely explanation is the enhanced rate of C—N bond fission that occurs in β-lactams as a result of the relief of ring strain (Proctor *et al.*, 1982).

Substituents at C(6) in penicillins affect both the carbonyl carbon and nitrogen of the β-lactam inductively, but their effect on C—N bond cleavage will be that predominantly of an acyl substituent. As described earlier, penicillins with an acylamido side chain at C(6) undergo degradation in acid solution to give a variety of products which must result from transformations involving the side chain amide. Electron-withdrawing substituents which cannot be involved in neighbouring group participation greatly retard the rate with a Hammett ρ_1-value of *ca* -4.0 to -5.0, depending upon the acidity (Proctor *et al.*, 1982). The effect of acyl substituents upon the rate of acid catalysed hydrolysis of acyclic amides is small, with electron-withdrawing substituents producing either a small increase or decrease in rate (Bruylants and Kezdy, 1960; Bolton and Jackson, 1969).

Electron-withdrawing substituents in the amine portion of the β-lactam decrease the rate of acid catalysed degradation of penicillins. The Brønsted β-value is *ca* 0.35 (Proctor *et al.*, 1982) compared with −0.26 for acyclic anilides and amides (Giffney and O'Connor, 1975). Although the effects of substituents are not large, they are significant and in the opposite direction for β-lactams compared with other amides which again is indicative of a different mechanism.

There is a large dependence of the rate of the acid catalysed degradation of penicillins upon the nature of the acylamido side chain. Electron-withdrawing substituents decrease the rate of degradation and for substituted phenylpenicillins the Hammett ρ-value is −1.60 (Yamana *et al.*, 1974b).

Because of the similarity of the rate constants for the degradation of benzylpenicillin and its methyl ester there is no evidence for neighbouring group participation of the carboxy-group in the fission of the β-lactam ring (Proctor *et al.*, 1982). However, penicillenic acid [37] is not formed significantly below pH 1 and is the major product only at *ca* pH 3 (Schwartz, 1965; Blaha *et al.*, 1976; Bundgaard, 1980). It has been suggested (Schwartz, 1965) that penicillenic acid is formed from the specific acid-catalysed reaction of ionised penicillin or the kinetically equivalent spontaneous rearrangement of the undissociated acid. This suggestion is based on the observation that the rate of formation of penicillenic acid reaches a plateau at pH *ca* 2, where the overall rate constant for degradation also shows an inflection point, thought to correspond to ionisation of the carboxy-group. However, benzylpenicillin methyl ester shows exactly the same behaviour and the rate of penicillenic acid formation follows (2) with $k_0 = 3.42 \times 10^{-4}\,s^{-1}$ and $K_a = 10^{-2.25}$ (Proctor *et al.*, 1982). The ionisation of a group with pK_a 2.25 obviously cannot correspond to the carboxy-group as suggested for benzylpenicillin itself. It seems doubtful, therefore, that the decreased formation of penicillenic acid from penicillin at low pH is due to protonation of the ionised carboxy-group.

$$k_{obs} = \frac{k_0[H^+]}{K_a + [H^+]} \qquad (2)$$

The acid catalysed degradation of C(6) acylamido penicillins show a rate of enhancement of *ca* 10^3 compared with that predicted from the Hammett plot for C(6) substituents (Proctor *et al.*, 1982). The mechanism of degradation must therefore incorporate the acylamido group in the rate-limiting step or in a pre-equilibrium step.

The important facts to be explained therefore are:

1. The rate enhancement observed for the acid catalysed degradation of benzylpenicillin indicates neighbouring group participation by the acylamido side chain.

2. Benzylpenicillenic acid [38] is not formed in acidic solution (pH < 1) where its rate of formation is pH-independent.

3. The majority of the penillic acid [40] and penicilloic acid [30] formed does not come from a mechanism involving D-incorporation at C(6), i.e. from penicillenic acid type intermediates (Degelaen *et al.*, 1979a; Kessler *et al.*, 1983).

4. The rate-limiting step for the acid catalysed degradation of all penicillins does not appear to involve water.

A mechanism compatible with these observations is shown in Scheme 5 (Proctor *et al.*, 1982). Reversible ring opening of the β-lactam gives an acylium ion which may be trapped by the intramolecular amido group rather than by water to give the protonated oxazolone-thiazolidine [42]. This can react with water to give penicilloic acid, undergo intramolecular nucleophilic attack of the thiazolidine nitrogen on the carbon of the protonated oxazolone to give penillic acid [40] or eliminate across C(5)—C(6) to give penicillenic acid [38]. The pK_a-value of the protonated oxazolone or that of the thiazolidine nitrogen could be the kinetically important one controlling penicillenic acid formation. The protonated oxazolone probably has a pK_a of *ca* 0 and furthermore there is no obvious chemical reason why protonation of the imine should inhibit elimination. For the thiazolidine nitrogen pK_a is estimated to be *ca* 3. Protonation of this nitrogen would inhibit penicillenic acid formation. The kinetically important ionisation is therefore attributed to that of the thiazolidine nitrogen.

There is a slight shadow cast over the neat picture for the acid catalysed degradation of β-lactams outlined in Scheme 5. The rate of penicillin degradation is *ca* 10^3 faster than that predicted from the σ-value for RCONH. The implication is that the acylamido-group participates in the rate-limiting step, which is acceptable if k_3 is rate-limiting (Scheme 5) for hydrolysis, i.e. the formation of the acylium ion is reversible. However, this then reintroduces water in the transition state for hydrolysis of the penicillins lacking an acylamido-side chain, and this is not indicated by the acidity dependence of the rate of reaction. An alternative explanation is that attack of the acylamido-group on the β-lactam carbonyl carbon is concerted with C—N bond fission, i.e. the reaction does not proceed via the acylium ion.

The acid hydrolysis of cephalosporins shows similar behaviour to that of the penicillins, but they are about 10^4-fold less reactive (Proctor *et al.*, 1982). Electron-withdrawing substituents at C(7) in cephalosporins decrease the rate of acid hydrolysis and, as for penicillins, the Hammett ρ_1-value is *ca* -5. There is no evidence for neighbouring group participation by the 7-acylamido-group as postulated for the penicillins. There seems no obvious explanation of the difference in behaviour between the cephalosporins and penicillins. Either attack of water on the acylium ion (Scheme 5) could be inhibited in penicillins relative to cephalosporins, and therefore trapping of the acylium ion by the acylamido-group is more effective in penicillins, or the latter process could be inhibited in cephalosporins. One difference is that the dihydrothiazine of cephalosporins has a less basic nitrogen than the thiazolidine of penicillins; enamine resonance lowers the pK_a-value by *ca* 2 units. This could account for the lower reactivity of 7-amino- and 7-chlorocephalosporanic acid compared with the analogous penicillin derivatives.

Similar to alkaline hydrolysis there is no evidence for the group at C(3′) in cephalosporins (acetate or pyridine) affecting the rate of reaction. In fact, the 3-methyl derivative is more reactive than the cephalosporins with acetate or pyridine at C(3′) which yet again indicates that expulsion of these groups is not important in the rate-limiting step.

At pHs above the pK_a of the 3-carboxyl group, the rate of degradation of penicillin is acid catalysed up to about pH 6. The pathway for hydrolysis could be an acid catalysed reaction of the penicillin with an ionised carboxylate or the kinetically equivalent mechanism, the spontaneous degradation of penicillin with an unionised carboxyl function as indicated in (3).

$$k_H[\mathrm{PenCO_2^-}][\mathrm{H^+}] = k_H K_a[\mathrm{PenCO_2H}] = k_0\,[\mathrm{PenCO_2H}] \qquad (3)$$

The calculated rate constant k_0 for benzylpenicillin at 30°C (Proctor *et al.*, 1982) is $6.30 \times 10^{-4}\,\mathrm{s^{-1}}$ which is about 6000 times greater than the spontaneous hydrolysis of benzylpenicillin with an ionised carboxylate at C(3).

Some investigators have favoured the spontaneous degradation of unionised penicillin mainly because of the lack of an appreciable ionic strength effect on k_H (Finholt *et al.*, 1965; Yamana *et al.*, 1974b) and an observed solvent isotope effect $k_0^{H_2O}/k_0^{D_2O}$ of 1.53 at 60°C (Yamana *et al.*, 1977). Penicillenic acid and methyl benzylpenicillenate are formed during the degradation of benzylpenicillin and benzylpenicillin methyl ester, respectively (Proctor *et al.*, 1982; Bundgaard, 1980). The mechanism of penicillin degradation could involve intramolecular participation by the C(3) carboxyl group or the 6-amido side chain.

The second-order rate constant k_H for the catalysed degradation of penicillin with an ionised carboxyl at C(3) is $0.354\,\mathrm{M^{-1}\,s^{-1}}$ at 30°C compared with $8.2 \times 10^{-2}\,\mathrm{M^{-1}\,s^{-1}}$ for benzylpenicillin methyl ester. This rate difference of 4-fold is insignificant. The calculated first-order rate constant k_0 for undissociated benzylpenicillin degradation is 10-fold greater than that for the methyl ester of benzylpenicillin. This is also probably too small to be indicative of intramolecular catalysis by the carboxyl group.

The 6-amido side chain also does not provide a significant rate enhancement in the degradation of penicillins between pH 3 and 6. The second-order rate constants for the acid catalysed degradation of 6-ammonium penicillanate and 6-aminopenicillanate are 3.5×10^{-3} and $0.675\,M^{-1}\,s^{-1}$, respectively, compared with $0.354\,M^{-1}\,s^{-1}$ for benzylpenicillin at 30°C (Proctor *et al.*, 1982). Replacement of the 6-amido side-chain by an amino group therefore leads to a slight rate enhancement. It appears that the amido side chain is not involved in the rate-limiting step of the degradation of penicillins from pH 3 to 6, but the mechanism of this degradation is still ambiguous.

6 Spontaneous hydrolysis

Penicillins undergo an acid and a base catalysed hydrolysis but there is no significant uncatalysed reaction. The pH minimum is at 7 (Fig. 9) and k_0, the apparent first-order rate constant for spontaneous or water-catalysed degradation, is $1 \times 10^{-7}\,s^{-1}$ at 30°C (Gensmantel *et al.*, 1978). By contrast, cephalosporins show a pH independent reaction between pH 3 and 7 with k_0 in the range 5×10^{-7} to $3 \times 10^{-6}\,s^{-1}$ at 30°C (Yamana and Tsuji, 1976; Yamana *et al.*, 1974a; Fujita and Koshiro, 1984; Berge *et al.*, 1983; Lumbreras *et al.*, 1982). It has been suggested that this pH-independent reaction involves intramolecular nucleophilic attack on the β-lactam by the 7-amido side chain (Yamana and Tsuji, 1976). There are several problems raised by this proposal. As outlined in the previous section, cephalosporins do not show neighbouring group participation by the 7-amido side chain in their acid catalysed degradation. It is difficult to understand why it would therefore occur in the uncatalysed reaction. Furthermore, 7-amido cephalosporins show a similar reactivity to 7-aminocephalosporanic acid in their spontaneous degradation. However, the deuterium solvent isotope effect $k_0^{H_2O}/k_0^{D_2O}$ is 0.93 at 35°C (Yamana and Tsuji, 1976) which is not typical of a water catalysed hydrolysis. Although cephaloridine [pyridine at C(3′)] is 6-fold more reactive towards hydroxide ion than cephalothin [acetate at C(3′)], it is 3-fold less reactive in its spontaneous degradation (Proctor *et al.*, 1982; Yamana and Tsuji, 1976).

Broad plateaus around neutral pH are observed for the hydrolysis of some penicillins, e.g. cloxacillin (Bundgaard and Ilver, 1970), phenethicillin (Schwartz *et al.*, 1962), cyclacillin (Yamana *et al.*, 1974b) and dicloxacillin (Pawelczyk *et al.*, 1981). The rate constants for the spontaneous hydrolysis of most penicillins are *ca* $10^{-7}\,s^{-1}$ at 30°C and there is little dependence of the rate upon the nature of the side chain (Yamana *et al.*, 1974b). Unlike the spontaneous degradation of cephalosporins, that of penicillins shows a significant solvent isotope effect, $k_0^{H_2O}/k_0^{D_2O}$ is 4.5 at 60°C (Yamana *et al.*, 1974b). The most likely mechanism of spontaneous hydrolysis therefore involves general base catalysis by water of nucleophilic attack by water on the β-lactam as indicated in [43], although a rate-limiting step involving breakdown of the intermediate is also possible.

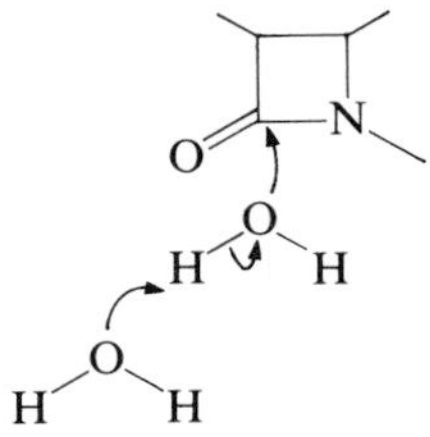

[43]

7 Buffer catalysed hydrolysis

The hydrolysis of both penicillins and cephalosporins are often catalysed by buffers. Catalysis by borate and phosphate buffers is usually found for the hydrolysis of some cephalosporins, but buffer catalytic effects are not always observed. Sometimes catalysis by phosphate monoanion is kinetically more important than that by the dianion (Lumbreras *et al.*, 1982).

TABLE 6

Summary of second-order rate constants for the reaction of benzylpenicillin with oxygen anions at 30°C; $I = 1.0$ M(KCl)

Conjugate acid of anion	pK_a	$k_{RO^-}/M^{-1}s^{-1}$
H_2O	13.83	1.54×10^{-1}
CH_3OH	15.54	15.96
$EtOCH_2CH_2OH$	14.98	3.45
$ClCH_2CH_2OH$	14.31	20.1
$CH \equiv CH—CH_2OH$	13.55	7.82×10^{-1}
CF_3CH_2OH	12.43	2.23×10^{-1}
$(CF_3)_2CHOH$	9.30	1.05×10^{-4}
HCO_3^-	9.71	1.25×10^{-4}
$(CF_3)_2C(OH)_2$	6.76	8.98×10^{-6}
$H_2PO_4^-$	6.51	1.08×10^{-5}
CH_3CO_2H	4.60	1.26×10^{-6}
HCO_2H	3.62	4.47×10^{-7}

Rate constants for the hydrolysis of benzylpenicillin catalysed by oxygen bases are given in Table 6. The corresponding Brønsted plot is shown in Fig. 10; its non-linearity is indicative of a change in mechanism. The Brønsted β-value for weak bases is 0.39 and probably represents general base catalysed hydrolysis. The reaction with alkoxide ions is nucleophilic. The

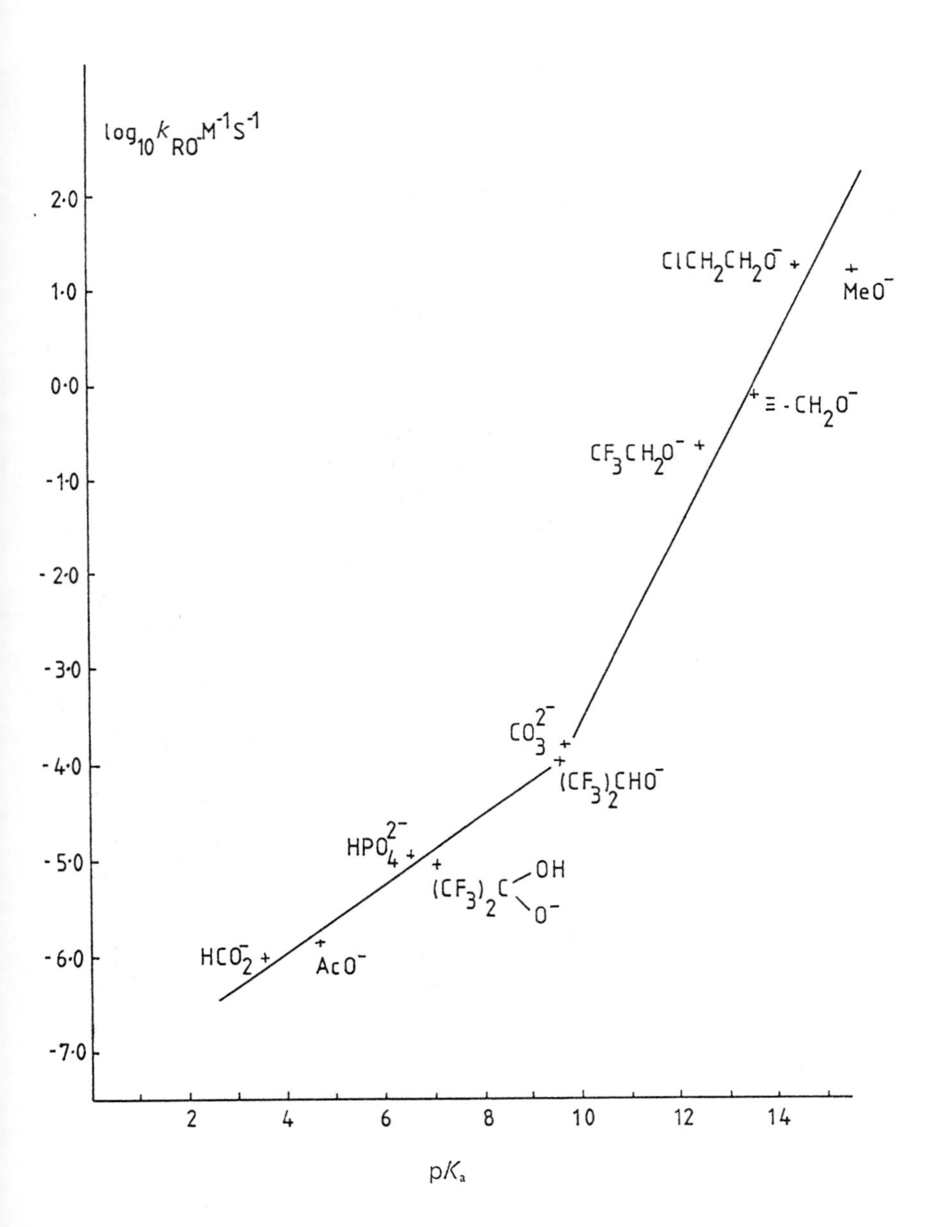

FIG. 10 Brønsted plot of the oxygen-anion catalysed hydrolysis of benzylpenicillin at 30°C. The steep slope represents nucleophilic catalysis whilst the less steep one for weakly basic anions represents general base catalysis

Brønsted β for bases whose conjugate acids have $pK_a > 7$ is 0.95 and represents nucleophilic catalysed hydrolysis (Proctor and Page, 1979). Evidence for an intermediate ester formed during the reaction has been obtained with alkoxide ions and phosphate dianion (Proctor and Page, 1979; Bundgaard and Hansen, 1981). Weakly basic catalysts probably act as general base catalysts. The reaction with acetate exhibits a solvent isotope effect $k_B^{H_2O}/k_B^{D_2O}$ of 2.1, as expected for general base catalysis but not for a nucleophilic reaction (Anderson *et al.*, 1961).

The reaction of benzylpenicillin with imidazole is dominated by imidazole buffer catalysis and hence the rate term which is first order in imidazole is difficult to determine (Bundgaard, 1976a, 1972). The rate constant reported for imidazole is similar to that for phosphate dianion which is surprising if they both represent nucleophilic catalysed hydrolysis.

The mechanism of the reaction with alkoxide ions is discussed in Section 13 but it can be noted here that the nucleophilic reaction probably proceeds by rate-limiting breakdown of the tetrahedral intermediate [44]. Fission of the β-lactam C—N bond may generate the amine anion or be catalysed by water acting as a general acid. Benzylpenicillin is about 1000-fold less reactive towards oxygen nucleophiles than is acetylimidazole which involves rate-limiting expulsion of the imidazole anion (Oakenfull and Jencks, 1971).

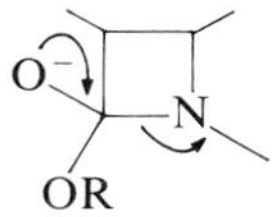

O^- N OR

[44]

8 Metal-ion catalysed hydrolysis

Transition-metal ions cause an enormous increase in the rate of hydrolysis of penicillins and cephalosporins (Gensmantel *et al.*, 1978, 1980; Cressman *et al.*, 1969). For example, copper(II) ions can enhance the rate of hydrolysis of benzylpenicillin 10^8-fold, a change in the half-life from 11 weeks to 0.1 seconds at pH 7. In the presence of excess metal ions, the observed apparent first-order rate constants for the hydrolysis of the β-lactam derivatives are first order in hydroxide ion but show a saturation phenomenon with respect to the concentration of metal ion which is indicative of the formation of an antibiotic/metal ion complex. A kinetic scheme is shown in (3), where M is

$$M + L \underset{}{\overset{K_1}{\rightleftharpoons}} ML \xrightarrow{k_2(OH)} \text{products} \qquad (3)$$

the metal ion and L is the β-lactam, and some relevant data are given in Tables 7 and 8. The rate of hydroxide-ion catalysed hydrolysis of benzylpenicillin bound to metal ion shows the following rate enhancements compared with the uncoordinated substrate: Cu(II), 8×10^7; Zn(II), 4×10^5; Ni(II), 4×10^4; Co(II), 3×10^4. The analogous data for cephaloridine are: Cu(II), 3×10^4, Zn(II), 2×10^3 (Gensmantel *et al.*, 1980).

TABLE 7

Summary of the rate and association constants for the copper(II)-catalysed hydrolysis of β-lactam derivatives in water at 30°C ($I = 0.5$ M)[a]

β-Lactam	$k_{OH}/M^{-1}s^{-1}$ [b]	K/M^{-1} [c]	$k_2^{OH}/M^{-1}s^{-1}$ [d]	k_2^{OH}/k_{OH}
Benzylpenicillin	0.154	187	1.22×10^7	8×10^7
Benzylpenicillin methyl ester	2.51	–	$<3 \times 10^4$ [e]	$<1.5 \times 10^4$
6-β-Aminopenicillanic acid	6.35×10^{-2}	232	5.15×10^6	8×10^7
Penicillanic acid	7.40×10^{-3}	120	1.58×10^6	2×10^7
Cephaloridine	0.526	2080	1.64×10^4	3×10^4
3-Methyl-7β-phenylacetamido ceph-3-em-4-carboxylic acid	2.41×10^{-2}	2400	1.56×10^3	7×10^4

[a] Gensmantel *et al.*, 1980
[b] Second-order rate constant for the hydroxide-ion catalysed hydrolysis
[c] Association constant for metal ion and β-lactam
[d] Second-order rate constant for the hydroxide-ion catalysed hydrolysis of metal-ion bound β-lactam
[e] $k_2^{OH}K$, saturation was not observed.

COORDINATION SITE

Copper(II) ion coordinates to the carboxylate group and the β-lactam nitrogen of benzylpenicillin as shown in [26] (Gensmantel *et al.*, 1980). Coordination occurring to the carboxylate group is indicated because esterification of this group decreases the rate enhancement by a factor of *ca* 5×10^3. Nonetheless, the rate of the hydroxide-ion catalysed hydrolysis of the copper(II)-bound methyl ester of benzylpenicillin is still 1.5×10^4-fold faster than the rate of hydrolysis of the uncoordinated ester, although the product of the reaction is not known.

TABLE 8

Summary of the rate and association constants for the metal-ion catalysed hydrolysis of benzylpenicillin and cephaloridine in water at 30°C ($I = 0.5$ M)[a]

	Benzylpenicillin		Cephaloridine	
Metal-ion	$k_2^{OH}/M^{-1}s^{-1}$	K/M^{-1}[c]	$k_2^{OH}/M^{-1}s^{-1}$[b]	K/M^{-1}[c]
Cu(II)	1.22×10^7	187	1.64×10^4	2080
Zn(II)	6.0×10^4	109	8.75×10^2	2181
Ni(II)	5.9×10^3	119		
Co(II)	4.1×10^3	178		

[a] Gensmantel *et al.*, 1980
[b] Second-order rate constant for the hydroxide-ion catalysed hydrolysis
[c] Association constant for metal ion and β-lactam

It has been suggested that copper(II) ions coordinate to the 6-acylamino side chain and the β-lactam carbonyl group (Cressman *et al.*, 1969). However, replacement of the acylamino side chain by the more basic amino group has little effect upon the binding constant and the rate enhancement for the hydroxide-ion catalysed hydrolysis for 6-aminopenicillanic acid [10] is very similar to that for benzylpenicillin. Furthermore, complete removal of the amido side chain as in penicillanic acid, also gives similar binding constants and rate enhancements. It is apparent from these observations that copper(II) ions do not bind to the amido side chain in penicillins, and that coordination probably occurs between the carboxylate oxygen and the β-lactam nitrogen [26] (Gensmantel *et al.*, 1980; Fazakerley *et al.*, 1976; Fazakerley and Jackson, 1977).

Model studies of the binding of metal ions to α-amido carboxylic acids indicate that coordination between the carboxylate group and the amide nitrogen does occur although the evidence is not unambiguous (Fazakerley and Jackson, 1975a,b; Weiss *et al.*, 1957; Weiss and Fallab, 1960; Eichelberger *et al.*, 1974; Ishidate *et al.*, 1960; Taguchi, 1960; Eichorn, 1973).

Copper(II) ions bind 10-fold more tightly to cephalosporins than to penicillins which would be surprising if the sites of coordination were similar. Molecular models indicate that one of the conformations of cephalosporins would be very suitable for metal-ion coordination between the carboxylate group and the β-lactam carbonyl oxygen [45]. The shortest distance between the carboxylate oxygen and the β-lactam nitrogen is similar (~2.7 Å) in penicillins and cephalosporins. However, the carboxylate oxygen–β-lactam oxygen shortest distance is much smaller (~2.7 Å) in cephalosporins than in penicillins (~4.6 Å). Precipitation of the β-lactam/metal-ion complex in the presence of excess ligand gives solids with interestingly different characteristics. Benzylpenicillin forms a 1 : 1 complex with both

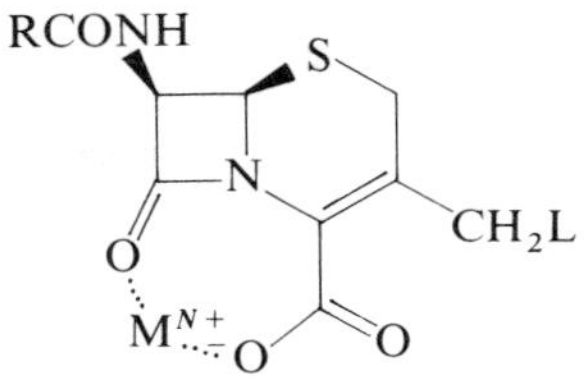

[45]

copper(II) and zinc(II) in which the asymmetric stretching frequencies of the β-lactam carbonyl and the carboxylate are decreased by *ca* 30 cm^{-1} compared with uncoordinated penicillin. The nmr spectrum of the zinc(II)/benzylpenicillin complex shows a downfield shift for the C(3) hydrogen, consistent with the proposed mode of binding (Gensmantel *et al.*, 1981; Fazakerley and Jackson 1975a,b; Asso *et al.*, 1984). However, solid ML_2 complexes of Mn(II), Pb(II) and penicillins have been reported (Chakrawarti *et al.*, 1982, 1984). Alkali-metal salts affect the infrared spectra of penicillin; the β-lactam carbonyl stretch is split in the potassium salt (1777 and 1757 cm^{-1}), and the amide I band is at 1669 cm^{-1} for the potassium salt and at 1700 cm^{-1} for the sodium salt (Zugara and Hidalgo, 1965). Thorium(IV) forms a complex with benzylpenicillin but not with its methyl ester (Ishidate *et al.*, 1960).

Cephalothin and 3-methyl-7β-phenylacetamidoceph-3-em-4-carboxylic acid form solid 2 : 1 complexes with transition-metal ions in which not only is the asymmetric stretching frequency of the carboxylate decreased, but also the β-lactam carbonyl stretching frequency by 10–30 cm^{-1}, depending upon the nature of the metal ion. The nmr spectrum of the zinc(II)/cephalothin complex shows a downfield shift of the C(7) hydrogen. The site of metal-ion coordination could thus be different for cephalosporins and involve the β-lactam carbonyl oxygen, although the situation in solution may be different. For example, the kinetic data indicate only a 1 : 1 complex and, of course, the thermodynamically favoured binding site is not necessarily the kinetically important one.

RATE ENHANCEMENT

The hydroxide-ion catalysed hydrolysis of benzylpenicillin probably proceeds by the formation of the tetrahedral intermediate [46]. The pK_a-value of the bridgehead nitrogen in [46] is estimated to be 8.0 (Page and Jencks, 1972b) so there is an enormous change in the basicity (> 12 pK_a units) of the β-lactam nitrogen as the reaction proceeds. The role of the metal ion in the hydroxide-ion catalysed hydrolysis is to stabilise the tetrahedral intermediate. An estimate of the binding constant of copper(II) ions to [46] can be made from a comparison with model compounds. Based on a Brønsted plot for the binding of model ligands such as thioproline, proline and glycine, the estimated association constant for copper(II) and [46] is $10^{7.2}$ M^{-1} (Gensmantel *et al.*, 1980).

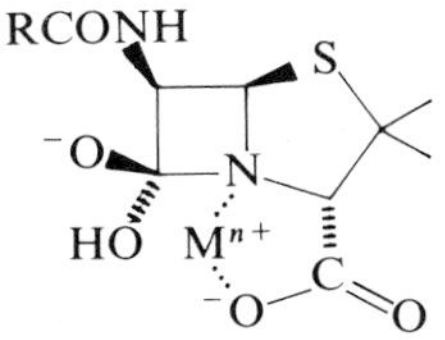

[46]

The rate of the hydroxide ion catalysed hydrolysis of copper(II)-bound benzylpenicillin is 8×10^7 faster than that of uncoordinated benzylpenicillin (Gensmantel *et al.*, 1978). A better estimate of the stabilisation of the transition state by the metal ion is from the comparison of the third-order rate constant, k_2K_1, for the metal and hydroxide-ion catalysed hydrolysis with the second-order rate constant for the hydroxide-ion catalysed hydrolysis. For copper(II) ions and benzylpenicillin this ratio is 1.2×10^{10} M. Copper(II) ion thus stabilises the transition state for hydroxide ion catalysis by 13.9 kcal mol^{-1} at 30°C compared with an estimated value of 9.8 kcal mol^{-1} for the stabilisation of the tetrahedral intermediate [46].

EFFECT OF TRANSITION METAL ION

The rate enhancements brought about by various metal ions for the hydrolysis of penicillins and cephalosporins are summarised in Tables 7 and 8. There is no correlation between the binding constant of the β-lactam antibiotic with the metal ion and the rate enhancement. The order of reactivity is that of the Irving–Williams series (Irving and Williams, 1953): Co(II) < Ni(II) < Cu(II) > Zn(II).

EFFECT OF β-LACTAM

Copper(II) ions bind *ca* 10-fold more tightly to cephalosporins than to penicillins (Table 7). This is at first surprising in view of the greater non-planarity of the penicillin molecule and the correspondingly greater basicity of the β-lactam nitrogen which is assumed. In cephalosporins the possibility of enamine type conjugation and the less favourable geometry [sp^2 C at C(3) and C(4)] for metal-ion coordination to the β-lactam nitrogen and the carboxylate group would be expected to hinder coordination. Nonetheless, the rate of hydroxide-ion catalysed hydrolysis of copper(II)-bound cephaloridine is *ca* 3×10^4-fold faster than that for the uncoordinated compound. This may be compared with a rate enhancement of

8×10^7 for benzylpenicillin. The ratio of the third-order rate constant, k_2K_1, for the copper(II) ion plus hydroxide ion catalysed hydrolysis of cephaloridine to the second-order rate constant for hydroxide-ion catalysed hydrolysis of the same substrate is 1.6×10^8 M. The corresponding ratio for benzylpenicillin is 1.2×10^{10} M. The transition state for cephaloridine hydrolysis is therefore stabilised by copper(II) ions *ca* 100-fold less than that for penicillin hydrolysis, but both transition states are greatly stabilised by the metal ion. Again, *ad hoc* explanations for this difference may be found in the lower basicity of the ring nitrogen in the tetrahedral intermediate formed from cephaloridine and/or a less favourable geometry.

Whether or not the group at C(3′) of the cephalosporin [2] is expelled or not makes little difference to the rate enhancement brought about by the metal ion. The 3-methyl derivative, has a similar association constant for binding of copper(II) ion to that for cephaloridine [11] (X = CH_2-pyridinium) (Table 7). The rate enhancement brought about by copper(II) ion is the same within a factor of 2 (Gensmantel *et al.*, 1980).

It has been suggested that a ternary complex is formed between benzylpenicillin, zinc(II) and tris buffers and that hydrolysis occurs by intramolecular nucleophilic attack of one of the coordinated buffer hydroxyl groups on the β-lactam (Schwartz, 1982; Tomida and Schwartz, 1983).

9 Micelle catalysed hydrolysis of penicillins

Non-functionalised micelles which catalyse reactions provide a simple illustration of the utilisation of the binding energy between a phase or macromolecule and the reactants to lower the free energy of activation (Jencks, 1975; Page, 1980b; Fendler and Fendler, 1975; Bunton, 1977, 1984; Cordes, 1978; Bunton and Savelli, 1986). Even if the rate constant for a bimolecular reaction within the micelle is the same as that in the bulk solvent, a rate enhancement may be observed if the reactants are confined to a smaller volume within the micelle (Jencks 1975; Martinek *et al.*, 1973, 1975). This requires the free energy of interaction between the reactants and the micelle to compensate for the loss of entropy resulting from the restriction of the reactants within the micelle. Micelles are of particular interest with respect to the hydrolysis of penicillins because they can provide different microenvironments for different parts of the reactant molecule. There is a nonpolar, hydrophobic core that can provide binding energy for similar groups on penicillin and a polar, usually charged, outer shell that can interact with the penicillin's polar groups.

Hydrophobic substrates and counterions are attracted to the micelle so that a cationic micelle should assist the reaction between a neutral molecule and anionic nucleophile while anionic micelles will inhibit such reactions.

The micelle catalysed hydrolysis of penicillins in alkaline solution is unusual because it involves the reaction between two anions, the hydroxide ion and the negatively charged benzylpenicillin (Gensmantel and Page, 1982a). The rate of the hydroxide-ion catalysed hydrolysis of benzylpenicillin decreases approximately three-fold in micellar solutions of itself (Hong and Kostenbauder, 1975).

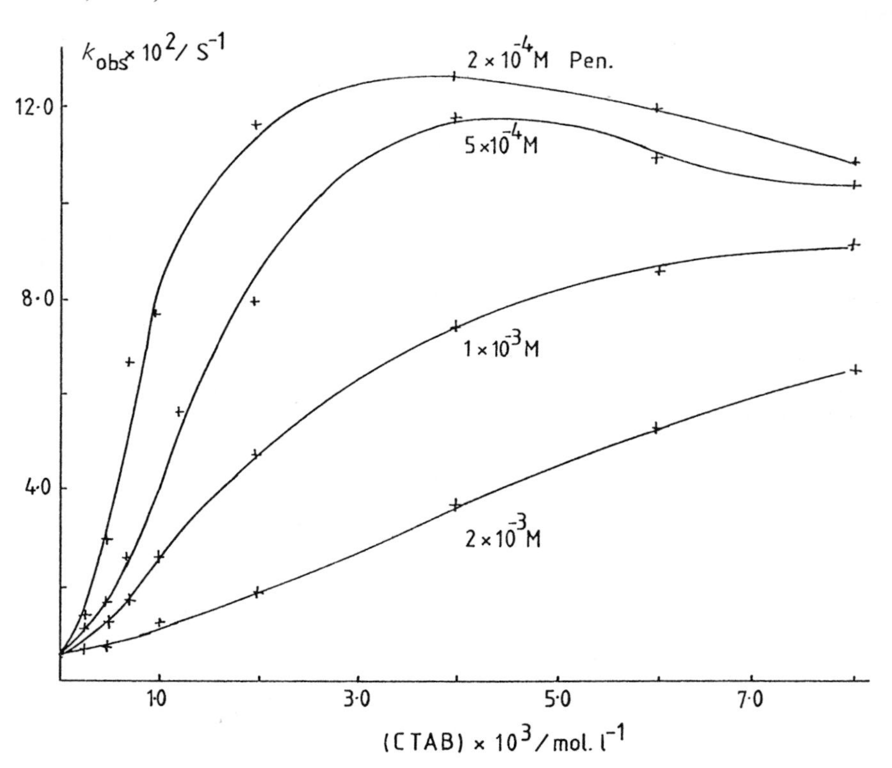

FIG. 11 Observed pseudo first-order rate constants for the hydrolysis of benzylpenicillin at the concentrations shown as a function of cetyltrimethylammonium bromide (CTAB) concentration at 30°C (Gensmantel and Page, 1982a)

The acid catalysed degradation of penicillins is inhibited in cationic micelles of cetyltrimethylammonium bromide (Tsuji *et al.*, 1982) and, as expected, neither anionic micelles of sodium dodecylsulphate nor polyoxyethylene lauryl ether promote the hydroxide-ion catalysed hydrolysis of benzylpenicillin (Gensmantel and Page, 1982a). In the presence of cetyltrimethylammonium bromide (CTAB) the pseudo first-order rate constants for the alkaline hydrolysis increase rapidly with surfactant concentration once above the critical micelle concentration (cmc) of the surfactant (Gensmantel and Page, 1982). Increasing the surfactant concentration eventually leads to a slow decrease in the observed rate (Fig. 11). This general shape of surfactant-rate profile has been found for many bimolecular reactions catalysed by cationic micelles (Menger and Portnoy, 1967; Yatsimirski *et al.*, 1971; Bunton *et al.*, 1970). However, unusually, the observed pseudo first-order rate constant is not independent of penicillin concentration. The binding constant between the micelle and substrate is unlikely to change significantly with concentration and yet the lower the concentration of benzylpenicillin the faster the rate increases and the greater the maximal rate obtained, the rate maximum shifting to a lower surfactant concentration. This observation could be explained if both hydroxide ion and benzylpenicillin compete for the same types of sites in the micelle and if benzylpenicillin binds better than hydroxide ion. Increasing the hydroxide ion concentration inhibits the rate of the micellar catalysed reaction while the rate in the bulk aqueous phase increases. The observed pseudo first-order rate constant for the micelle catalysed hydrolysis does not increase linearly with increasing hydroxide-ion concentration at constant surfactant concentration but reaches a maximum value (Gensmantel and Page, 1982a).

The kinetic evidence implies that there must be some binding between the benzylpenicillin anion and the micelles of CTAB and this has been shown spectroscopically (Chaimovich *et al.*, 1985). The maximum rate acceleration in the alkaline hydrolysis of benzylpenicillin by CTAB micelles is about 50.

The rate increase observed for many reactions upon the addition of detergents above the cmc has been explained on the basis of Scheme 6

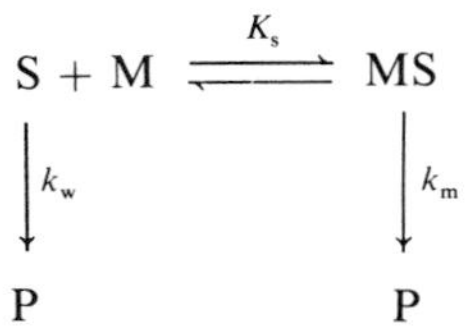

$$\begin{array}{ccc} S + M & \overset{K_s}{\rightleftharpoons} & MS \\ \downarrow k_w & & \downarrow k_m \\ P & & P \end{array}$$

Scheme 6

(Menger and Portnoy, 1967). The substrate, S, associates with the micelle, M, to form a substrate–micelle complex MS with an equilibrium constant K_s. The substrate and substrate-micelle complex form the product P with rate constants k_w and k_m referring to bulk aqueous and micellar phases, respectively. The observed first-order constant is given by (4) in which C_M is the concentration of micelles. Equation (4) has been used to explain catalysis

$$k_{obs} = \frac{k_w + k_m K_S C_M}{1 + K_S C_M} \tag{4}$$

in the presence of surfactant and, since it takes a similar form to the Michaelis–Menten equation for enzyme catalysed reactions, the rate and equilibrium constants can be evaluated by the procedure of double reciprocal plotting (Martinek *et al.*, 1977). However, (4) cannot explain many experimental observations for bimolecular or higher order reactions when two species or reactants compete for the same vacant "sites" in the micelles.

In recent years development of the kinetic theory for micellar reactions of molecularity greater than one have led to two general approaches to micellar catalysis. Equation (5) has been derived (Martinek *et al.*, 1973,

$$K_{obs} = \frac{k_m P_A P_B C_D V + K_b(1 - C_D V)}{[1 + (P_A - 1) C_D V][1 + (P_B - 1) C_D V]} \tag{5}$$

1975) for the reaction between two unchanged molecules and quantitatively explains surfactant-rate profiles for these bimolecular reactions; P_A and P_B are the partition coefficients of molecules A and B between aqueous and micellar phase, V is the molar volume of the micelle and C_D is the concentration of surfactant.

Romsted (1977) has suggested an expression, (6), which considers the effects of ions present in solution as well as substrate binding to the micelle.

$$k_{obs} = \frac{k_m \beta S K_s(C_D\text{-}CMC)}{[K_s(C_D\text{-}CMC) + 1][C_{It} + C_{Xt}K_i]} + \frac{k_w}{[K_s(C_D\text{-}CMC) + 1]} \tag{6}$$

The major assumption in deriving (6) is that the total number of counterions bound to a micelle is constant, allowing evaluation of the rate constant associated with the substrate–micelle complex. Exchange constant K_i refers to equilibrium (7) where I is the reactive, and X the unreactive counterion,

$$I_m + X_w \overset{K_i}{\rightleftharpoons} I_w + X_m \tag{7}$$

(total concentrations C_{It}, C_{Xt}), and the subscripts m and w refer to the micelle and bulk phases, respectively. The constant β is the degree of binding of the counterions to the Stern layer, S is the molar density of the micellar phase and C_D is the surfactant concentration.

The inhibitory effect of increasing benzylpenicillin concentration can be rationalised by the pseudo-phase ion-exchange model, but as the number of molecules of the antibiotic bound to the micelle increases (> 10 for concentrations $> 2 \times 10^{-4}$ M) the behaviour of a micelle covered with benzylpenicillin is probably different from a typical CTAB micelle (Chaimovich *et al.*, 1985).

Catalysis by micelles of the hydroxide-ion catalysed hydrolysis of substrates appears to be qualitatively understood on the basis of a concentration effect of reactant on, or around, the micelle surface and need not necessarily involve a difference in the free energies of activation in the micelle and bulk phase. That is not to say that the cationic micelles could not and do not cause electrostatic stabilisation of the transition state. The cationic micelle surface can act as an electrostatic sink for the anionic intermediate leading to its stabilisation, but a rate enhancement requires preferential stabilisation of this intermediate compared with the reactant. The small rate enhancement of the micelle catalysed reaction, about 50-fold, is equally well explained by considering that the increased concentration of reactants at the micelle surface leads to a higher observed rate. Incorporation of the reactants into a limited volume decreases the entropy loss that is associated with bringing reactants together in the transition state and this leads to an increase in the pseudo first-order rate constants in the presence of surfactant micelles. Cationic micelles of CTAB have also been shown to facilitate the alkaline hydrolysis of the cephalosporin, cephalexin (Yatsuhara *et al.*, 1977).

Added salts decrease the rate of the CTAB micelle catalysed alkaline hydrolysis of benzylpenicillin (Gensmantel and Page, 1982a). The salt effect can be considered to be due to competitive binding of the anions with the micelle. Increasing the unreactive anion concentration displaces hydroxide ion bound in the Stern layer leading to a reduction in the observed rate.

If the assumption is made that the inhibition is competitive but only between the added anion and hydroxide ion, then the equation derived by Romsted can be applied to the kinetic data. The association constant, K_s, between benzylpenicillin and micelle has been estimated to be 300 M^{-1} and the exchange constant, K_i, for bromide relative to hydroxide ion 25 M^{-1}. The relative degree of anion inhibition is that the larger the anion the lower its charge density and the larger the inhibition. The hydrolysis reaction is also inhibited by the addition of the hydrolysis product, benzylpenicilloate [30], a dianion which appears to bind no more tightly to the micelle than does benzylpenicillin itself. In benzylpenicilloate there are two carboxylate anions, yet the inhibition which results from increasing its concentration is similar to that caused by increasing the benzylpenicillin concentration. The effect of acetate ions is not large and it appears that carboxylate anions are rather ineffective anions in terms of binding with the CTAB micelle relative to simple inorganic anions. The attraction of organic molecules into a micelle can be due to both electrostatic and hydrophobic interactions. The rate of the hydroxide-ion catalysed hydrolysis of benzylpenicillin in the presence of micelles of CTAB is sensitive to electrolytes (Gensmantel and Page, 1982a), supporting the idea of electrostatic interactions between substrates and the micelle surface. The importance of other effects has been demonstrated by modifying the 6-β-side chain of penicillin to increase the substrate lipophilicity and hence the micelle–substrate hydrophobic interaction (Gensmantel and Page, 1982b).

The second-order rate constants for the hydroxide-ion catalysed hydrolysis of 6-substituted penicillins are given in Table 9, and are independent of the alkyl substituent. Conversion of the 6-amino substituent to an acylamido group increases the rate constant *ca* 3-fold. The rate maximum in the rate-surfactant concentration profiles for the base catalysed hydrolysis of alkylpenicillins in the presence of CTAB moves to a lower surfactant concentration with increasing substrate lipophilicity, and the rate "maximum" is dependent on the penicillin concentration. Increasing the 6-β-acylamino chain length increases the lipophilic character of the substrate and increases the binding constant K_s and the rate enhancement (Gensmantel and Page, 1982b).

Increasing the 6-acylamino chain length of the penicillin substrate not only decreases the surfactant concentration at which the maximum rate is observed but also results in a slightly increased maximal rate. The first effect may be rationalised on the basis of increasing affinity of the substrate for the micelle phase brought about by the increased hydrophobic interaction when the 6-acylamino side chain is increased in length. The second aspect, that of different rate maxima, must be more subtle. Increasing the surfactant concentration should eventually lead to all the substrate being associated with the micelle and, since the substrates hydrolyse in water with similar second-order rate constants, then the same rate maximum would be expected for each substrate, if, as generally accepted the rate constant within the micelle is similar to that in the aqueous phase. For compounds with lower affinities for the micelle, it is necessary to use higher concentrations of surfactant to incorporate all the penicillin substrate. Increasing the surfactant concentration also increases the concentration of unreactive counterion, and it is probably the displacement of reactants from the micelle surface by bromide ion that causes different rate maxima to be achieved for different substrates.

Catalysis occurs below the cmc of CTAB and is most marked for the more lipophilic substrates, suggesting that induced micelle formation may be occurring (Gensmantel and Page, 1982b).

The CTAB catalysed hydrolysis of penicillin derivatives appears to exhibit some degree of specificity. Increasing the hydrophobicity of the 6-β-side chain increases micellar catalysis. The association of the penicillin substrate with the micelle is presumably the result of interactions similar to those that give micelles stability relative to their monomeric form in aqueous solution; hence the not unexpected increase in substrate binding with increased lipophilicity of the molecule. It appears that once the 6-β-side has been

TABLE 9

Summary of the data for the hydroxide-ion catalysed hydrolysis of penicillin derivatives in the presence and absence of micelles of cetyltrimethylammonium bromide at 30°C[a]

Derivative	$k_w^{OH}/M^{-1}s^{-1}$ [b]	$k_m^{OH}/M^{-1}s^{-1}$ [c]	K_s/M^{-1} [d]	$k_m^{OH}K_s/M^{-2}s^{-1}$	k_m^{OH}/k_w^{OH}	$k_m^{OH}K_s/k_w^{OH}/M^{-1}$
6β-Aminopenicillanic acid	0.039	0.11	10	1.1	2.8	28
6-Methylpenicillin	0.138	0.72	40	28.8	5.2	208
6-Ethylpenicillin	0.116	1.05	90	94.5	9.1	819
6-Propylpenicillin	0.109	1.42	180	256	13.0	2340
6-Butylpenicillin	0.116	1.68	250	420	14.5	3625
6-Pentylpenicillin	0.121	1.92	280	538	15.9	4452
6-Heptylpenicillin	0.126	2.55	320	816	20.2	6464
6-Nonylpenicillin	0.132	2.32	330	766	17.6	5808
6-Undecylpenicillin	0.115	2.06	350	721	17.9	6265
6-Benzylpenicillin	0.137	2.20	300	660	16.1	4830
6-Benzylpenicillin methyl ester	3.69	48.0	145	6660	13.0	1885

[a] Gensmantel and Page, 1982b
[b] Second-order rate constant for hydrolysis in the absence of micelles, ionic strength $I = 0.05$ M
[c] Apparent second-order rate constant for hydrolysis in the presence of CTAB micelles, 0.05 M-NaOH and 2×10^{-4} M penicillin
[d] Apparent binding constant of penicillin derivative to micelle

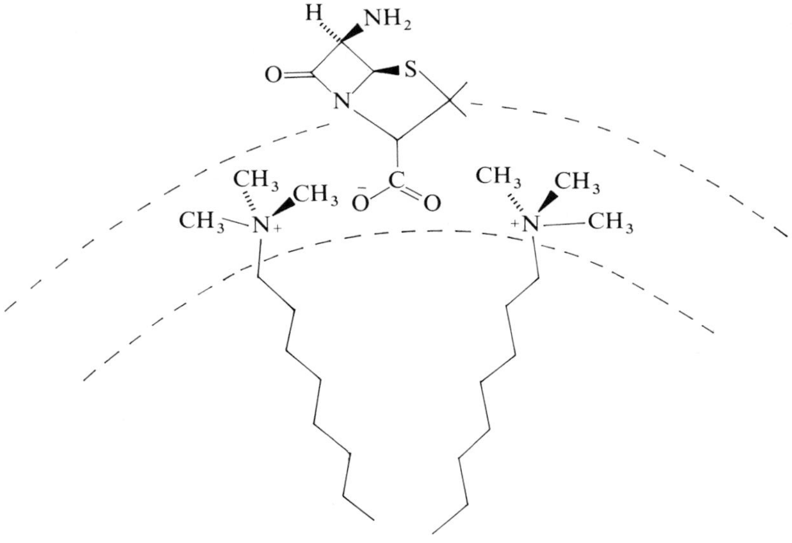

FIG. 12 Hypothetical orientation of 6-β-aminopenicillanic acid bound to micelles of CTAB (Gensmantel and Page, 1982b)

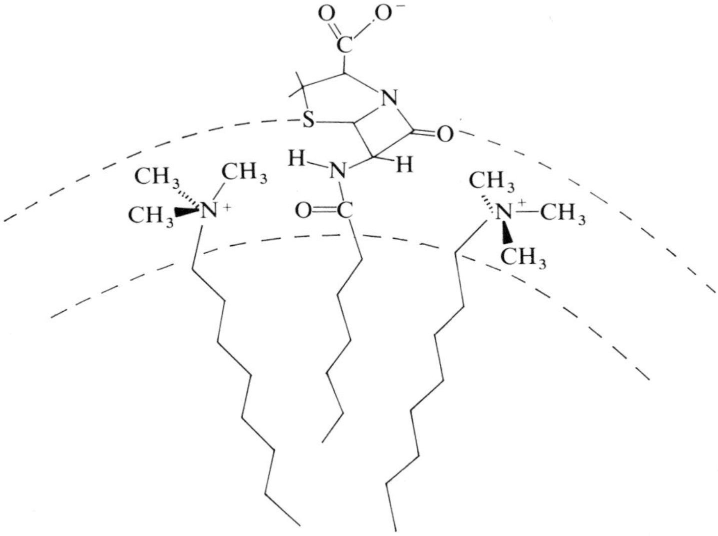

FIG. 13 Hypothetical orientation of alkylpenicillin bound to micelles of CTAB (Gensmantel and Page, 1982b)

extended to $CH_3(CH_2)_4CONH$-, further extension does not significantly increase the binding constant. It is interesting to note that there is no evidence of the longer chain compounds pulling the whole penicillin molecule into the interior of the micelle. The polar compound 6-β-aminopenicillanic acid is only weakly bound to the micelle, and electrostatic interactions may be all that exist between the substrate and micelle. Figure 13 illustrates schematically how increasing the length of the 6-β-acylamino side chain increases the hydrophobic interaction between the substrate and micelle and may thus alter the major orientation compared with that adopted by 6-APA as shown by comparison of Figs. 12 and 13. The β-lactam carbon is suitably situated for reaction with hydroxide ion and the carbonyl oxygen of the 6-amide linkage is positioned to allow some electrostatic interaction between the electron density on the carbonyl oxygen and the micelle surface. Figure 13 also explains why forming the methyl ester of benzylpenicillin leads to only a small reduction in the micelle-substrate binding constant relative to that for benzylpenicillin. Because the carboxylate anion points away from the micelle surface, it can have only a weak electrostatic interaction with the micelle surface.

There have been relatively few studies of the micellar catalysed hydrolysis of amides and the effects are small (Gani and Viout, 1978; Broxton and Duddy, 1979; Anoardi and Tonellato, 1977; O'Connor and Tan, 1980).

From the mechanism of micellar catalysis outlined in Scheme 6 the ratio k_m/k_w gives the difference between the free energy of activation in the micellar phase and in the bulk aqueous phase. For bimolecular reactions an apparent rate enhancement of 10^3 to 10^4 can result from the higher concentration of reactants in the smaller volume of micelles given by $\boldsymbol{R}T \ln V_m/V_w$, where V_m and V_w are the respective volumes of micelle and aqueous phases. This acceleration can occur even if the rate constants within the two phases are identical. To observe this maximum rate enhancement resulting from a simple concentration effect, the free energy of transfer of the reactant from the aqueous to the micellar phase must be more than enough to offset the loss of entropy from its restriction to a smaller volume within the micelle.

A comparison of the constant k_mK_s with k_w is dependent upon the choice of standard state because the micelle catalysed reaction is a higher order process. For a bimolecular reaction within the micelle, k_mK_s has units of concentration^{-2} time^{-1} and represents the free energy difference between the reactants and micelle in the bulk aqueous phase and the transition state in the micellar phase. The analogous reaction in the absence of micelles proceeds with a rate constant k_w with units of concentration^{-1} time^{-1}. The ratio k_mK_s/k_w has units of concentration and represents the free energy of transfer of the transition state from the aqueous phase to the micellar phase.

Relative values of k_m give the relative free energies of binding the substituent to the micelle in the ground state and transition state; relative values of K_s give the free energy of transfer of the substituent from the aqueous to the micellar phase in the ground state and relative values of $k_m K_s$ give the free energy of transfer of the substituent from the aqueous phase in the ground state to the micellar phase in the transition state.

The comparison of the micelle and non-micelle catalysed hydrolysis of penicillin derivatives is given in Table 9. The data refer to 0.05 M sodium hydroxide but the apparent rate enhancements would be greater at lower hydroxide-ion concentration (Gensmantel and Page, 1982a). The logarithm of the binding constants K_s show a non-linear dependence upon the Hansch π-substituent constant for the 6-alkyl side chain. This non-linear relationship is reflected in an apparent decrease in the free energy of transfer of a methylene unit from water to the micelle with a maximum value of 0.48 kcal mol^{-1} for transfer in the ground state and a maximum of 0.71 kcal mol^{-1} for transfer in the transition state (Gensmantel and Page, 1982b). A saturation phenomenon with respect to increasing alkyl hydrophobicity of the substrate is not always observed and depends upon the structure of the rest of the substrate (Gensmantel and Page, 1982b). It has been estimated (Molyneux *et al.*, 1965) that the free energy change for the complete transfer of a single methylene unit from water to the micellar phase is 0.65 kcal mol^{-1} which corresponds to a maximum rate or equilibrium difference of 3 at 25°C. The free energy of transfer of a methylene group from water to a non-polar liquid is about 1.0 kcal mol^{-1} (Nelson and DeLigny, 1968) and that to an enzyme from 2.1 to 3.8 kcal mol^{-1} (Page, 1976, 1977). The smaller value for transfer to micelles compared with enzymes presumably results from the "loose" interactions between the micelle, composed of several molecules of surfactant separated by their van der Waals radii, and the substrate compared with the "tight" interactions available from the substrate molecule and one molecule of enzyme, composed of many atoms closely packed together (Page, 1984b).

10 Cycloheptaamylose catalysed hydrolysis

Cycloamyloses (cyclic α-1,4-linked oligomers of D-glucose) have a toroidal or "doughnut"-shaped structure. The primary hydroxy groups are located on one side of the torus while the secondary ones lie on the other side. Relative to water the interior of the cycloamylose torus is apolar. The catalytic properties of cycloamyloses depend on the formation of inclusion complexes with the substrate and subsequent catalysis by either the hydroxy, or other groups, located around the circumference of the cavity (Komiyama and Bender, 1984; Page and Crombie, 1984).

Under mildly alkaline conditions and in the presence of excess cycloheptaamylose the rate of degradation of penicillin is increased 20–90-fold compared with the rate of alkaline hydrolysis (Tutt and Schwartz, 1971). Michaelis–Menten kinetics are observed which are indicative of complex formation. The apparent binding constant of 6-substituted penicillins varies little with the length of the alkyl side chain although it is increased about 10-fold for diphenylmethyl penicillin. The reaction is catalytic and hydrolysis proceeds by the formation of a penicilloyl-β-cyclodextrin covalent intermediate, i.e. ester formation, by nucleophilic attack of a carbohydrate hydroxyl on the β-lactam.

11 The aminolysis of β-lactam antibiotics

The reaction of amines with penicillins to give penicilloyl amides (Scheme 7) is of interest because the major antigenic determinant of penicillin allergy is the penicilloyl group bound by an amide linkage to ε-amino-groups of lysine residues in proteins (Levine and Ovary, 1961; DeWeck and Bulm, 1965; Parker *et al.*, 1962). The formation of the penicilloyl haptenic groups could conceivably occur by the direct aminolysis of penicillin (Schneider and DeWeck, 1966; Batchelor *et al.*, 1965) or by the aminolysis of penicillenic acid formed from a rearrangement of penicillin (Levine, 1961; Bundgaard, 1980; De Weck, 1962; Schwartz, 1969) or by the reaction of amines with the ketene [47] (Gensmantel *et al.*, 1978) formed by an elimination mechanism (Scheme 7).

The aminolysis of penicillin is also worthy of study because the reaction is an amide exchange, a normally difficult process but one which occurs readily with β-lactams (Blackburn and Plackett, 1973). Carbon—nitrogen bond fission in amides usually requires protonation of the nitrogen to avoid expulsion of the unstable anion, but in β-lactams this process is accompanied by a large release of strain energy which modifies the requirements for catalysis compared with normal amides. Another important difference between carbon—nitrogen bond fission in β-lactams compared with that in amides is that the latter may be accompanied by a more favourable entropy change as the molecule fragments into two separate entities (Page, 1973).

Because of the rigidity and shape of the penicillin molecule it is a suitable substrate to study the effectiveness of intramolecular catalysis and, in particular, to elucidate any preferred direction of nucleophilic attack upon the β-lactam carbonyl group (Martin *et al.*, 1978).

There have been several studies on the self aminolysis of penicillins containing amino groups which leads to dimerisation and polymerisation products (Bundgaard, 1977a,b; Larsen and Bundgaard, 1977, 1978a,b). The

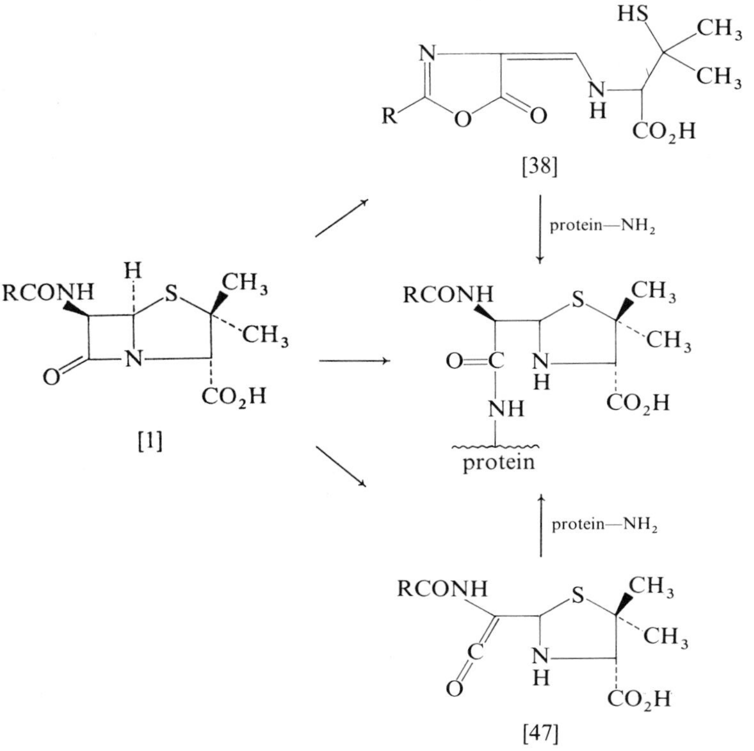

[38]

protein—NH_2

[1]

protein

protein—NH_2

[47]

Scheme 7

kinetics of the aminolysis of penicillins with protein has also been reported (Bundgaard and Buur, 1983).

INTERMOLECULAR GENERAL BASE CATALYSIS

The aminolysis of penicillin is a substitution reaction in which an acyl group is transferred from one amino group to another. This reaction requires at least two proton transfers, proton removal from the attacking amine and proton addition to the leaving amino group. These proton transfers are facilitated by buffers (Morris and Page, 1980a), and the kinetic importance of such catalysis is usually related to the observation that "catalysis occurs where it is most needed". Buffer catalysis is needed in the aminolysis of penicillin because covalent bond formation and fission between heavy atoms is accompanied by large changes in the acidity and basicity of the reacting groups (Jencks, 1976). If a proton is not removed from the attacking amine at some stage during the reaction the acidity of the NH group would change by *ca* 40 pK units, from *ca* 30 in the reactant to *ca* −10 in the hypothetical N-protonated amide [48]. Similarly, fission of the β-lactam C—N bond causes a change of *ca* 35 pK units in the basicity of the leaving amino group.

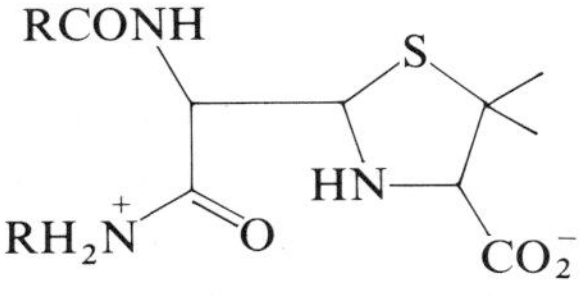

[48]

These large changes in pK can give rise to unstable intermediates such as [49] and [50], and buffer catalysis is observed because it can increase the rate of the reaction by either trapping such unstable intermediates or by stabilising or bypassing the transition states leading to their formation (Jencks, 1976).

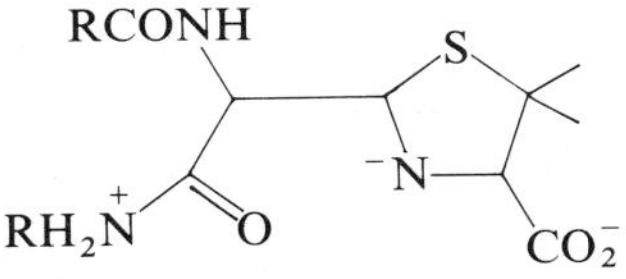

[49]

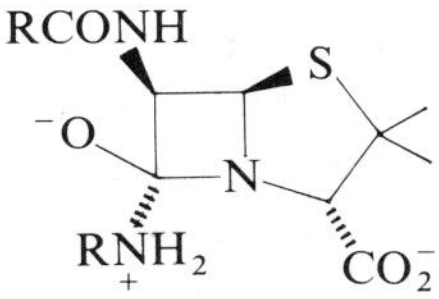

[50]

It is generally accepted that acyl-transfer reactions involve the intermediate formation of tetrahedral addition compounds, such as [50], i.e. bond formation to the attacking group occurs before bond fission to the leaving group. The bond-breaking process may occur after, during, or even before (Page and Jencks, 1972a) the rate-limiting step. Further complications in the elucidation of the detailed mechanism of acyl-transfer reactions arise from the problem of the timing of the proton-transfer steps, i.e. are they concerted with, or separate processes from, covalent bond changes between heavy atoms?

The aminolysis of penicillins and cephalosporins is a stepwise process catalysed predominantly by bases which remove a proton from the attacking amine. The evidence for the *reversible* formation of a tetrahedral intermediate is kinetic and based on linear free-energy relationships (Page, 1984a).

The aminolysis of benzylpenicillin at 30°C in aqueous solutions of the amine follows the rate law (8), where k_{obs} is the observed pseudo first-order rate constant for the disappearance of penicillin and k_0 is the second-order

$$\frac{\text{Rate}}{[\text{Pen}]} = k_{obs} = k_0[\text{OH}^-] + k_u[\text{RNH}_2] + k_b[\text{RNH}_2]^2 + k_{OH}[\text{RNH}_2][\text{OH}^-] \quad (8)$$

rate constant for the hydrolysis reaction (Tsuji *et al.*, 1975; Bundgaard, 1976d; Morris and Page, 1980a). The general acid-catalysed aminolysis of penicillin makes a negligible contribution to the observed rate. The dominant form of buffer catalysis in the aminolysis is general base catalysis. The relative importance of the terms in (8) depends on the basicity and the concentration of the amine and the pH. For strongly basic amines the amine catalysed (k_b) and the hydroxide-ion catalysed (k_{OH}) terms contribute most to the observed rate with the k_b term, of course, being more important with increasing concentration of amine. Consequently, the rate constants k_u for the uncatalysed reactions of basic monoamines are not of high precision. For the more weakly basic amines aminolysis occurs mainly through the uncatalysed (k_u) and amine catalysed (k_b) pathways because of the low concentration of hydroxide ion. The hydroxide-ion catalysed term (k_{OH}) makes a negligible contribution to the observed rate of aminolysis in buffers of amines with $pK_a < ca$ 9, and can only be determined in solutions of sodium hydroxide (Morris and Page, 1980a).

Linear-free energy relationships have been determined by varying independently the reactivity of the amine nucleophile and the catalyst. A plot of k_b for the general base catalysed aminolysis of benzylpenicillin for a series of primary monoamines against the pK_a values of the amines gives a straight line, the slope of which, the Brønsted β-value, is 1.09 ± 0.09 (Morris and Page, 1980a). This means that the reaction behaves as if, in the transition state, approximately unit positive charge is developed and is distributed between the nucleophilic and the catalysing amine molecules. A Brønsted β-value of 0.68 at 60°C has been reported for this reaction, but this was based on a series of diverse amines including imidazole (Tsuji *et al.*, 1975). Another reported value of 0.82 at 35°C was derived from results that included data for glycine ethyl ester which is known to hydrolyse under the reaction conditions (Bundgaard, 1976d). The Brønsted β-value of *ca* unity is indicative of a transition state in which full covalent bond formation has taken place between the nitrogen of the attacking amine and the carbonyl carbon and which carries the positive charge on either the nitrogen of the nucleophilic amine or on the catalytic amine molecule. The simplest mechanism that is consistent with this observation is shown in Scheme 8. The first step involves nucleophilic attack of the amine to form the tetrahedral intermediate $T^{\pm}$, for which there is independent kinetic evidence (see later). However, the intermediate $T^{\pm}$ breaks down rapidly to starting materials by expulsion of the attacking amine (k_{-1}). Catalysis of the reaction occurs by the formation of an encounter complex between $T^{\pm}$ and the basic catalyst B (Scheme 8). For general base catalysis by amines, B is an amine. Subsequent proton transfer from $T^{\pm}$ to B forms T^{-} which then breaks down to products.

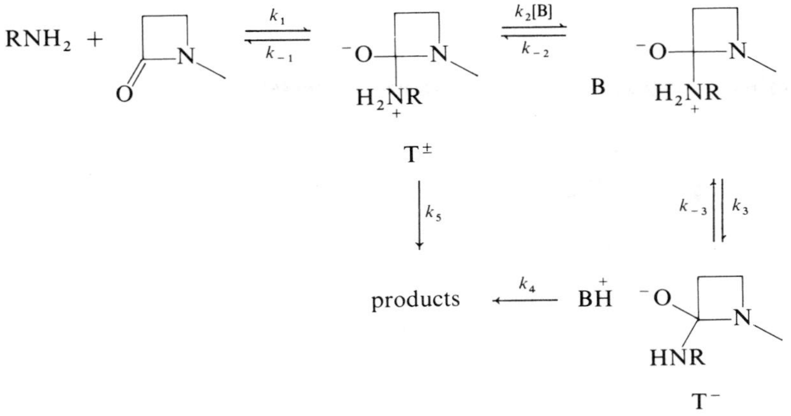

$$RNH_2 + \text{β-lactam} \underset{k_{-1}}{\overset{k_1}{\rightleftharpoons}} T^{\pm} \underset{k_{-2}}{\overset{k_2[B]}{\rightleftharpoons}} B \cdot T^{\pm} \underset{k_{-3}}{\overset{k_3}{\rightleftharpoons}} BH^{+}\, T^{-} \xrightarrow{k_4} \text{products}; \quad T^{\pm} \xrightarrow{k_5} \text{products}$$

Scheme 8

The Brønsted β_{nuc}-value for the hydroxide-ion catalysed aminolysis of benzylpenicillin (k_{OH}) for a series of primary monoamines is 0.96 (Morris and Page, 1980a). This value also indicates that the reaction behaves as if a unit positive charge is developed on the attacking amine in the transition state. The assignment of charge density is unambiguous, unlike the case for the general base catalysed reaction, and the simplest interpretation of the β_{nuc}-value is that the attacking amine resembles its conjugate acid, i.e. is fully protonated, in the transition state. This is compatible with the mechanism of Scheme 8 in which the bond between the attacking amine and the carbonyl carbon is fully formed. The β_{nuc}-value indicates the location of the proton in

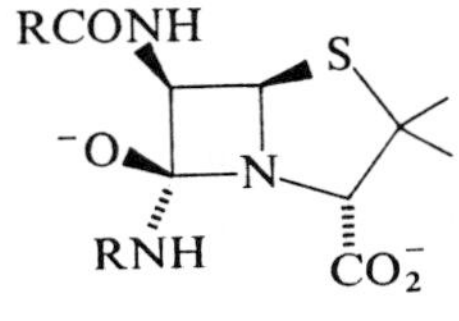

[51]

the transition state, for there can be little or no proton transfer from the attacking amine to the hydroxide-ion catalyst. It is consistent with rate-limiting diffusion-controlled encounter of the tetrahedral intermediate $T^{\pm}$ and hydroxide ion giving [51].

Hydrazine shows an enhanced nucleophilic reactivity towards penicillin compared with amines of similar basicity which is attributed to the α-effect. This has allowed a study of the effect of varying the basicity of the catalyst with a constant nucleophile even in the presence of strongly basic catalysts.

For example, catalysis of the reaction of hydrazine with benzylpenicillin occurs even with the strongly basic amine propylamine (Morris and Page, 1980b). There is a non-linear dependence of the rate of hydrazinolysis of benzylpenicillin upon the basicity of both oxygen and nitrogen base catalysts. For strongly basic catalysts there is little dependence of the rate constants upon basicity and the Brønsted β-value is $\leqslant 0.2$. However, catalysis by weak bases shows a much stronger dependence upon the basicity of the catalyst with $\beta \geqslant 0.8$. A curved or non-linear Brønsted plot is required to describe the behaviour of both oxygen and nitrogen bases.

The large sensitivity of the rate constants to base strength for weakly basic catalysts indicates that the catalyst resembles its conjugate acid in the transition state, i.e. there is a large amount of, or complete, proton transfer to the catalyst in the transition state. For strongly basic catalysts the small sensitivity of the rate constants upon base strength suggests that the catalyst resembles its free unprotonated basic form in the transition state.

The simplest mechanism compatible with the observation involves the formation of an unstable dipolar tetrahedral addition intermediate $T^{\pm}$ which rapidly reverts to the starting materials by expulsion of the attacking amine, (k_{-1}). Reaction only proceeds if the intermediate is trapped by an encounter with a base that results in proton transfer to form the anionic intermediate T^{-}, which rapidly breaks down to products (Scheme 8). Proton transfer between electronegative atoms is thought to occur by a stepwise process involving the diffusion-controlled encounter of the proton donor and acceptor, followed by proton transfer itself and then diffusion apart (Eigen, 1964). Proton transfer itself (k_3) is not usually rate-limiting. The application of these suggestions to the mechanism of aminolysis of penicillin provides an explanation for the non-linear Brønsted plot. When the tetrahedral intermediate, $T^{\pm}$, is a stronger acid than the conjugate acid of the basic catalyst, proton transfer is thermodynamically favourable. The rate-limiting step will therefore be the diffusion-controlled encounter of $T^{\pm}$ and the catalyst (k_2) and the observed rate will be independent of the basicity of the catalyst (Jencks, 1976). However, for weakly basic catalysts proton transfer is thermodynamically unfavourable and the rate-limiting step changes to the diffusion apart of the deprotonated intermediate, T^{-}, and the protonated catalyst (k_4). This kinetic scheme explains the large dependence of the rate upon the basicity of the catalyst for weakly basic catalysts and its insensitivity for strongly basic catalysts (Morris and Page, 1980b).

In addition to this change in rate-limiting step deduced from non-linear free energy relationships by changing the basicity of the catalyst, another change has been observed directly from the kinetics of the hydroxide-ion catalysed aminolysis of benzylpenicillin (Gensmantel and Page, 1979a). In aqueous sodium hydroxide the aminolysis occurs largely by the k_{OH} pathway

(8). There is a non-linear dependence of the apparent second-order rate constants upon the concentration of hydroxide ion. At low concentrations of hydroxide ion the rate is first-order in hydroxide ion and the initial slopes give values of k_{OH} which agree well with those determined at lower pH in buffer solutions (Gensmantel and Page, 1979a; Morris and Page, 1980a). At high concentrations of hydroxide ion the rate becomes independent of the concentration of hydroxide ion. This change in the kinetic dependence on hydroxide ion is indicative of a change in the rate-limiting step of the reaction which, in turn, requires that there be at least two sequential steps in the reaction. One of these steps is rate-limiting at low concentrations of hydroxide ion and the transition state for this step contains hydroxide ion, or its kinetic equivalent. The other step is rate-limiting at high concentrations of hydroxide ion but the transition state for this step does not contain hydroxide ion. The existence of two sequential steps demands that there be an intermediate in the reaction which is probably the tetrahedral intermediate $T^{\pm}$.

The mechanism of Scheme 8 is compatible with this observation. At low concentrations of hydroxide ion the rate of collapse of the tetrahedral intermediate to reactants must be faster than its reaction with hydroxide ion ($k_{-1} \gg k_2[OH^-]$); the observed rate constant is dependent upon the concentration of hydroxide ion with k_2, the diffusion-controlled step, being rate-limiting. The calculated pK_a-values for the protonated amine of the tetrahedral intermediates are well below that for water. Proton transfer from the tetrahedral intermediate to hydroxide ion is therefore in the thermodynamically favourable direction and it is to be expected that the rate-limiting step for this process is the diffusion-controlled encounter of the proton donor and acceptor.

At high concentrations of hydroxide ion the tetrahedral intermediate and hydroxide ion diffuse together faster than the intermediate collapses back to reactants ($k_2[OH^-] \gg k_{-1}$). Under these conditions the observed rate constant is independent of hydroxide ion concentration and k_1, the rate of formation of the tetrahedral intermediate, is rate-limiting (Gensmantel and Page, 1979a). Values of k_1 thus determined for a series of amines yield a Brønsted β_{nuc} of 0.3. This indicates that the reaction behaves as if there is a development of a charge of *ca* + 0.3 on the attacking amine nitrogen in the transition state [52], which must therefore occur early along the reaction coordinate with little C—N bond formation.

Assuming that the diffusion controlled step k_2 has a value of 10^{10} $M^{-1} s^{-1}$, values of k_{-1} and the equilibrium constants for the formation of the tetrahedral intermediates have been obtained (Gensmantel and Page, 1979a). The rates of expulsion of the attacking amine from the tetrahedral intermediate to regenerate the reactants (k_{-1}) are very rapid, *ca*

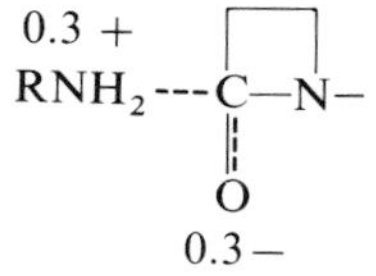

$$\begin{array}{c} 0.3+ \\ RNH_2\text{---}C\text{—}N\text{—} \\ \| \\ O \\ 0.3- \end{array}$$

[52]

10^9–$10^{10}s^{-1}$. Although these rate constants are very large they are of the order of magnitude that have been postulated for the breakdown of tetrahedral intermediates formed in acyl transfer reactions.

Similarly, the equilibrium constants for the formation of the tetrahedral intermediates, $T^{\pm}$, have been obtained. These vary substantially with the basicity of the amine from $4 \times 10^{-11} M^{-1}$ for 2-cyanoethylamine to $9 \times 10^{-9} M^{-1}$ for propylamine. The Brønsted β_{nuc}-value for the equilibrium is 0.9. This provides experimental support for the Brønsted β-value of 1.0 that is often postulated for the formation of the tetrahedral intermediate from amines and carbonyl groups in which the amine nitrogen develops a unit positive charge, and presumably resembles the conjugate acid of the amine in structure and in its stability dependence upon substituents. Similar observations have been made in the aminolysis of cephalosporins (Page and Proctor, 1984), and these are discussed in Section 12.

The aminolysis of cephalosporins follows a similar mechanism to that for penicillins (Proctor and Page, 1984). Although it has been suggested that the hydroxide-ion catalysed aminolysis involves proton transfer concerted with nucleophilic attack (Bundgaard, 1975), the Brønsted β_{nuc}-values of *ca* 1.0 are consistent with the stepwise mechanism. This is also supported by the non-linear dependence of the rate of aminolysis of cephalosporins upon hydroxide ion concentration (Proctor and Page, 1984) (see Section 12).

The partitioning of the tetrahedral intermediate ($T^{\pm}$, Scheme 8) formed by nucleophilic attack upon a β-lactam is controlled by the ease of exocyclic versus endocyclic bond fission. Endocyclic C—N bond fission is favoured by the release of the strain energy of the four-membered ring. Expulsion of the attacking nucleophile by exocyclic bond cleavage is accompanied by a relatively favourable entropy change because two molecules are generated from one (Page, 1973).

Except in the presence of strongly acidic catalysts, general acid catalysed breakdown of the tetrahedral intermediate to products by proton transfer to the β-lactam nitrogen is a relatively unimportant pathway in aminolysis. This is expected because of the weakly basic β-lactam nitrogen in the tetrahedral intermediate (Morris and Page, 1980a).

That expulsion of the attacking amine nucleophile from [50] occurs more readily than fission of the β-lactam C—N bond is confirmed by the observation that 2-azetidinylideneammonium salts [53] react with hydroxide to give 2-azetidinones [54] presumably through the intermediate formation of [55] (Poortere *et al.*, 1974; Agathocleous *et al.*, 1985). Similarly, several synthetic reactions have been reported in which the exocyclic β-lactam oxygen is exchanged without fission of the β-lactam ring (Gilpin *et al.*, 1981; Wojtkowski *et al.*, 1975). The rate law for hydrolysis of [53] shows a third order term which is first order in substrate, hydroxide and carbonate ions and indicates that the rate-limiting step must involve breakdown of the tetrahedral intermediate [55] and that formation of [55] must be reversible. The simplest explanation is that C—N cleavage occurs by general acid (HCO_3^-) catalysed breakdown of the anion of the tetrahedral intermediate, [55] (Page *et al.*, 1987).

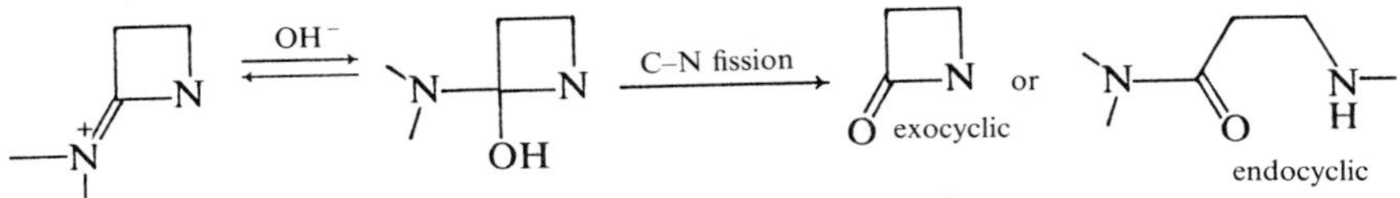

OH^- ; C–N fission ; or ; exocyclic ; endocyclic

[53] [55] [54]

INTRAMOLECULAR GENERAL BASE CATALYSIS

Intermolecular general base catalysis in the aminolysis of penicillins is a major pathway for product formation (p. 234). It is not surprising therefore that intramolecular general base catalysed aminolysis has been observed (Schwartz, 1968).

The rate constant k_u for the reaction of 1,2-diaminoethane with benzylpenicillin is *ca* 30-fold greater than that predicted for a monoamine of the same basicity from the Brønsted plot (Martin *et al.*, 1976; Morris and Page, 1980a). The rate enhancement is interpreted as evidence for intramolecular general base catalysis of aminolysis by the second amino group in 1,2-diaminoethane [56]. Proton transfer occurs from the amino group that acts as the nucleophile to the terminal amino group acting as a general base.

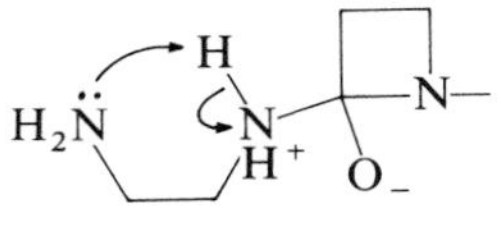

[56]

Most of this rate enhancement is a result of the greater basicity of the amino group compared with water. Very little of the rate enhancement is attributable to intramolecularity, with the catalyst being covalently linked to the nucleophile. This is evident from the effective molarity of *ca* 1 mol l^{-1} for the reaction which is obtained by dividing the second-order rate constant, k_u, for the reaction of 1,2-diaminoethane with penicillin by the third-order rate constant, k_b, for intermolecular catalysis of aminolysis by a second molecule of amine of similar basicity. The effective molarity is the concentration of catalysing amine required to give the same rate of reaction as the diamine. Similar, small effective molarities have been observed for intramolecular general acid base catalysed reactions (Page, 1973; Kirby, 1980). Intramolecular catalysis is observed because of the importance of general base catalysis in these reactions compared with uncatalysed aminolysis.

By analogy with the intermolecular general base catalysed reaction, the rate-limiting step in the intramolecular reaction is probably a conformational change. It appears that the dominant contribution to the low effective molarity is the "loose" transition state of the intermolecular reaction. The rate-limiting step of the intermolecular general base catalysed aminolysis of penicillin is the diffusion-controlled encounter of the tetrahedral intermediate, $T^{\pm}$, with the base (Scheme 8). The transition state is thus very "loose" and the bimolecular step, k_B in Scheme 8, will be associated with a small entropy change giving rise to the low effective concentration of the intramolecular reaction (Page, 1977; Morris and Page, 1980a).

A suitably placed amino group within the penicillin molecule, as opposed to one in the attacking amine nucleophile, could conceivably also act as an intramolecular general base catalyst for aminolysis. In fact, the aminolysis of 6-β-aminopenicillanic acid [10] occurs predominantly by an uncatalysed pathway. The term k_B in rate law (8), which is second order in amine and predominant for the aminolysis of penicillin, is of minor significance for the reaction of monoamines with 6-β-aminopenicillanic acid. This has been interpreted as evidence for intramolecular general base catalysis [57] (Schwartz and Wu, 1966).

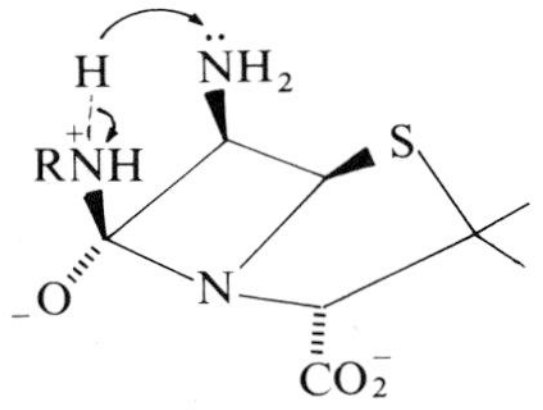

[57]

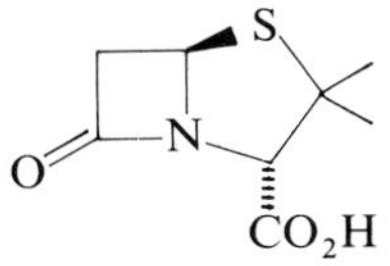

[58]

However, the second-order rate constants, k_u, for the reaction of monoamines with benzylpenicillin and 6-β-aminopenicillanic acid are similar; there is no rate enhancement. Furthermore, the aminolysis of penicillanic acid [58] also does not show a significant general base catalysed term in the rate law (Gensmantel and Page, 1979b). Clearly, this predominance of the uncatalysed aminolysis pathway cannot be attributed to intramolecular general base catalysis.

The aminolysis of 6-β-aminopenicillanic acid [10] in solutions of sodium hydroxide has enabled the rate constants k_1 and k_{-1} and the equilibrium constant K (Scheme 8) to be deduced. The values of k_1 for the formation of the tetrahedral intermediates from 6-β-aminopenicillanic acid and from benzylpenicillin are similar, which is again indicative of the lack of intramolecular general base catalysis by the neighbouring amino group in [10] (Gensmantel and Page, 1979a).

The reason for the absence of significant intramolecular general base catalysis in the aminolysis of 6-β-APA and penicillanic acid is discussed on p. 244.

INTRAMOLECULAR GENERAL ACID CATALYSIS AND THE DIRECTION OF NUCLEOPHILIC ATTACK

The rate constant k_u for the reaction of penicillin with the monocation of 1,2-diaminoethane is *ca* 100-fold greater than that predicted from the Brønsted plot for a monoamine of the same basicity. The rate enhancement is attributed to intramolecular general acid catalysis of aminolysis by the protonated amine (Morris and Page, 1980a; Martin *et al.*, 1979). Breakdown of the tetrahedral intermediate, $T^{\pm}$, is facilitated by proton donation from the terminal protonated amino group to the β-lactam nitrogen [59]. It is not known whether proton transfer and carbon—nitrogen bond fission are concerted or occur by a stepwise process.

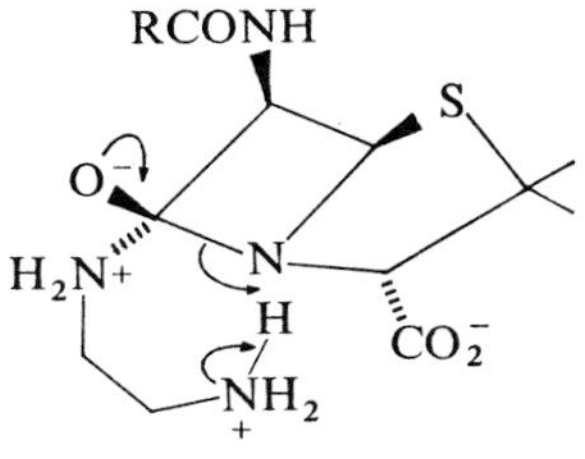

RCONH O⁻ H_2N^+ N H NH_2^+ S CO_2^-

[59]

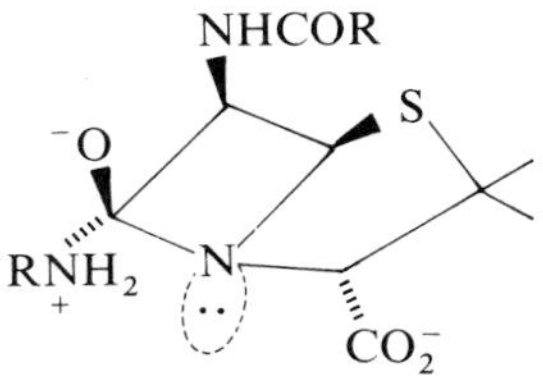

NHCOR ⁻O $R\overset{+}{N}H_2$ N S CO_2^-

[60]

According to the theory of stereoelectric control of Deslongchamps (1975), the breakdown of tetrahedral intermediates is facilitated by the lone pairs of the heteroatoms attached to the incipient carbonyl carbon being *anti*periplanar to the leaving group. Application of this theory to the microscopic reverse steps predicts that the direction of nucleophilic attack on the carbonyl carbon be such that the lone pairs on the heteroatoms will be *anti*periplanar to the attacking group.

Penicillins have a fairly rigid structure because of the fusion of the β-lactam and the thiazolidine rings giving a V-shaped molecule. A consequence of the non-planarity of the fused bicyclic ring system is that the electron density of the lone pair of the β-lactam nitrogen will be concentrated heavily on the α-face of the penicillin molecule [17] and particularly of the tetrahedral intermediate [60]. According to the theory of stereoelectronic control nucleophilic attack on penicillins should, therefore, take place from the β-side. However, this face is sterically hindered and it has been suggested that nucleophilic attack may therefore take place from the less hindered α-side (Martin *et al.*, 1979; Gensmantel and Page, 1979a).

The observation of intramolecular general acid catalysis in the reaction with the monocation of 1,2-diaminoethane gives an indication of the direction of nucleophilic attack upon penicillin. In order that ready proton transfer takes place from the protonated amine to the β-lactam nitrogen, it is essential that the tetrahedral intermediate has the geometry shown [60]. Although intramolecular general acid catalysis could conceivably take place if the amine attacked from the β-face [61], this would involve considerable non-bonded interactions and/or the proton transfer taking place through one or more water molecules. Further evidence for nucleophilic attack taking place from the α-face comes from the *absence* of intramolecular

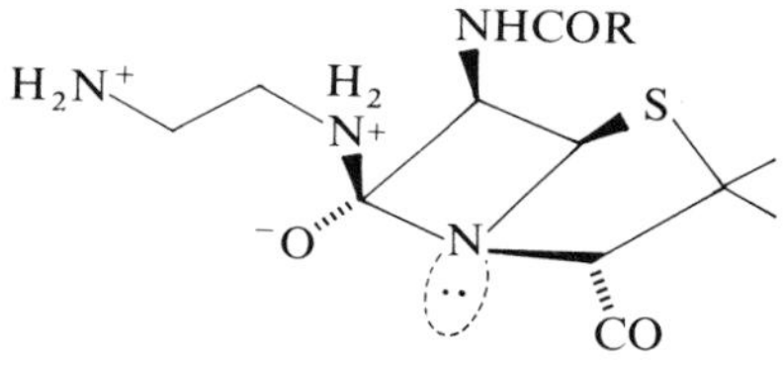

NHCOR
H_2N^+ H_2 N^+ S
^-O N
CO

[61]

general base catalysis in the aminolysis of 6-β-aminopenicillanic acid (p. 242). That the lone pair on the β-lactam nitrogen takes up the geometry with respect to the carboxy-group shown in [60] is supported by the observation that copper(II) ions catalyse the aminolysis of penicillin by coordination to the β-lactam nitrogen and the carboxy-group, thus stabilising the tetrahedral intermediate (Gensmantel *et al.*, 1978). The low effective molarity of intramolecular aminolysis from the β-side is also consistent with this side being sterically unfavourable (p. 249). Thus nucleophilic attack on penicillins, at least by amines, appears to take place from the least hindered α-side in disagreement with the prediction of the theory of stereoelectronic control. It is unlikely that α-attack would give the stereoisomer predicted by stereoelectronic control because this would introduce a highly strained *trans*-fused bicyclic system.

UNCATALYSED AMINOLYSIS

The Brønsted β_{nuc}-value for the uncatalysed aminolysis of penicillins is 1.0 (Morris and Page, 1980a), which indicates that the reaction behaves as if a unit positive charge is developed on nitrogen in the transition state. The uncatalysed pathway characterised by k_u could represent either a purely uncatalysed reaction of amine and penicillin or solvent catalysis with water acting either as a general base, removing a proton from the attacking amine, or as a general acid, donating a proton to the β-lactam nitrogen. The k_u pathway cannot represent rate-limiting formation of the tetrahedral intermediate because the β_{nuc}-value for this is known to be 0.3, and the rate constants k_1 (Scheme 8) are known to be much greater than the observed k_u-values (Gensmantel and Page, 1979a). The rate constant k_u for hydrazine divided by the concentration of water, 55 M, gives a third-order rate constant with a large positive deviation from the Brønsted plot for general base catalysed hydrazinolysis (Morris and Page, 1980b); this indicates that water is not acting as a proton acceptor. It is not easy to distinguish between uncatalysed rate-limiting breakdown of the tetrahedral intermediate, $T^{\pm}$, [62] and breakdown of the same intermediate general acid catalysed by water [63].

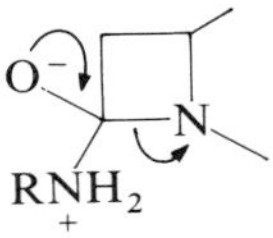

O$^-$ N RNH$_2$ +

[62]

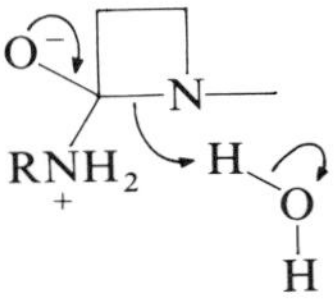

O$^-$ N H O H RNH$_2$ +

[63]

The equilibrium constant K_1 for the formation of $T^{\pm}$ from penicillin and propylamine is known to be 8.86×10^{-9} l mol^{-1} (Gensmantel and Page, 1979a). Because the rate constant for the uncatalysed reaction, k_u, is given by k_5K_1 (Scheme 8), k_5, the rate constant for the uncatalysed or water-catalysed breakdown of the tetrahedral intermediate, is $k_u/K_1 = 1.5 \times 10^6 s^{-1}$. The rate of protonation of the β-lactam nitrogen of $T^{\pm}$ by water may be estimated from its pK_a-value of 5.2 to be *ca* $10^3 s^{-1}$. This means that the uncatalysed breakdown of $T^{\pm}$ cannot proceed by stepwise proton transfer from water to the β-lactam nitrogen. However, it could occur either by a concerted mechanism – proton transfer from water occurring synchronously with carbon—nitrogen bond fission – or uncatalysed expulsion of the nitrogen anion. The rate of expulsion of the imidazolyl anion from the tetrahedral intermediate formed in the aminolysis of acetylimidazole is $\geqslant 10^6 s^{-1}$ (Page and Jencks, 1972b). If the strain energy of the β-lactam ring of *ca* 120 kJ mol^{-1} is relieved upon ring opening, then expulsion of the nitrogen as the anion should be treated as a leaving group of pK_a-value of *ca* 10 rather than the normal value of 30 for ordinary amines. It is conceivable therefore that carbon—nitrogen bond fission occurs without protonation of the β-lactam nitrogen [62]. The uncatalysed aminolysis of cephalosporins is discussed in Section 12.

The reason for the absence of a significant general base catalysed term in the rate law for the aminolysis of 6-β-aminopenicillanic acid [10] and penicillanic acid [58] is that the rate constants for the uncatalysed breakdown of the tetrahedral intermediates, k_5 in Scheme 8, are 13- and 50-fold, respectively, greater than that for benzylpenicillin (Gensmantel and Page, 1979b). The observed contributions of general base catalysed and uncatalysed aminolysis to the rate depend on the ratio of k_5 to $k_2\,[RNH_2]$ (Scheme 8). For [10] and [58] $k_2\,[RNH_2]$ does not become greater than k_5 until the amine concentration is greater than 0.5 M.

METAL-ION CATALYSED AMINOLYSIS

In aqueous solution in the presence of copper(II) ion penicillin reacts with amines to form the corresponding penicilloyl amide. The kinetics of this reaction show a saturation phenomenon with the concentration of metal ion but are complicated by the complexation of copper(II) ions with the amine (Gensmantel *et al.*, 1978).

A kinetic scheme compatible with the observed data is given in Scheme 9 and where B represents the amine, M copper(II) ion, P penicillin, MP is the penicillin–copper(II) complex and MB is the amine–copper(II) complex.

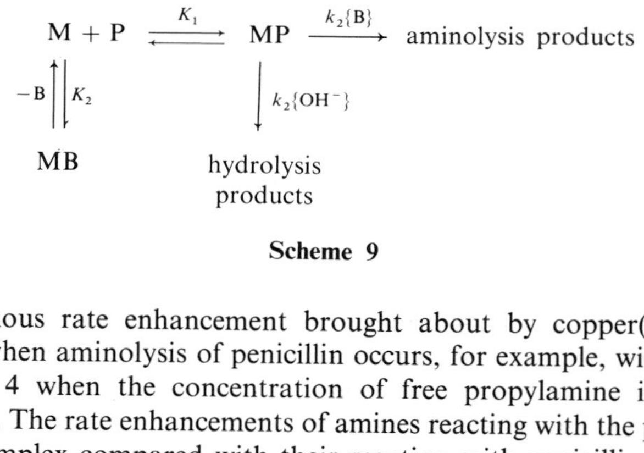

$$\begin{array}{ccccc} M + P & \underset{}{\overset{K_1}{\rightleftharpoons}} & MP & \xrightarrow{k_2\{B\}} & \text{aminolysis products} \\ -B \Big\updownarrow K_2 & & \Big\downarrow k_2\{OH^-\} & & \\ MB & & \text{hydrolysis products} & & \end{array}$$

Scheme 9

The enormous rate enhancement brought about by copper(II) ion is appreciated when aminolysis of penicillin occurs, for example, with propylamine at pH 4 when the concentration of free propylamine is only *ca* 10^{-7}–10^{-8}M. The rate enhancements of amines reacting with the penicillin–copper(II) complex compared with their reaction with penicillin alone are *ca* 4×10^6 and 10^7 respectively. These rate enhancements are similar to that for hydroxide ion, *ca* 9×10^7, described earlier (Section 9).

These large rate enhancements are attributable to the copper(II) ion complexing with penicillin [26] because methylation of the free carboxylate group in benzylpenicillin reduces the rate of the copper(II) ion-trifluoroethylamine reaction by *ca* 10^3. There is no kinetic dependence upon the concentration of trifluoroethylamine which indicates that aminolysis does not occur in this reaction. This observation may be rationalised by metal-ion coordination to the carboxy-group in penicillin which is reduced or does not occur in the methyl ester. However, the rate of reaction of the methyl ester is increased by *ca* 10^3 in the presence of 2×10^{-3}M copper(II) ions and the rate is apparently first order in metal ion. This may be due either to weak coordination of the metal ion to the β-lactam nitrogen and the methoxy-carbonyl group compared with penicillin itself, or to binding at another site in the penicillin molecule. The product of the reaction of the methyl ester in the presence of copper(II) ions is not known (Gensmantel *et al.*, 1978).

The dependence of the rate constants for the attack of amine on the metal ion–penicillin complex upon the basicity of the attacking amine gives a Brønsted β-value of 0.87. This is taken to indicate that there is approximately a unit positive charge on the amine nitrogen in the transition state. A mechanism consistent with this involves the rate-limiting breakdown of the tetrahedral intermediate [64].

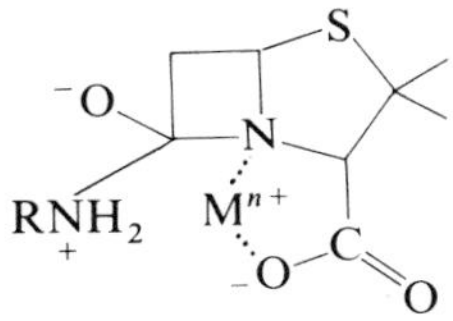

[64]

In the rate law for the metal-ion catalysed reaction there is no evidence of a term second order in amine which, if present, would indicate general base catalysis by a second molecule of amine. The coordination of the metal ion to penicillin apparently makes this normally dominant mode of catalysis unnecessary.

If the mechanisms of the reactions of penicillin involves the intermediate formation of a keten [47] or penicillenic acid [38], then the products of the reaction should show deuterium incorporation at C(6) if the reactions are carried out in D_2O. The nmr spectra of benzylpenicilloic acid and penicilloyl amides obtained from the hydrolysis and aminolysis of benzylpenicillin in D_2O in the presence of copper(II) ion shows no incorporation of deuterium at C(6). This indicates that the elimination-addition mechanism is not a major pathway for either of these reactions (Gensmantel *et al.*, 1978).

The presence of copper(II) ion has little effect upon the rate of formation of penicillenic acid from benzylpenicillin. Furthermore, the rates of hydrolysis and aminolysis of benzylpenicillenic acid are retarded in the presence of copper(II) ion. Since the observed rates of hydrolysis and aminolysis of penicillin in the presence of copper(II) ion are at least 10^3 times faster than the rate of formation of penicillenic acid from penicillin these reactions cannot occur through the intermediate formation of penicillenic acid (Gensmantel *et al.*, 1978).

Zinc(II) and tris-buffers are effective catalysts for the aminolysis of benzylpenicillin. It is suggested that this is due to formation of a ternary complex in which the metal ion binds both penicillin and tris. Nucleophilic attack of the ionised hydroxyl on bound tris forms a penicilloyl ester which may then react with tris to form a penicilloyl amide (Schwartz, 1982; Tomida and Schwartz, 1983). A kinetically equivalent mechanism, however, would simply involve nucleophilic attack of tris on the zinc–penicillin complex.

IMIDAZOLE CATALYSED ISOMERISATION OF PENICILLINS

Unlike other amines, imidazole catalyses the isomerisation of benzylpenicillin to benzylpenicillenic acid [38] (Bundgaard, 1971a,b, 1972, 1976a). The rate law shows that the nucleophilic reaction with imidazole to give the intermediate penicilloylimidazole [65] is general base catalysed by another molecule of imidazole and general acid catalysed by the conjugate acid of imidazole.

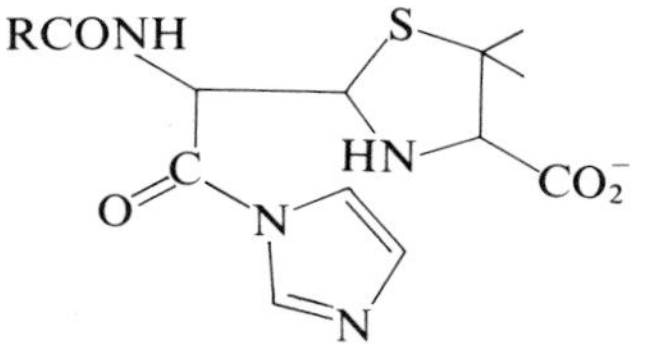

RCONH S HN CO_2^- O C N N

[65]

The mechanisms originally proposed for the formation of penicilloyl imidazole involved a concerted reaction in which nucleophilic attack by imidazole occurred simultaneously with proton transfer (Bundgaard, 1972; Yamana *et al.*, 1975). More recently it has been suggested (Butler *et al.*, 1982) that the mechanism is similar to the stepwise process proposed for the aminolysis of penicillins (Morris and Page, 1980a).

Imidazole also catalyses the aminolysis of penicillins (Yamana *et al.*, 1975; Bundgaard, 1976a). The formation of penicilloyl amides could occur by acyl transfer from the intermediate penicilloyl imidazole [65] to the amine or by aminolysis of penicillenic acid. The aminolysis of acylimidazoles is well known (Page and Jencks, 1972b), but it is claimed that intramolecular attack by the 6-side chain amido group to displace imidazole and to give the oxazolinone-thiazolidine and then penicillenic acid will be faster than intermolecular attack of amine (Yamana *et al.*, 1975). Imidazole catalysed aminolysis of penicillin was therefore suggested to occur exclusively through the reaction of the amine with penicillenic acid. Oxazolinone formation from

N-benzoylglycinate esters occurs when there is a good leaving group as in phenyl esters (Williams, 1975). Acetylimidazole is more reactive than phenyl acetate (Oakenfull and Jencks, 1971) and so displacement of the imidazole by the intramolecular amido group is expected.

However, penicillins incapable of forming penicillenic acid undergo an imidazole catalysed aminolysis, presumably via the intermediate formation of an N-penicilloylimidazole. Furthermore, penicillenic acid reacts with imidazole to form N-penicilloylimidazole suggesting that the latter may be the acylating agent for aminolysis (Bundgaard, 1976a).

INTRAMOLECULAR AMINOLYSIS

There have been many reports of a suitably placed intramolecular amino group attacking the β-lactam of cephalosporins (Tsuji *et al.*, 1981, 1983; Bundgaard, 1977b; Dinner 1977). Cephalosporins which have an α-amino group in the 7-amido side chain, e.g. cephalexin, form piperazine-2,5-diones by such a pathway [66], whilst analogous penicillins do not (Indelicato *et al.*,

[66]

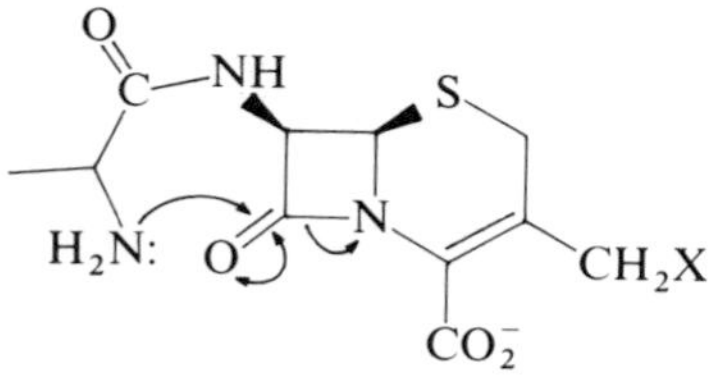

1974). Ring closure is predominant at neutral pH and is subject to hydroxide ion and general acid-base catalysis. There is a non-linear dependence of the rate upon buffer concentration which is indicative of a change in rate-limiting step and evidence for the formation of an intermediate. At 35°C the rate constant for the uncatalysed intramolecular aminolysis of cephaloglycin is $6.3 \times 10^{-5} s^{-1}$ and the second-order rate constant for the hydroxide ion catalysed reaction is $0.22\,M^{-1}s^{-1}$ (Bundgaard, 1976b). The estimated rate constants for the equivalent intermolecular reactions for an amine of $pK_a = 7$ are $2 \times 10^{-6} M^{-1}s^{-1}$ and $6 \times 10^{-3} M^{-2}s^{-1}$ respectively (Page and Proctor, 1984). This gives an effective molarity of only about 35 M for the intramolecular amino group which is very low compared with other intramolecular aminolysis reactions (Kirby, 1980; Page, 1973). Such a low value is probably due to the introduction of unfavourable steric strain in the intramolecular reaction and problems with keeping the side chain amido group coplanar. Nucleophilic attack on the β-lactam of penicillins takes place from the α-side (p. 243) and intramolecular attack from the sterically hindered β-side is unfavourable, particularly because of interference from the 2-β-methyl group. It is not surprising therefore that intramolecular aminolysis by amino groups in the 6-β-side chain of penicillins does not occur (Indelicato *et al.*, 1974) and that intermolecular aminolysis of cephalosporins is competitive with intramolecular aminolysis (Bundgaard, 1976c).

6-(N-Phenylureido)penicillanic acids undergo a rapid cyclisation to the isomeric 3-phenylhydantoin-thiazolidines by intramolecular nucleophilic attack of the ureido nitrogen anion on the β-lactam carbonyl carbon. The second-order rate constant for the hydroxide ion catalysed reaction is over 10^3-fold greater than that for hydroxide ion catalysed hydrolysis (Bundgaard, 1973). The rate enhancement is consistent with neighbouring group participation.

12 The stepwise mechanism for expulsion of C(3′)-leaving groups in cephalosporins

That expulsion of the leaving group at C(3′) is not generally concerted with C—N bond fission of the β-lactam when cephalosporins react with nucleophiles was described in Section 4. Despite experimental evidence (Hamilton-Miller *et al.*, 1970b; Page and Proctor, 1984; Agathocleous *et al.*, 1985; Grabowski *et al.*, 1985; Faraci and Pratt, 1984, 1985) that this is the case for both enzymic and non-enzymic reactions, the claim persists (Boyd, 1985) that the mechanism is concerted. The recent experimental evidence will therefore be briefly reviewed.

The stepwise process for β-lactam cleavage prior to loss of the C(3) leaving group generates the enamine [36] before the conjugated imine [37]. ^{13}C nmr observations of the ammonolysis of cephamycins in liquid ammonia at $-50°C$ are consistent with the intermediate formation of the enamine [36] (Grabowski *et al.*, 1985).

The β-lactamase catalysed hydrolysis of cephalosporins shows spectral changes in the ultraviolet which are consistent with the formation of [36] prior to expulsion of the leaving group at C(3′) (Faraci and Pratt, 1984, 1985; Agathocleous *et al.*, 1984).

There is a non-linear dependence of the rate of aminolysis of cephalosporins upon hydroxide ion concentration (Page and Proctor, 1984) which is consistent with a change in rate-limiting step and hence formation of an intermediate as described in Section 11.

The uncatalysed pathway k_u in the rate law (8) could represent either a purely uncatalysed reaction of amine and cephalosporin, or solvent catalysis with water acting as a general base to remove a proton from the attacking amine, or as a general acid to donate a proton to the β-lactam nitrogen. The k_u pathway cannot represent rate-limiting formation of the tetrahedral intermediate because the observed rate constants are much smaller than the miscroscopic rate constants calculated for this step at high hydroxide ion concentration. This is substantiated by a Brønsted β_{nuc} of 1.05 for the uncatalysed aminolysis of cephaloridine which indicates that the reaction behaves as if a unit positive charge is developed on the amine nitrogen in the transition state. The uncatalysed pathway therefore represents a rate-limiting breakdown of the tetrahedral intermediate [62] or general acid catalysed breakdown of the same intermediate by water [63].

The observed rate constant k_u for the uncatalysed aminolysis must be given by $k_5K = k_5k_1/k_{-1}$ (Scheme 8). Since the values of K, the equilibrium constant for formation of the tetrahedral intermediate, are known, the values of k_5 the rate constant for the breakdown of the intermediate to products, can be calculated. These are all about $10^6 s^{-1}$ (Page and Proctor, 1984) whether or not a leaving group is expelled at C(3′). This therefore argues against expulsion of the leaving group at C(3′) being concerted with fission of the carbon—nitrogen bond of the β-lactam [67]. At least this is true

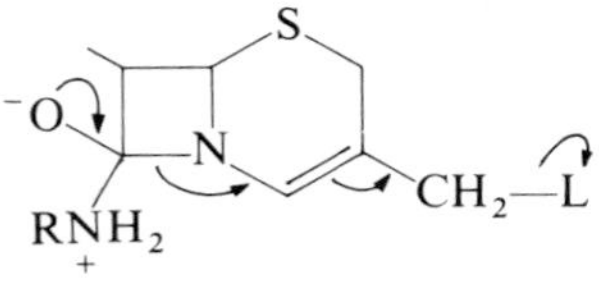

[67]

in the sense that there can be no significant coupling of the processes so that there is a significant lowering of the activation energy for carbon—nitrogen bond fission in the β-lactam *because* a group at C(3′) is expelled. It could be the case that the intermediate formed by β-lactam carbonyl carbon—nitrogen bond fission is so unstable and that its lifetime is so short as to preclude its existence and therefore the breakdown of the tetrahedral intermediate is enforced to be concerted. However, the important conclusion is that *expulsion of a leaving group at C(3′) does not significantly enhance the rate of carbon–nitrogen fission in the β-lactam.*

Finally, the enamine [36] and conjugated imine [37] are in equilibrium. The imine [37] can be generated from the β-lactamase catalysed hydrolysis of a variety of cephalosporins in aqueous solution. The addition of thiols generates the enamine [36] which can also be generated by the hydrolysis of the corresponding cephalosporin with the same thiol group at C(3′). The equilibrium between [36] and [37] is pH-dependent with the enamine favoured at low pH, (Buckwell *et al.*, 1986). The equilibrium constant between [36] and [37] varies with the pK_a-value of the thiol. The equilibrium [36][RS^-][H^+]/[37] shows a Brønsted β_{lg} of 0.7, with, for example, that for butanethiol being $2 \times 10^{-15} M^2$ at 30°C (Buckwell and Page, 1986).

Thiols presumably add to the conjugated imine [37] by a type of Michael addition reaction. It is not inconceivable that this could be an important pathway for inactivation of enzymes by cephalosporins using a suitably placed nucleophilic group on the enzyme.

13 Reaction with alcohols and other oxygen nucleophiles

Several bacterial penicillin-binding proteins have been shown to be serine enzymes (Section 2). β-Lactamases are efficient and clinically important enzymes which play an important part in bacterial resistance to the normally lethal action of β-lactam antibiotics. A major class of β-lactamases are also serine enzymes that function by covalent catalysis with the intermediate formation of an acyl-enzyme (Knott-Hunziker *et al.*, 1982; Cohen and Pratt, 1980; Fisher *et al.*, 1980; Anderson and Pratt, 1983; Cartwright and Fink, 1982, Joris *et al.*, 1984). The β-lactamase catalysed hydrolysis of a penicillin thus proceeds by formation of an acyl-enzyme which is an α-penicilloyl ester of a serine residue. The mechanism of reaction of penicillin with alcohols is therefore of obvious relevance, but in addition acyl transfer from nitrogen to oxygen nucleophiles is of current interest.

The observed pseudo first-order rate constant for the degradation of benzylpenicillin in water in the presence of alcohols is given by (9), where K_a

$$k_{obs} = k_0[OH^-] + k_n \cdot \frac{K_a}{[H^+] + K_a} \cdot [ROH]_T = k_0[OH^-] + k_n[RO^-] \qquad (9)$$

is the ionisation constant of the alcohol, k_o and k_n are the second-order rate constants for alkaline hydrolysis and the reaction of the alkoxide anion with penicillin, respectively. The values of k_n given in Table 6 form the Brønsted plots of Figs 10 and 14 (Davis *et al.*, 1987).

Oxygen anions can catalyse the hydrolysis of penicillins by either acting as general bases or as nucleophiles. Several observations indicate that basic anions act as nucleophilic catalysts forming an intermediate penicilloyl ester [68] whereas weakly basic ones act as general base catalysts.

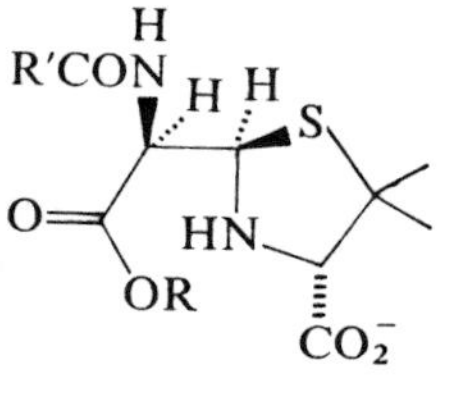

H
R′CON H H S
O HN
OR
CO_2^-

[68]

$^-$O
N
RO

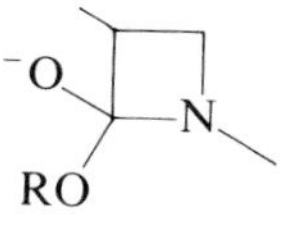

[69]

The non-linear Brønsted plot (Fig. 10) is indicative of a change in reaction mechanism. Weakly basic oxygen anions probably act as general base catalysts for the hydrolysis of penicillins as discussed earlier (Section 7). The

Brønsted β-value of 0.39 is typical for this sort of mechanism (Oakenfull and Jencks, 1971).

Basic oxygen anions of pK_a > 9 act as nucleophilic catalysts for hydrolysis with the intermediate formation of a penicilloyl ester [68]. The reaction of

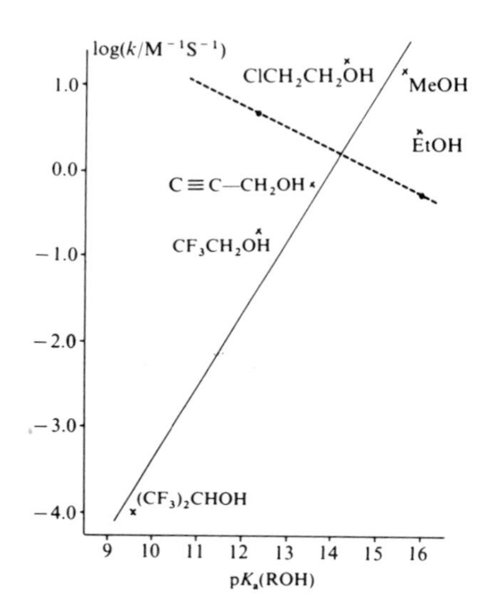

FIG. 14 Brønsted plot (solid line) for the second-order rate constants for the alcoholysis of benzylpenicillin against pK_a for the alcohol (+). The broken line is for the second-order rate constant for the alkaline hydrolysis of α-penicilloyl esters [68]

various phenolate anions, carbohydrates and other hydroxyl-containing compounds with penicillins has been shown to proceed *via* the unstable ester by the penamaldate assay (Schwartz and Delduce, 1969; Schneider and DeWeck, 1968; Bundgaard, 1976d; Larsen and Bundgaard, 1978b; Yamana *et al.*, 1977; Schwartz and Pflug, 1967; Tutt and Schwartz, 1971). The phosphate catalysed hydrolysis of benzylpenicillin is thought to occur by the nucleophilic pathway (Bundgaard and Hansen, 1981). Carbohydrates such as glucose and fructose form intermediate penicilloyl esters in their catalysed hydrolysis of penicillin and cephalosporins (Bundgaard and Larsen, 1983).

It has been shown that the addition of amine nucleophiles to the β-lactam carbonyl carbon to form a tetrahedral intermediate is reversible. The rate-limiting step in the aminolysis of β-lactam antibiotics involves proton removal from the attacking amine for base catalysed reactions and β-lactam C—N bond fission for the uncatalysed reaction (Page, 1984a; Section 11). The rate-limiting step for the hydroxide ion catalysed hydrolysis of β-lactams appears to be formation of the tetrahedral intermediate (p. 199).

The addition of an alkoxide ion to β-lactams generates the tetrahedral intermediate [69]. The rate-limiting step in the alcoholysis of penicillins depends on the relative rates of alkoxide ion expulsion from [69] compared with the subsequent step leading to β-lactam C—N fission. In the aminolysis of penicillins, expulsion of the attacking amine is faster than the subsequent step (p. 234). The relative rates of bond cleavage expelling alkoxide ions, compared with amines of equal basicity, are not easy to predict. Amines are expelled 10^5-fold *faster* than alkoxide ions from phthalimidium cation adducts (Gravitz and Jencks, 1974). However, the expulsion of trimethylamine from formaldehyde carbinolamine zwitterion (Hine and Kokesh, 1970) is about 10^4-fold *slower* than the expulsion of alkoxide ions, of similar basicity, from formaldehyde hemiacetal anion (Funderburk *et al.*, 1978). In the forward direction the breakdown of the aminolysis tetrahedral intermediate [50] to products occurs with a rate constant of *ca* $10^6\ s^{-1}$ (Gensmantel and Page, 1979a). However, this generates an unstable N-protonated amide and it is therefore expected that the equivalent reaction of [69] to generate an ester [68] will be faster because of the "push" provided by the developing conjugation from the oxygen lone pair. It would not be surprising therefore, if the rate-limiting step for the alcoholysis of β-lactams is formation of the tetrahedral intermediate. However, the large Brønsted β_{nuc} of 0.95 for alkoxide ions is indicative of rate-limiting breakdown of the tetrahedral intermediate. A β_{nuc}-value of 1.0 has also been reported for the reaction of benzylpenicillin with phenolate anions (Bundgaard, 1976d; Yamana *et al.*, 1977). The Brønsted β_{nuc} for the formation of the tetrahedral intermediate for amines adding to the β-lactam carbonyl is 0.3 (Gensmantel and Page, 1979a). The β_{nuc}-values for rate-limiting breakdown of tetrahedral intermediates formed from oxygen anions and esters, thioesters and acetylimidazole are 1.4, 1.4 and 1.3 respectively (Hupe and Jencks, 1977; Oakenfull and Jencks, 1971). Although the value of 1.0 is perhaps a little low compared with these values and the maximum of 1.7, it is consistent with rate-limiting breakdown and is therefore yet another illustration that β-lactam C—N bond fission is not a facile process.

If the reaction of benzylpenicillin with alkoxide ions proceeded with rate-limiting attack, basic alkoxide ions would be expected to show a negative deviation from the Brønsted plot which is usually curved for such reactions (Hupe and Jencks, 1977; Jencks *et al.*, 1982).

As for the uncatalysed aminolysis of penicillins the reaction with oxygen anions could involve β-lactam C—N bond fission with or without assistance from general acid catalysis by water. The solvent isotope effect $k^{RO^-}{}_{H_2O}/k^{RO^-}{}_{D_2O}$ for trifluoroethoxide ion reacting with benzylpenicillin is 3.9 (Davis *et al.*, 1987). Although there are difficulties in the quantitative interpretation of solvent isotope effects (Gold and Grist, 1972) the values of

0.5–0.7 for the attack of methoxide ion on phenyl acetates and methyl phenyl carbonates in MeOD are typical for rate-limiting formation of tetrahedral intermediates (Mitton *et al.*, 1969). It seems likely therefore that the mechanism for the formation of penicilloyl esters involves general acid catalysis by the solvent as shown in [70]. No term for buffer catalysis is observed in the rate law for alcoholysis of penicillin (Davis *et al.*, 1987); this would be expected if buffers could act as general acid catalysts.

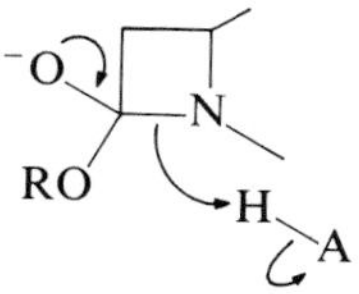

$^{-}$O, N, RO, H, A

[70]

It is interesting to note that the hydroxide ion catalysed hydrolysis of benzylpenicillin involves rate-limiting attack whereas alkoxide ions react with benzylpenicillin with rate-limiting breakdown of the tetrahedral intermediate. This could result from the intermediate formed by hydroxide ion attack either breaking down to reactants or products slower or faster, respectively, than [69].

The product of the alkoxide-ion catalysed hydrolysis of benzylpenicillin is benzylpenicilloic acid (see, however, Section 14). The hydrolysis of the intermediate α-penicilloyl esters [68] can be studied independently (Davis *et al.*, 1986). The Brønsted β_{lg}-value for the hydroxide-ion catalysed hydrolysis of the esters is -0.3, electron withdrawing substituents increasing the rate of hydrolysis (Fig. 14). Consequently, the oxygen anion catalysed hydrolysis of benzylpenicillin does not give a detectable ester intermediate for alcohols of $pK_a < 12$. Conversely, for alcohols of $pK_a > 14$ the hydrolysis of the ester intermediate is slower than its rate of formation. It is interesting to note that the rates of hydrolysis of the ester intermediates show no evidence of steric inhibition by the penicilloyl residue and are similar to those of simple alkyl esters (Davis *et al.*, 1986; Gensmantel *et al.*, 1981). This suggests that, if slow hydrolysis of the penicilloyl enzyme formed during antibiosis is responsible for enzyme inhibition, it is not the result of intrinsic stabilisation but the result of interaction with the protein preventing access to the ester function.

Penicilloyl esters of basic alcohols undergo reactions in addition to hydrolysis (Davis and Page, 1985) and these are discussed below.

THIAZOLIDINE RING OPENING

The first chemical step in the antibacterial activity of penicillins is thought to be the ring opening of the β-lactam by a serine hydroxyl of a transpeptidase enzyme to form an intermediate which is an ester of penicilloic acid [68; R = Enz] (Waxman and Strominger, 1980). It is not known if it is the stability of this intermediate which causes enzyme inhibition and why the ester is not hydrolysed to regenerate the enzyme and penicilloic acid. It is conceivable that a subsequent reaction of the ester produces an electrophilic entity which is ultimately responsible for enzyme inhibition by irreversibly reacting with a nucleophilic group on the enzyme.

Methyl 5*R*, 6*R*-benzylpenicilloate [68; R = Me] in water undergoes reactions other than simple ester hydrolysis and produces intermediates which may be of relevance to enzyme inhibition. The alkaline hydrolysis of methyl 5*R*, 6*R*-benzylpenicilloate shows an optical density increase followed by a decrease at 280 nm with the absorbance maximum increasing with concentration of hydroxide ion. The observed pseudo first-order rate coefficients k_{obs} for both of these two phases show non-linear dependences upon hydroxide ion. For the first phase k_{obs} is pH-independent up to 0.07 M sodium hydroxide but at higher concentrations it becomes first-order in hydroxide ion. By contrast, k_{obs} for the second phase, corresponding to a decrease in optical density, changes from being hydroxide-ion dependent to independent with increasing pH. These kinetic observations are compatible with Scheme 10 where SH = ester [68; R = Me]. The only observable products are penicilloic acid [30] and methanol (Davis and Page, 1985). The ester [68; R = Me] undergoes base-catalysed reversible ionisation in a pH-dependent equilibrium and at a rate which is faster than the hydroxide-ion catalysed hydrolysis of the ester. If ionisation corresponds to a simple deprotonation of [68] then it would have an apparent pK_a of 12.9 and k_2, the second-order rate constant, is 0.55 $M^{-1}s^{-1}$, as expected for hydrolysis of a methyl ester.

$$\mathrm{OH^-} + \mathrm{SH} \underset{k_{-1}}{\overset{k_1}{\rightleftharpoons}} \mathrm{S^-}$$

$$\downarrow k_2$$

Products

Scheme 10

The simplest interpretation of these observations is that methyl 5*R*, 6*R*-benzylpenicilloate [68] undergoes reversible elimination across C(6)–C(5) and ring opening of the thiazolidine to give the enamine tautomer of methyl penamaldate [71] (Scheme 11). The amount of enamine formed increases with increasing hydroxide concentration which explains the increasing absorbance at 280 nm, the wavelength of the absorption maxima of penamaldates. The thiol anion can be trapped with Ellman's reagent. Hydrolysis

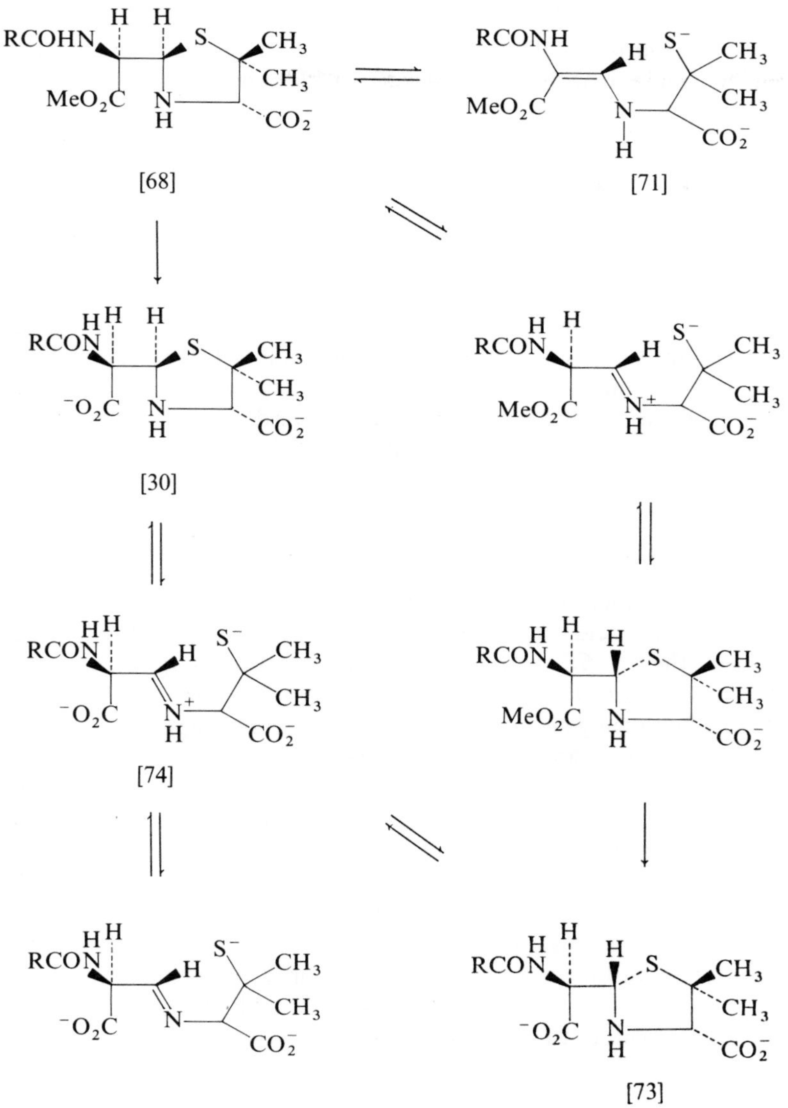

[68] [71] [30] [74] [73]

Scheme 11

of the ester occurs at low concentrations of hydroxide ion by the normal base-catalysed mechanism but becomes pH-independent at high concentrations of hydroxide because the substrate exists predominantly as the enamine tautomer [71], which is presumably less reactive towards ester hydrolysis. Hydrolysis of the ester must occur through [68], the concentration of which decreases with increasing pH (Davis and Page, 1985).

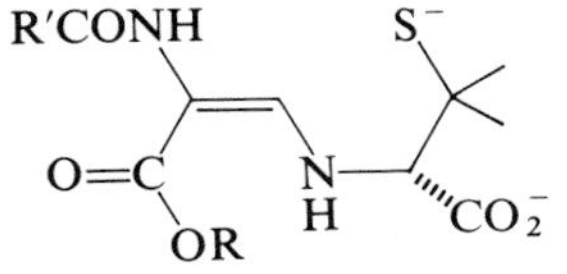

[71] 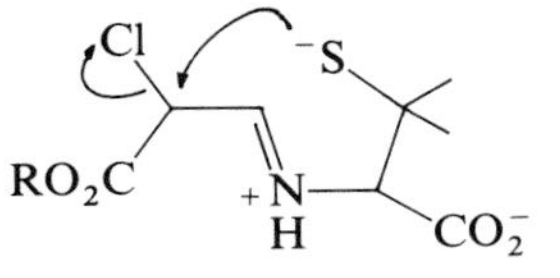 [72]

Similar observations have been made with 6-α-halopenicillanic acids but in this case the thiol anion in the analogous enamine [71] displaces the halide at C(6) *via* the imine tautomer [72] (Gensmantel *et al.*, 1981). If the enamine [71] is produced in the transpeptidase-catalysed reaction with penicillins, then inactivation of the enzyme could occur by a nucleophilic group on the enzyme attacking C(5) in a Michael-type addition reaction.

14 Epimerisation of penicillin derivatives

The initial product of alkaline hydrolysis of benzylpenicillin is 5*R*, 6*R*-benzylpenicilloic acid. However, epimerisation then occurs at C(5) to give a mixture of the 5*R*, 6*R*- and 5*S*, 6*R*- penicilloic acids [73] (Davis and Page, 1985; Carroll *et al.*, 1977; Busson *et al.*, 1976b; Kessler *et al.*, 1983; Bird *et al.*, 1983). This change in configuration at C(5) is accompanied by a decrease in pK_a of the protonated thiazolidine from 5.3 to 4.8, a change in the nmr chemical shifts of the protons at C(5) and C(6) and those on the α and β methyl groups at C(2), a decrease in the coupling constants between the C(5)–C(6) hydrogens and a change in specific rotation. Epimerisation does not occur at C(6).

The equilibrium constant for the ratio of the 5*S*, 6*R*-epimer to that of the 5*R*, 6*R*-benzylpenicilloate is 4. The observed polarimetric pseudo first-order rate constants for this process are pH- and buffer-independent from pH 6 to pH 12.5 but become first order in hydroxide ion at higher pH. Epimerisation in D_2O occurs without D-incorporation at C(5) or C(6) over most of the pH range (Davis and Page, 1985).

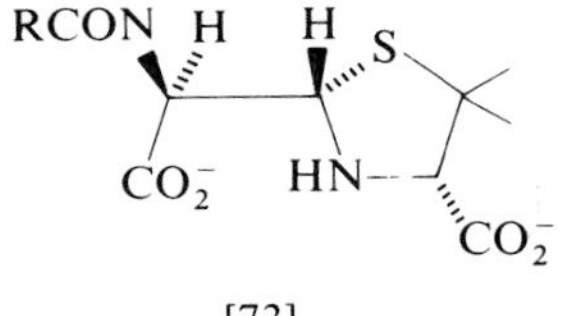

[73] 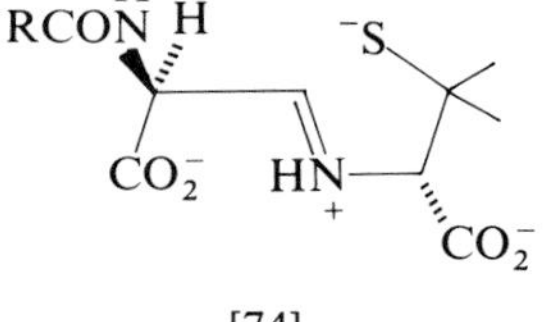 [74]

The pH-independent epimerisation of penicilloic acid at C(5) probably occurs by unimolecular ring opening of the thiazolidine to form the iminium ion tautomer of penamaldic acid [74]. There are two possible explanations for the base-catalysed reaction. Above pH 12 deprotonation of the iminium ion could become faster than ring closure, and the rate of epimerisation would increase because the steady state concentration of the ring-opened thiazolidine is increased. Alternatively, epimerisation could occur by the penamaldate mechanism followed by penicilloyl esters (Davis and Page, 1985).

At hydroxide ion concentrations above 1 M elimination across C(6)–C(5) and thiazolidine ring-opening occurs to give the enamine, similar to [71]. This is suggested by the observation of the appearance of the chromophore at 280 nm and deuterium-exchange at C(6) (Davis and Page, 1986).

Around neutral pH α-penicilloyl esters [68] epimerise at C(5) faster than the ester function is hydrolysed (Davis and Page, 1985). If this occurs with penicilloyl enzymes from penicillins and the serine hydroxyl at the active site of the enzyme, the ring-opening of the thiazolidine generates electrophilic sites capable of irreversibly inactivating the enzyme.

Penicillins with common amido side chains at C(6) are hydrolysed initially to 3*S*, 5*R*, 6*R*-penicilloic acid without D-incorporation or epimerisation at C(3), C(5) or C(6). However, penicillins without an ionisable NH on the C(6) side chain can undergo epimerisation from 6-β- to 6-α-substituted penicillins faster than ring-opening of the β-lactam (Wolfe and Lee, 1968; Johnson and Mania, 1969). The epimerisation of hetacillins to epihetacillin and that of mecillinam is hydroxide-ion catalysed but shows no general base catalysis (Tsuji *et al.*, 1977; Baltzer *et al.*, 1979).

There have been several unusual observations related to proton abstraction at C(6) in penicillins which indicate that the carbanion formed is exceptional. Treatment of the penicillin Schiff base derivative [75] with phenyllithium in tetrahydrofuran appears to lead to abstraction of the C(6) proton but reprotonation regenerates [75] despite the 6-α-epimer being thermodynamically more stable (Firestone *et al.*, 1972, 1974). Even more strange is the observation that reprotonation with D_2O/CD_3CO_2D does not give the 6-α-deuterio Schiff base [75]. However, epimerisation by triethylamine in acetonitrile containing D_2O is accompanied by deuteriation of the

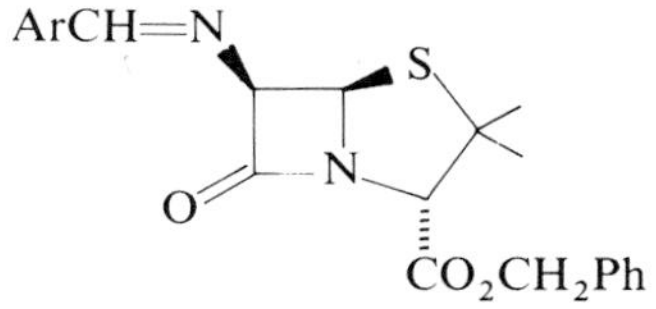

ArCH=N
CO_2CH_2Ph

[75]

X

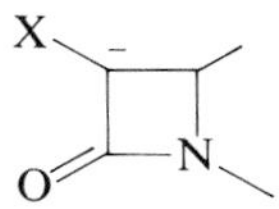

[76]

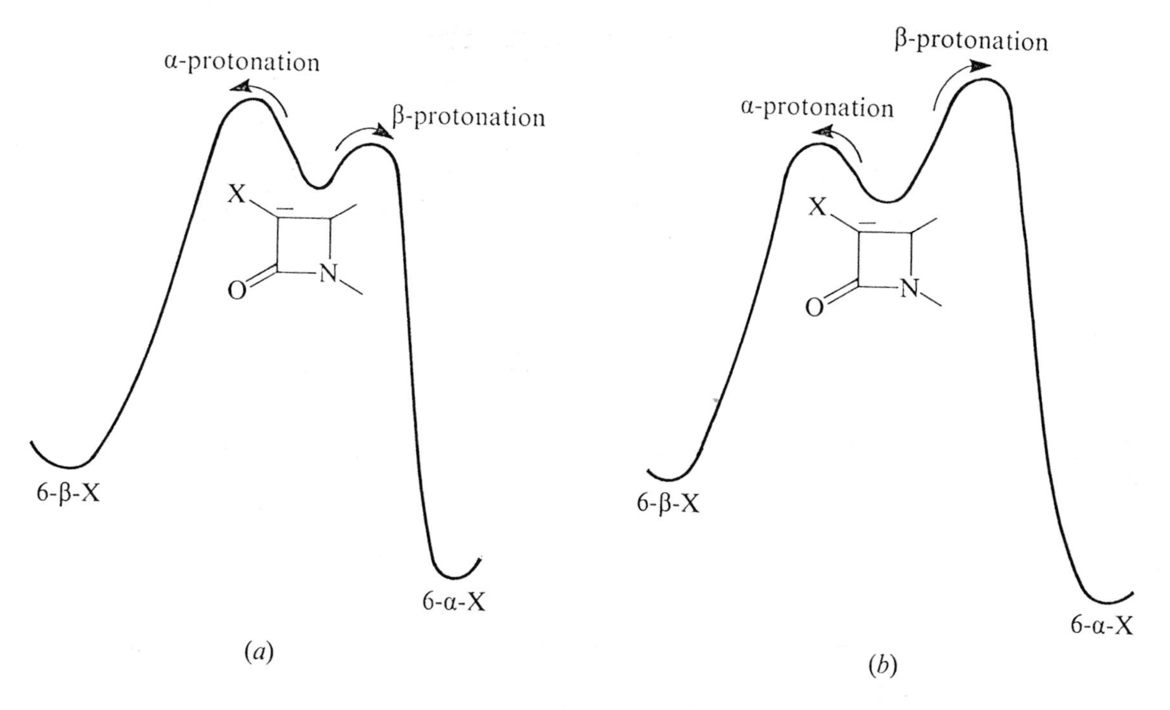

FIG. 15 Energy-profiles for abstraction of the proton on C(6) in 6-X-substituted penicillins and reprotonation of the intermediate carbanion

presumed carbanion intermediate [76] and occurs preferentially from the least hindered α-face to give the less stable β-epimer. Deuterium exchange at C(6) of the 6-β epimer occurs faster than epimerisation at C(6) (Firestone and Christensen, 1977). The calculated isotope effect k_H/k_D for protonation of the carbanion is 12.7 which is ascribed to proton tunnelling.

By contrast, 6-α-chloropenicillanic acid undergoes D-exchange at C(6) in $NaOD/D_2O$ without epimerisation and at a rate faster than β-lactam ring-opening. 6-β-Epimers appear to undergo epimerisation and deuteriation at C(6) at the same rate in water (Clayton *et al.*, 1969; Gensmantel *et al.*, 1981).

If it is assumed that epimerisation and D-exchange at C(6) occur *via* the same carbanion [76], two schemes can be envisaged (Fig. 15). In (*a*) β-protonation of the carbanion is faster than α-protonation whereas the reverse is true in (*b*). β-Protonation would have a lower activation energy if either the transition state is "late" and the factors stabilising the α-epimer in the product are reflected in the transition state [presumably less unfavourable steric interactions between the C(6) substituent and the C(2) methyl] or an "early" transition state but a non-planar carbanion [76] with a preponderance of electron density on the β-face (Gensmantel *et al.*, 1981). For a planar carbanion [76] an "early" transition state for protonation would favour the sterically favourable *exo* α-direction. It is conceivable that there is a change from (*a*) to (*b*) (Fig. 15) on going from aqueous to non-polar solvents.

As described in the previous sections protonation of the enamine [71] appears to occur stereospecifically since the stereochemistry at C(6) is preserved when [71] ring closes to regenerate [68].

There have been several reports of the rates and mechanism of epimerisation of asymmetric centres in side chains of β-lactam antibiotics. These have usually involved an asymmetric centre adjacent to the carbonyl carbon of the penicillin C(6) or cephalosporin C(7) amide side chain (Bird *et al.*, 1984; Hashimoto and Tanaka, 1985).

References

Abraham, E. P., Chain, E., Baker, W. and Robinson, R. (1943). *Pen Report* No. 103

Abrahamsson, S., Hodgkin, D. C. and Maslen, E. N. (1963). *Biochem. J.* **86**, 514

Adriaens, P., Meesschaert, B., Frère, J. M., Vanderhaeghe, H., Degelaen, J., Ghuysen, J. M. and Eyssen, H. (1978). *J. Biol. Chem.* **253**, 3660.

Agathocleous, D., Buckwell, S., Proctor, P. and Page, M. I. (1985). *In* "Recent Advances in the Chemistry of β-Lactam Antibiotics" (eds A. G. Brown and S. M. Roberts) p. 18. Royal Society of Chemistry, London

Allinger, M. L., Tribble, M. T. and Miller, M. A. (1972). *Tetrahedron* **28**, 1173

Anderson, B. M., Cordes, E. H. and Jencks, W. P. (1961). *J. Biol. Chem.* **236**, 455

Anderson, E. G. and Pratt, R. F. (1983). *J. Biol. Chem.* **258**, 13120

Anoardi, L. and Tonellato, V. (1977). *J. Chem. Soc. Chem. Commun.* 401

Applegate, H. E., Kolfini, J. E., Puar, M. S., Slusarchyk, W. A., Toeplitz, B. and Gougoutas, J. Z. (1974). *J. Org. Chem.* **39**, 2794

Aroney, M. J., Le Fère, R. J. W., Radom, L. and Ritchie, G. L. D. (1968). *J. Chem. Soc. B* 507

Asso, M., Panossian, R. and Guiliano, M. (1984). *Spectroscopy Lett.* **17**, 271

Baltzer, B., Lund, F. and Rastrup-Andersen, N. (1979). *J. Pharm. Sci.* **68**, 1207

Batchelor, F. R., Dewdney, J. M. and Gazzard, D. (1965). *Nature* **206**, 362

Bender, M. L. and Thomas, R. J. (1961). *J. Am. Chem. Soc.* **83**, 4183

Berge, S. M., Henderson, N. L. and Frank, M. J. (1983). *J. Pharm. Sci.* **72**, 59

Berger, S. A. (1984). *New England J. Med.* **312**, 1395.

Bird, A. E., Cutmore, E. A., Jennings, K. R. and Marshall, A. C. (1983). *J. Pharm. Pharmacol.* **35**, 138

Bird, A. E., Charsley, C.-H., Jennings, K. R. and Marshall, A. C. (1984). *Analyst* **109**, 1209

Blackburn, G. M. and Dodds, H. L. H. (1974). *J. Chem. Soc. Perkin Trans 2* 377

Blackburn, G. M. and Plackett, J. D. (1972). *J. Chem. Soc., Perkin Trans 2* 1366

Blackburn, G. M. and Plackett, J. D. (1973). *J. Chem. Soc., Perkin Trans 2* 981

Blaha, J. M., Knevel, A. M., Kessler, D. P., Mincy, J. W. and Hern, S. L. (1976). *J. Pharm. Sci.* **65**, 1165

Bolton, P. D. and Jackson, G. L. (1969). *Austral. J. Chem.* **22**, 527

Bose, A. K. and Srinivasan, P. R. (1979). *Org. Mag. Res.* **12**, 34

Bottini, A. T. and Nash, C. P. (1962). *J. Am. Chem. Soc.*, **84**, 734

Bouchoux, G. and Houriet, R. (1984). *Tetrahedron Lett.* 5755

Boyd, D. B. (1972). *J. Am. Chem. Soc.* **94**, 6513

Boyd, D. B. (1982). *In* "Chemistry and Biology of β-Lactam Antibiotics", (eds R. B. Morin and M. Gorman) Vol. 1, p. 437 Academic Press, New York

Boyd, D. B. (1983). *J. Med. Chem.* **26**, 1010

Boyd, D. B. (1985). *J. Org. Chem.* **50**, 886

Boyd, D. B. and Lunn, W. H. W. (1979). *J. Med. Chem* **22**, 778

Boyd, D. B., Hermann, R. B., Presti, D. E. and Marsh, M. M. (1975). *J. Med. Chem.* **18**, 408

Boyd, D. B., Yeh, C.-Y. and Richardson, F. S. (1976). *J. Am. Chem. Soc.* **98**, 6100

Boyd, D. B., Herron, D. K., Lunn, W. H. W. and Spitzer, W. A. (1980). *J. Am. Chem. Soc.* **102**, 1812.

Brown, H. C. and Ichikawa, K. (1957). *Tetrahedron* **1**, 221

Broxton, T. J. and Duddy, N. W. (1979). *Austral. J. Chem.* **32**, 1717

Bruylants, A. and Kezdy, F. (1960). *Rec. Chem. Progr.* **21**, 213

Buckwell, S. and Page, M. I. (1985). Unpublished observations

Buckwell, S., Page, M. I. and Longridge, J. L. (1986). *J. Chem. Soc. Chem. Commun.* 1039

Bucourt, R., Heymès, R., Lutz, A., Pénasse, L. and Perronet, J. (1978). *Tetrahedron* **34**, 2233

Bundgaard, H. (1971a). *J. Pharm. Sci.* **60**, 1273

Bundgaard, H. (1971b). *Tetrahedron Lett.* 4613

Bundgaard, H. (1972). *Dansk. Tidsskr. Farm.* **46**, 29, 85

Bundgaard, H. (1973). *Acta Pharm. Suecica* **10**, 309

Bundgaard, H. (1975). *Arch. Pharm. Chemi. Sci. Ed.* **3**, 94

Bundgaard, H. (1976a). *J. Pharm. Pharmacol.* **28**, 725

Bundgaard, H. (1976b). *Arch. Pharm. Chemi. Sci. Ed.* **4**, 25

Bundgaard, H. (1976c). *Acta. Pharm. Suecica* **13**, 299

Bundgaard, H. (1976d). *Arch. Pharm. Chemi. Sci. Ed.* **4**, 91
Bundgaard, H. (1977a). *Acta Pharm. Suecica* **14**, 47, 67
Bundgaard, H. (1977b). *Arch. Pharm. Chemi. Sci. Ed.* **5**, 141, 149
Bundgaard, H. (1980). *Arch. Pharm. Chemi. Sci. Ed.* **8**, 161
Bundgaard, H. and Angelo, H. R. (1974). *Tetrahedron Lett.* 3001
Bundgaard, H. and Buur, A. (1983). *Arch. Pharm. Chem. Sci. Ed.* **11**, 15
Bundgaard, H. and Hansen, J. (1981). *Int. J. Pharm.* **9**, 273
Bundgaard, H. and Ilver, K. (1970). *Dansk. Tidsskr. Farm.* **44**, 365
Bundgaard, H. and Larsen, C. (1983). *Int. J. Pharm.* **16**, 319
Bunton, C. A. (1977). *Pure Appl. Chem.* **49**, 969
Bunton, C. A. (1984). *In* "The Chemistry of Enzyme Action" (ed. M. I. Page) p. 461. Elsevier, Amsterdam
Bunton, C. A. and Savelli, G. (1986), *Adv. Phys. Org. Chem.* **22**, 213
Bunton, C. A., Robinson, L. and Stam, M. (1970). *J. Am. Chem. Soc.* **92**, 7393
Buss, V. Gleiter, R. and Schleyer, P. von R. (1971). *J. Am. Chem. Soc.* **93**, 3927
Busson, R. and Vanderhaeghe, H. (1978). *J. Org. Chem.* **43**, 4438
Busson, R., Claes, P. J. and Vanderhaeghe, H. (1976a). *J. Org. Chem.* **41**, 2556
Busson, R., Vanderhaeghe, H. and Toppet, S. (1976b). *J. Org. Chem.* **41**, 3054
Busson, R., Roets, E. and Vanderhaeghe, H. (1978). *J. Org. Chem.* **43**, 4434
Butler, A. R., Robinson, I. R. and Wright, D. E. (1982). *J. Chem. Soc. Perkin Trans 2* 827
Carlsen, N. R., Radom, L. and Riggs, N. V. (1979). *J. Am. Chem. Soc.* **101**, 2233
Carroll, R. D., Jung, S. and Sklavounos, C. G. (1977). *J. Heterocycl. Chem.* **14**, 503
Cartwright, S. J. and Fink, A. L. (1982). *FEBS Lett.* **137**, 186
Chaimovich, H., Correia, V. R., Araujo, P. S., Aleixo, M. V. and Cuccovia, I. M. (1985). *J. Chem. Soc. Perkin Trans 2* 925
Chakrawarti, P. B., Tiwari, A. and Sharma, H. N. (1982). *Indian J. Chem.* **21A**, 200
Chakrawarti, P. B., Tiwari, C. P., Tiwari, A. and Sharma, H. N. (1984). *J. Indian Chem. Soc.* **61**, 705
Chambers, R. and Doedens, R. J. (1980). *Acta Crystallogr. Sect. B.* **36**, 1507
Clayton, J. P., Nayler, J. H. C., Southgate, R. and Stove, E. R. (1969). *J. Chem. Soc. Chem. Commun.* 129
Coene, B., Schanck, A., Dereppe, J.-M. and Van Meerssche, M. (1984). *J. Med. Chem.* **27**, 694
Cohen, S. A. and Pratt, R. F. (1980). *Biochemistry* **19**, 3996
Collings, A. J., Jackson, P. F. and Morgan, K. J. (1970). *J. Chem. Soc. B* 581
Cordes, E. H. (1978). *Pure Appl. Chem.* **49**, 969
Cressman, W. A., Sugita, E. T., Coluisio, J. T. and Niebergall, P. J. (1969). *J. Pharm. Sci.* **58**, 1471
Crowfoot, D., Bunn, C. W., Rogers-Low, B. W. and Turner-Jones, A. (1949). *In* "The Chemistry of Penicillin" (eds H. T. Clarke, J. R. Johnson and R. Robinson) p. 310. Princeton University Press, Princeton, New Jersey
Davis, A. M. and Page, M. I. (1985). *J. Chem. Soc. Chem. Commun.* 1702
Davis, A. M. and Page, M. I. (1986). Unpublished observations
Davis, A. M., Proctor, P. and Page, M. I. (1987). *J. Chem. Soc. Perkin Trans 2* In press
Degelaen, J., Loukas, S. L., Feeney, J., Roberts, G. C. K. and Burgen, A. S. V. (1979a). *J. Chem. Soc. Perkin Trans 2* 86
Degelaen, J., Feeney, J., Roberts, G. C. K., Burgen, A. S. V., Frère, J. M. and Ghuysen, J. M. (1979b). *FEBS Lett.* **98**, 53

Demarco, P. V. and Nagarajan, R. (1973). *In* "Cephalosporins and Penicillins" (ed. E. H. Flynn) Ch. 8. Academic Press, New York
Dennen, D. W. and Davis, W. W. (1962). *Antimicrob. Ag. Chemother.* **1**, 531
Dereppe, J. M., Schanck, A., Coene, B., Moreau, C. and Van Meerssche, M. (1978). *Org. Magn. Resonance* **11**, 638
Deslongchamps, P. (1975). *Tetrahedron* **31**, 2463
DeWeck, A. L. (1962). *Int. Arch. Allergy Appl. Immunol.* **21**, 20
DeWeck, A. L. and Bulm, G. (1965). *Int. Arch. Allergy Appl. Immunol.* **27**, 221
Dewhirst, K. C. and Cram, D. J. (1958). *J. Am. Chem. Soc.* **80**, 3115
Dhami, K. S. and Stothers, J. B. (1964). *Tetrahedron Lett.* 631
Dinner, A. (1977). *J. Med. Chem.* **20**, 963
Doering, W. von E., Birladeanu, L., Andrews, D. W. and Pagnotta, M. (1984). *J. Am. Chem. Soc.* **107**, 428
Doyle, F. P. and Nayler, J. H. C. (1964). *In* "Advances in Drug Research" (eds N. J. Harper and A. B. Simmonds) Vol. 1, p. 1. Academic Press, New York
Dusart, J., Marquet, A., Ghuysen, J. M., Frère, J. M., Moreno, R., Leyh-Bouille, M., Johnson, K., Lucchi, C., Perkins, H. R. and Nieto, M. (1973). *Antimicrob. Agents Chemo. Ther.* **3**, 181
Earl, H. A., Marshall, D. R. and Stirling, C. J. M. (1983). *J. Chem. Soc. Chem. Commun.* 779
Eichelberger, H. R., Mayeste, R. J. and Good, M. L. (1974). *16th Proc. Int. Conf. Coord. Chem.* **1**, 21
Eichorn, G. (1973). *Inorg. Biochem.* **1**, 79
Eigen, M. (1964). *Angew. Chem. [Int. Edn]* **3**, 1
Faraci, W. S. and Pratt, R. F. (1984). *J. Am. Chem. Soc.* **106**, 1489
Faraci, W. S. and Pratt, R. F. (1985). *Biochemistry* **24**, 903
Fazakerley, G. V. and Jackson, G. E. (1975a). *J. Chem. Soc. Perkin Trans 2* 567
Fazakerley, G. V. and Jackson, G. E. (1975b). *J. Inorg. Nucl. Chem.* **37**, 2371
Fazakerley, G. V. and Jackson, G. E. (1977). *J. Pharm. Sci.* **66**, 533
Fazakerley, G. V. Jackson, G. E. and Linder, P. W. (1976). *J. Inorg. Nucl. Chem.* **38**, 1397
Fendler, J. H. and Fendler, E. J. (1975). "Catalysis in Micellar and Macromolecular Systems". Academic Press, New York
Fersht, A. R. and Requena, Y. (1971). *J. Am. Chem. Soc.* **93**, 3499, 3502
Finholt, P., Jurgensen, G. and Kristiansen, H. (1965). *J. Pharm. Sci.* **54**, 387
Firestone, R. A. and Christensen, B. G. (1977). *J. Chem. Soc. Chem. Commun.* 288
Firestone, R. A., Schelechow, N., Johnston, D. B. R. and Christensen, B. G. (1972). *Tetrahedron Lett.* 375
Firestone, R. A., Maciejewicz, N. S., Ratcliffe, R. W. and Christensen, B. G. (1974). *J. Org. Chem.* **39**, 437
Firestone, R. A., Fahey, J. L., Maciejewicz, N. S., Patel, G. S. and Christensen, B. G. (1977). *J. Med. Chem.* **20**, 551
Fisher, J., Belasco, J. G., Khosla, S. and Knowles, J. R. (1980). *Biochemistry* **19**, 2985
Fleming, A. (1929). *Br. J. Exp. Pathol.* **10**, 226
Flynn, E. H. (1972). "Cephalosporins and Penicillins: Chemistry and Biology" Academic Press, New York
Frère, J. M. and Joris, B. (1985). *CRC Crit. Revs. Microbiol.* **11**, 289
Frère, J. M., Leyh-Bouille, M., Ghuysen, J. M. and Perkins, H. R. (1974). *Eur. J. Biochem.* **50**, 203
Frère, J. M., Ghuysen, J. M. and Iwatsubo, M. (1975a). *Eur. J. Biochem.* **57**, 343

Frère, J. M., Ghuysen, J. M., Degelaen, J., Loffet, A. and Perkins, H. R. (1975b). *Nature (London)* **258**, 168
Frère, J. M., Ghuysen, J. M., Vanderhaeghe, H., Adriaens, P., Degelaen, J. and De Graeve, J. (1976a). *Nature (London)* **260**, 451
Frère, J. M., Duez, C., Ghuysen, J. M. and Vandekerkhove, J. (1976b). *FEBS Lett.* **70**, 727
Frère, J. M., Ghuysen, J. M. and De Graeve, J. (1978). *FEBS Lett.* **88**, 147
Frère, J. M., Kelly, J. A., Klein, D., Ghuysen, J. M. Claes, P. J. and Vanderhaeghe, H. (1982). *Biochem. J.* **203**, 223
Frère, J. M., Klein, D., Kelly, J. A. and Ghuysen, J. M. (1984). *FEMS Microbiol. Lett.* **21** 213
Fujita, T. and Koshiro, A. (1984). *Chem. Pharm. Bull.* **32**, 3651
Funderburk, L., Aldwin, L. and Jencks, W. P. (1978). *J. Am. Chem. Soc.* **100**, 5444
Gani, V. and Viout, P. (1978). *Tetrahedron* **34**, 1337
Gassman, P. G., Cryberg, R. L. and Shudo, L. (1972). *J. Am. Chem. Soc.,* **94**, 7600
Gensmantel, N. P. and Page, M. I. (1979a). *J. Chem. Soc. Perkin Trans 2* **137**
Gensmantel, N. P. and Page, M. I. (1979b). Unpublished observations
Gensmantel, N. P. and Page, M. I. (1982a). *J. Chem. Soc. Perkin Trans 2* 147
Gensmantel, N. P. and Page, M. I. (1982b). *J. Chem. Soc. Perkin Trans 2* 155
Gensmantel, N. P., Gowling, E. W. and Page, M. I. (1978). *J. Chem. Soc. Perkin Trans 2* 335
Gensmantel, N. P., Proctor, P. and Page, M. I. (1980). *J. Chem. Soc. Perkin Trans 2* 1725
Gensmantel, N. P., McLellan, D., Morris, J. J., Page, M. I., Proctor, P. and Randahawa, G. (1981). *In* "Recent Advances in the Chemistry of β-Lactam Antibiotics" (ed. G. Gregory) p. 227. Royal Society of Chemistry, London
Georgopapadakou, N. H., Smith, S. A. and Cimarusti, C. M. (1981). *Eur. J. Biochem.* **115**, 53
Ghuysen, J.-M. (1977). *Cell Surfaces Rev.* **4**, 463
Ghuysen, J.-M., Frère, J.-M., Leyh-Bouille, M., Nguyen-Distèche, M., Coyette, J., Dusart, J., Joris, B., Duez, C., Dideberg, O., Charlier, P., Dive, G. and Lamotte-Brasseur, J. (1985). *Scand. J. Infect. Dis. Suppl.* **42**, 17
Giffney, C. J. and O'Connor, C. J. (1975). *J. Chem. Soc. Perkin Trans 2* 1357
Gilpin, M. L., Harbridge, J. B., Howarth, T. F. and King, T. J. (1981). *J. Chem. Soc. Chem. Commun.* 929
Glidewell, C. and Mollison, G. S. M. (1981). *J. Mol. Struct.* **72**, 203
Gold, V. and Grist, S. (1972). *J. Chem. Soc. Perkin Trans 2* 89
Grabowski, E. J. J., Douglas, A. W. and Smith, G. B. (1985). *J. Am. Chem. Soc.* **107**, 267
Gravitz, N. and Jencks, W. P. (1974). *J. Am. Chem. Soc.* **96**, 499
Green, G. G. F. H., Page, J. E. and Straniforth, S. E. (1965). *J. Chem. Soc.* 1595
Hamilton-Miller, J. M. T., Richards, E. and Abraham, E. P. (1970a). *Biochem. J.* **96**, 739
Hamilton-Miller, J. M. T., Richards, E. and Abraham, E. P. (1970b). *Biochem. J.* **116**, 385
Hammond, G. S. (1956). *In* "Steric Effects in Organic Chemistry" (Ed. M. S. Newman), Ch. 9. John Wiley and Sons, New York
Hashimoto, N. and Tanaka, H. (1985). *J. Pharm. Sci.* **74**, 68
Hehre, W. J., Radom, L. and Pople, J. A. (1972). *J. Am. Chem. Soc.* **94**, 1496
Higgins, C. E., Hamill, R. L., Sands, T. H., Hoehn, M. M., Davis, N. E., Nagarajan, R. and Boeck, L. D. (1974). *J. Antibiot.* **27**, 298

Hine, J. and Kokesh, F. C. (1970). *J. Am. Chem. Soc.* **92**, 4383
Ho, P. P. K., Towner, R. D., Indelicato, J. M., Wilham, W. L., Spitzer, W. A. and Koppel, G. A. (1973). *J. Antibiot.* **36**, 313
Hong, J. T. and Kostenbauder, H. B. (1975). *J. Pharm. Sci.* **64**, 1378
Hupe, D. J. and Jencks, W. P. (1977). *J. Am. Chem. Soc.* **99**, 451
Indelicato, J. M. and Wilham, W. L. (1974). *J. Med. Chem.* **17**, 528
Indelicato, J. M., Norvilas, T. T., Pfeiffer, R. R., Wheeler, W. J. and Wilham, W. L. (1974). *J. Med. Chem.* **17**, 523
Irving, H. and Williams, R. J. P. (1953). *J. Chem. Soc.* 3192
Ishidate, H., Sakaguchi, T., Taguchi, K. and Kanao, S. (1960). *Anal. Chim. Acta* **22**, 452
James, M. N. G., Hall, D. and Hodgkin, D. C. (1968). *Nature (London)* **220**, 168
Jeffrey, G. A., Ruble, J. R., McMullen, R. K., DeFrees, D. J., Binkley, J. S. and Pople, J. A. (1980). *Acta Cryst.* **B36**, 2242
Jencks, W. P. (1975). *Adv. Enzymol,* **43**, 219
Jencks, W. P. (1976). *Accounts Chem. Res.* **9**, 425
Jencks, W. P. and Page, M. I. (1972). *Proc. Eighth FEBS Meeting, Amsterdam,* **29**, 45
Jencks, W. P., Brant, S. R., Gandler, J. R., Fendrich, G. and Nakamura, C. (1982). *J. Am. Chem. Soc.* **104**, 7045
Johnson, D. A. and Mania, D. (1969). *Tetrahedron Lett.* 1779
Joris, B., Dusart, J., Frère, J.-M., van Beeumen, J., Emanuel, E. L., Petursson, S., Gagnon, J. and Waley, S. G. (1984). *Biochem. J.* **233**, 271
Kelly, J. A., Frère, J.-M., Klein, D. and Ghuysen, J. M. (1981). *Biochem. J.* **199**, 129
Kelly, J. A., Knox, J. R., Moews, P. C., Hite, G. J., Bartolone, J. B., Haiching, Z., Joris, B., Frère, J.-M. and Ghuysen, J.-M. (1985). *J. Biol. Chem.* **260**, 6449
Kessler, H. (1970). *Angew. Chem. [Int. Edn]* **9**, 219
Kessler, D. P., Cushman, M., Ghebre-Sellassie, I., Knevel, A. M. and Hem, S. L. (1983). *J. Chem. Soc. Perkin Trans 2* 1699
Kirby, A. J. (1980). *Adv. Phys. Org. Chem.* **17**, 183
Knott-Hunziker, V., Petursson, S., Waley, S. G., Jaurin, B. and Grundstrom, T. (1982). *Biochem. J.* **207**, 315
Komiyama, M. and Bender, M. L. (1984). *In* "The Chemistry of Enzyme Action" (ed. M. I. Page) p. 505. Elsevier, Amsterdam
Krabbenhoft, H. A., Wiseman, J. R. and Quinn, C. B. (1974). *J. Am. Chem. Soc.* **96**, 258
Kresge, A. J., Fitzgerald, P. H. and Chiang, Y. (1974). *J. Am. Chem. Soc.* **96**, 4698
Larsen, C. and Bundgaard, H. (1977). *Arch. Pharm. Chemi. Sci. Ed.* **5**, 201
Larsen, C. and Bundgaard, H. (1978a). *Arch. Pharm. Chemi. Sci. Ed.* **6** 33
Larsen, C. and Bundgaard, H. (1978b). *J. Chromat.* **147**, 143
Laurent, G., Durant, F., Frère, J. M., Klein, D. and Ghuysen, J. M. (1984). *Biochem. J.* **218**, 933
Levine, B. B. (1961). *Arch. Biochem. Biophys.* **93**, 50
Levine, B. B. and Ovary, Z. (1961). *J. Exp. Med.* **114**, 875
Levy, G. C. and Nelson, G. L. (1972a). "Carbon-13 Nuclear Magnetic Resonance for Organic Chemists". Wiley, Interscience, New York
Levy, G. C. and Nelson, G. L. (1972b). *J. Am. Chem. Soc.* **94**, 4897
Lichter, R. L. and Dorman, D. E. (1976). *J. Org. Chem.* **41**, 582
Liler, M. (1969). *J. Chem. Soc. B.* 385
Liler, M. (1975). *Adv. Phys. Org. Chem.* **11**, 328
Longridge, J. L. and Timms, D. (1971). *J. Chem. Soc. B.* 852

Luche, J. L., Kagan, H. B., Parthasarathy, R., Tsoucaris, G., De Rango, C. and Zeliver, C. (1968). *Tetrahedron* **24**, 1275
Lumbreras, J. M., Fernandez, G. M. and Ordonez, D. (1982). *An. Real Acad. Farm.* **48**, 51
Malihowski, E. R., Manhas, M. S., Goldberg, M. and Fanelli, V. (1974). *J. Mol. Struct.* **23**, 321
Manhas, M. S., Jeng, S. and Bose, A. K. (1968). *Tetrahedron* **24**, 1237
Martin, A. F., Morris, J. J. and Page, M. I. (1976). *J. Chem. Soc. Chem. Commun.*, 495
Martin, A. F., Morris, J. J. and Page, M. I. (1979). *J. Chem. Soc. Chem. Commun.*, 298
Martinek, K., Yatsimirski, A. K., Osipov, A. P. and Berezin, I. V. (1973). *Tetrahedron* **29**, 963
Martinek, K., Yatsimirski, A. K., Osipov, A. P. and Berezin, I. V. (1975). *Tetrahedron* **31**, 709
Martinek, K., Yatsimirski, A. K., Levashov, A. V. and Berezin, I. V. (1977). *In* "Micellisation, Solubilisation and Microemulsions" (ed. K. L. Mittal,) Vol. 2. Plenum Press, New York
McLelland, R. A. and Reynolds, W. F. (1974). *J. Chem. Soc. Chem. Commun.* 824
McLelland, R. A. and Reynolds, W. F. (1976). *Can. J. Chem.* **54**, 718
Menger, F. M. and Portnoy, C. E. (1967). *J. Am. Chem. Soc.* **89**, 4698
Meyer, W. P. and Martin, J. C. (1976). *J. Am. Chem. Soc.* **98**, 1231
Miller, I. M., Stapley, E. O. and Chaiet, L. (1962). *Bact. Proc.* **A49**, 32
Minhas, H. S. and Page, M. I. (1982). Unpublished observations
Mitton, C. G., Gresser, M. and Schowen, R. L. (1969). *J. Am. Chem. Soc.* **91**, 2045
Modro, T. A., Beaufays, F. and Yates, K. (1977). *Can. J. Chem.* **55**, 3050
Moll, F. (1968). *Arch. Pharm.* **301**, 272
Molyneux, P., Rhodes, C. T. and Swarbick, J. (1965). *Trans. Faraday Soc.*, **61**, 1043
Mondelli, R. and Ventura, P. (1977). *J. Chem. Soc. Perkin Trans 2* 1749
Morin, R. B., Jackson, B. G., Mueller, R. A., Lavagnino, E. R., Scanlon W. B. and Andrews, S. L. (1969). *J. Am. Chem. Soc.* **91**, 1401
Morris, R. B. and Jackson, B. G. (1970). *Fortsch. Chem. Org. Naturst.* **28**, 343
Morris, J. J. and Page, M. I. (1978). Unpublished observations
Morris, J. J. and Page, M. I. (1980a). *J. Chem. Soc. Perkin Trans 2* 212
Morris, J. J. and Page, M. I. (1980b). *J. Chem. Soc. Perkin Trans 2* 220
Morris, J. J. and Page, M. I. (1980c). *J. Chem. Soc. Perkin Trans 2* 679, 685
Murakami, K., Takasuka, M., Motokawa, K. and Yoshida, T. (1981). *J. Med. Chem.* **24**, 88
Murakami, K. and Yoshida, T. (1982). *Antimicrob. Agents Chemother.* **21**, 254
Nagarajan, R., Boeck, L. D., Gorman, M., Hamill, R. L., Higgens, C. E., Hoehn, M. M., Stark, W. M. and Whitney, J. G. (1971). *J. Am. Chem. Soc.* **93**, 2308
Narisada, M., Onoue, H. and Nagata, W. (1977). *Heterocycles* **7**, 389
Narisada, M., Yoshida, T., Ohtani, M., Ezumi, K. and Takasuka, M. (1983). *J. Med. Chem.* **26**, 1577
Nelson, H. D. and DeLigny, D. L. (1968). *Recl. Trav. Chim. Pays-Bas* **87**, 623
Newton, G. G. F. and Abraham, E. P. (1955). *Nature (London)* **175**, 548
Nikaido, H. (1981). *In* "β-Lactam Antibiotics" (eds M. Salton and G. D. Shockman) pp. 249–60. Academic Press, New York
Nishikawa, J. and Tori, K. (1981). *J. Antibiotics* **34**, 1641, 1645
Nishikawa, J. and Tori, K. (1984). *J. Med. Chem.* **27**, 1657

Nishikawa, J., Tori, K., Takasuka, M., Onoue, H. and Narisada, M. (1982). *J. Antibiotics* **35**, 1724
Oakenfull, D. G. and Jencks, W. P. (1971). *J. Am. Chem. Soc.* **93**, 178
O'Callaghan, C. H. O., Kirby, S. M., Morris, A., Waller, E. R. and Duncombe, R. E. (1972). *J. Bacteriol* **110**, 988
O'Connor, C. J. (1970). *Quart. Rev.* **24**, 553
O'Connor, C. J. and Tan, A.-L. (1980). *Austral. J. Chem.* **33**, 747
Page, M. I. (1973). *Chem. Soc. Rev.* 295
Page, M. I. (1976). *Biochem. Biophys. Res. Commun.* **72**, 456
Page, M. I. (1977). *Angew. Chem. [Int. Edn]* **16**, 449
Page, M. I. (1980a). *Int. J. Biochem.* **11**, 331
Page, M. I. (1980b). *In* "Macromolecular Chemistry", p. 397. Royal Society of Chemistry, London
Page, M. I. (1981). *Chem. Ind.* 144
Page, M. I. (1984a). *Accounts Chem. Res.* **17**, 144
Page, M. I. (1984b). *In* "The Chemistry of Enzyme Action" (ed. M. I. Page) p. 1. Elsevier, Amsterdam
Page, M. I. and Jencks, W. P. (1972a). *J. Am. Chem. Soc.* **94**, 3263
Page, M. I. and Jencks, W. P. (1972b). *J. Am. Chem. Soc.* **94**, 8828
Page, M. I. and Proctor, P. (1984). *J. Am. Chem. Soc.* **106**, 3820
Page, M. I. and Crombie, D. A. (1984). *In* "Macromolecular Chemistry", Vol 3, p. 351. Royal Society of Chemistry, London
Parker, C. W., Shapiro, J., Kern, M. and Eisen, H. N. (1962). *J. Exp. Med.* **115**, 821
Parthasarathy, R. (1970). *Acta Crystallogr.* **B26**, 1283
Paschal, J. W., Dorman, D. E., Srinivasan, P. R. and Lichter, R. L. (1978). *J. Org. Chem.* **43**, 2013
Paukstelis, J. V. and Kim, M. (1974). *J. Org. Chem.* **39**, 1503
Pawelczyk, E., Zajac, M. and Knitter, B. (1981). *Pol. J. Pharmacol. Pharm.* **33**, 241
Petrongolo, C., Prescatori, E., Ranghino, G. and Scordamaglia, S. (1980). *Chem. Phys.* **45**, 291
Pfaendler, H. R., Gosteli, J., Woodward, R. B. and Rihs, G. (1981). *J. Am. Chem. Soc.* **103**, 4526
Pitt, G. J. (1952). *Acta Cryst.* **5**, 770.
Poortere, M. De., Marchand-Brynaert, J. and Ghosez, L. (1974). *Angew. Chem. [Int. Edn]* **13**, 267
Pracejus, H., Kehlen, M., Kehlen, H. and Matschiner, H. (1965). *Tetrahedron* **21**, 2257
Pratt, R. F., Surh, Y. S. and Shaskus, J. J. (1983). *J. Am. Chem. Soc.* **105**, 1006
Proctor, P. and Page, M. I. (1979). Unpublished observations
Proctor, P. and Page, M. I. (1984). *J. Am. Chem. Soc.* **106**, 3820
Proctor, P., Gensmantel, N. P. and Page, M. I. (1982). *J. Chem. Soc. 2* 1185
Radom, L. and Riggs, N. V. (1980). *Austral. J. Chem.* **33**, 249
Ree, B. and Martin, J. C. (1970). *J. Am. Chem. Soc.* **92**, 1660
Reinicke, B., Blumel, P., Labischinski, H. and Giesbrecht, P. (1985). *Arch. Microbiol.* **141**, 309
Richardson, F. S., Yeh, C.-Y., Troxell, T. C. and Boyd, D. B. (1977). *Tetrahedron* **33**, 711
Romsted, L. S. (1977). *In* "Micellisation, Solubilisation and Microemulsions" (ed. K. L. Mittal) Vol. 1, p. 509. Plenum Press, New York
Schanck, A., Coene, B., Van Meerssche, M. and Dereppe, J. M. (1979). *Org. Magn. Reson.* **12** 337

Schanck, A., Coene, B., Dereppe, J. M. and Van Meersch, M. (1983). *Bull. Soc. Chim. Belg.* **92**, 81
Schneider, C. H. and DeWeck, A. L. (1966). *Helv. Chim. Acta* **49**, 1695, 1707
Schneider, C. H. and DeWeck, A. L. (1968). *Biochim. Biophys. Acta* **169**, 27
Schowen, R. L., Jayaraman, H. and Kershner, L. (1966). *J. Am. Chem. Soc.* **88**, 3373
Schrinner, E., Limbert, M. Pénasse, L. and Lutz, A. (1980). *J. Antimicrob. Chemother.* **6**, (Suppl. 1) 25
Schwartz, M. A. Granatek, A. P. and Buckwalter, F. H. (1962). *J. Pharm. Sci.* **51**, 523
Schwartz, M. A. (1965). *J. Pharm. Sci.* **54**, 472
Schwartz, M. A. and Delduce, A. J. (1969). *J. Pharm. Sci.* **58**, 1137
Schwartz, M. A. and Wu, G.-M. (1966). *J. Pharm. Sci.* **55**, 550
Schwartz, M. A. (1968). *J. Pharm. Sci.* **57**, 1209
Schwartz, M. A. (1969). *J. Pharm. Sci.* **58**, 643
Schwartz, M. A. (1982). *Biorg. Chem.* **11**, 4
Schwartz, M. A. and Pflug, G. R. (1967). *J. Pharm. Sci.* **56**, 1459
Simon, G. L., Morin, R. B. and Dahl, L. F. (1972). *J. Am. Chem. Soc.* **94**, 8557
Smith, C. R. and Yates, K. (1972). *J. Am. Chem. Soc.* **94**, 8811
Spratt, B. G. (1975). *Proc. Natl. Acad. Sci. USA* **72**, 2999
Spratt, B. G. (1977). *Eur. J. Biochem.* **72**, 341
Spratt, B. G. (1980). *Phil. Trans. Roy. Soc.* **289**, 27
Stackhouse, J., Bagchler, R. D. and Mislow, K. (1971). *Tetrahedron Lett.* 3437
Stewart, W. E. and Siddall, III, T. H. (1970). *Chem. Rev.* **70** 517
Stirling, C. J. M. (1979). *Accounts Chem. Res.* **12**, 198
Strominger, J. L. (1967). *Antibiotics* **1**, 706
Sweet, R. M. and Dahl, L. F. (1970). *J. Am. Chem. Soc.* **92**, 5489
Sweet, R. M. (1973). *In* "Cephalosporins and Penicillins: Chemistry and Biology" (ed. E. H. Flynn,) p. 280. Academic Press, New York
Sykes, R. B., Parker, W. L. and Wells, J. S. (1985). *Trends in Antibiot. Res.*, 115
Taguchi, K. (1960). *Chem. Pharm. Bull. (Tokyo)* **8**, 205
Takasuka, M., Nishikawa, J. and Tori, K. (1982). *J. Antibiot.* **35**, 1729
Tipper, D. J. (1970). *Int. J. Syst. Bacteriol.* **26**, 361
Tipper, D. J. (1985). *Pharmac. Ther.* **27**, 1
Tipper, D. J. and Strominger, J. L. (1965). *Proc. Natl. Acad. Sci. USA* **54**, 1133
Tomasz, A. (1979). *Ann. Rev. Microbiol.* **33**, 113
Tomasz, A. (1983). *In* "Handbook of Experimental Pharmacology (eds A. L. Demain and N. A. Solomon) Vol. 67/1, pp. 15–95. Springer-Verlag, Berlin
Tomida, H. and Schwartz, M. A. (1983). *J. Pharm. Sci.* **72**, 331
Tori, K., Nishikawa, J. and Takeuchi, Y. (1981). *Tetrahedron Lett.* 2793
Tsuji, A., Itatani, Y. and Yamana, T. (1977). *J. Pharm. Sci.* **66**, 1004
Tsuji, A., Yamana, T., Miyamoto, E. and Kiya, E. (1975). *J. Pharm. Pharmacol.* **27**, 580
Tsuji, A., Nakashima, E., Deguchi, Y., Nishide, K., Shimizu, T., Horiuchi, S., Ishikawa, K. and Yamana, T. (1981). *J. Pharm. Sci.* **70**, 1120
Tsuji, A., Miyamoto, E., Matsuda, M., Nishimura, K. and Yamana, T. (1982). *J. Pharm. Sci.* **71**, 1313
Tsuji, A., Nakashima, E., Nishide, K., Deguchi, Y., Hamano, S. and Yamana, T. (1983). *Chem. Pharm. Bull.* **31**, 4057
Tutt, D. E. and Schwartz, M. A. (1971). *J. Am. Chem. Soc.* **93**, 767
Vijayan, K., Anderson, B. F. and Hodgkin, D. C. (1973). *J. Chem. Soc. Perkin Trans 1* 484

Vishveshwara, S. and Rao, V. S. R. (1983). *J. Mol. Struct.* **92**, 19
Wan, P., Modro, T. A. and Yates, K. (1980). *Can. J. Chem.* **58**, 2423
Waxman, D. J. and Strominger, J. L. (1980). *J. Biol. Chem.* **255**, 3964
Waxman, D. J. and Strominger, J. L. (1983). *Ann. Rev. Biochem.* **52**, 825
Wei, C.-C., Borgese, J. and Weigele, M. (1983). *Tetrahedron Lett.* 1875
Weiss, A., Fallab, S. and Garlenmeyer, H. (1957). *Helv. Chim. Acta* **50**, 576
Weiss, A. and Fallab, S. (1960). *Helv. Chim. Acta* **40**, 61
Wheland, G. W. (1955). *In* "Resonance in Organic Chemistry", pp. 367, 508. John Wiley and Sons, New York
Williams, A., (1975). *J. Chem. Soc. Perkin Trans 2*, 947
Williams, A. (1976). *J. Am. Chem. Soc.* **98**, 5645
Williamson, K. L. and Roberts, J. D. (1976). *J. Am. Chem. Soc.* **98**, 5082
Wittig, G. and Steinhoff, G. (1964). *Justus Liebigs Ann. Chem.* **676**, 21
Wojtkowski, P. W., Dolfini, J. E., Kocy, O. and Cimarusti, C. M. (1975). *J. Am. Chem. Soc.* **97**, 5628
Wolfe, S. and Lee, W. S. (1968). *J. Chem. Soc. Chem. Commun.* 242
Wolfe, S., Godfrey, J. C., Holdrege, C. T. and Perron, Y. G. (1963). *J. Am. Chem. Soc.* **85**, 643
Wolfe, S., Godfrey, J. C., Holdrege, C. T. and Perron, Y. G. (1968). *Can. J. Chem.* **46**, 2549
Wolfe, S., Ducep, J.-B., Tin, K.-C. and Lee, S.-L. (1974). *Can. J. Chem.* **52**, 3996
Woodward, R. B. (1949). *In* "The Chemistry of Penicillin" (eds H. T. Clarke, J. R. Johnson and R. Robinson) p. 443. Princeton University Press, Princeton, New Jersey
Woodward, R. B. (1980). *Phil. Trans. Roy. Soc. London. Ser. B* **289**, 239
Yamana, T. and Tsuji, A. (1976). *J. Pharm. Sci.* **65**, 1563
Yamana, T., Tsuji, A., Kanayama, K. and Nakano, O. (1974a). *J. Antibiot.* **27**, 1000
Yamana, T., Tsuji, A. and Mizukami, Y. (1974b). *Chem. Pharm. Bull.* **22**, 1186
Yamana, T., Tsuji, A., Miyamoto, E. and Kiya, E. (1975). *J. Pharm. Pharmacol.* **27**, 283.
Yamana, T., Tsuji, A., Kiya, E. and Miyamoto, E. (1977). *J. Pharm. Sci.* **66**, 861
Yates, K. and Stevens, J. B. (1965). *Can. J. Chem.* **43**, 529
Yatsimirski, A. K., Martinek, K. and Berezin, I. V. (1971). *Tetrahedron* **27**, 2855
Yatsuhara, M., Sato, F., Kimura, T., Muranishi, S. and Sezaki, H. (1977). *J. Pharm. Pharmac.* **29**, 638
Zugara, A. and Hidalgo, A. (1965). *Rev. Real Acad. Cienc. Exact. Fis. Nat. Madrid* **59**, 221

Free Radical Chain Processes in Aliphatic Systems involving an Electron Transfer Reaction

GLEN A. RUSSELL

Department of Chemistry, Iowa State University, Ames, Iowa 50011

1 Introduction

There are only a limited number of elementary reactions that have been incorporated into free radical chain processes. The major propagation reactions of free radicals not involving electron transfer are:

(1) addition to an unsaturated or coordinatively unsaturated system in an inter- or intra-molecular fashion;

ADVANCES IN PHYSICAL ORGANIC CHEMISTRY
VOLUME 23 ISBN 0 12 033523 9

(2) elimination of the α, β, γ or ε-type;
(3) S_H2 atom or group transfer processes.

Chain reactions not involving electron transfer involve appropriate combinations of (1)–(3) plus the required initiation and termination processes.

Metal cations capable of existing in several oxidation states can participate in cyclic processes involving electron transfer wherein the metal ion cycles between two oxidation states. Oxidation-reduction couples such as Fe(II)/Fe(III), Cu(0)/Cu(I), Cu(I)/Cu(II) are effective in a number of such processes. In benzenediazonium ion chemistry, the Sandmeyer (Waters, 1942; Nonhebel and Waters, 1957), Meerwein (Koelsch and Boekelheide, 1944; Dickerman and Weiss, 1957; Dickerman *et al.*, 1956, 1958; Kochi, 1957, 1967), and Cohen (Cohen *et al.*, 1968) processes have been formulated as shown in Schemes 1–3 in homogeneous acetone solutions where $CuCl_2$ is reduced to CuCl.

$$ArN_2^+ + Cu^n \longrightarrow Cu^{n+1} + ArN_2\cdot \longrightarrow Ar\cdot + N_2$$
$$Ar\cdot + A^- \longrightarrow ArA\cdot^-$$
$$ArA\cdot^- + Cu^{n+1} \longrightarrow ArA + Cu^n$$

Scheme 1

$$ArN_2^+ + CuCl \longrightarrow CuCl^+ + ArN_2\cdot \longrightarrow Ar\cdot + N_2$$
$$Ar\cdot + CH_2{=}CHPh \longrightarrow ArCH_2CH(Ph)\cdot$$
$$ArCH_2CH(Ph)\cdot + CuCl^+ \longrightarrow CuCl + ArCH_2CH(Ph)^+ \xrightarrow{Cl^-} ArCH_2CH(Ph)Cl$$

Scheme 2

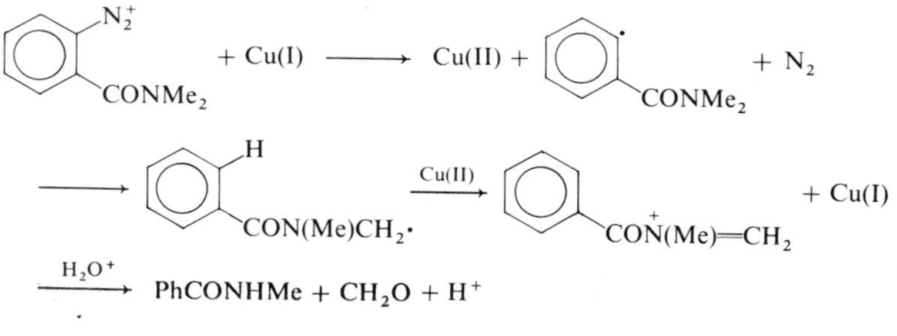

$$o\text{-}(N_2^+)C_6H_4CONMe_2 + Cu(I) \longrightarrow Cu(II) + o\text{-}(\cdot)C_6H_4CONMe_2 + N_2$$
$$\longrightarrow o\text{-}(H)C_6H_4CON(Me)CH_2\cdot \xrightarrow{Cu(II)} C_6H_5CO\overset{+}{N}(Me){=}CH_2 + Cu(I)$$
$$\xrightarrow{H_2O^+} PhCONHMe + CH_2O + H^+$$

Scheme 3

Peroxides, alkyl peresters or peracids can be forced to react in similar processes such as the Kharasch (Schemes 4, 5) (Kharasch *et al.*, 1953;

Kharasch and Sosnovsky, 1958) and Minisci (Schemes 6, 7) (Citterio *et al.*, 1977) substitution processes

$$PhC(O)OOCMe_3 + Cu(I) \longrightarrow Cu(II) + PhCO_2^- + Me_3CO\cdot$$
$$Me_3CO\cdot + RH \longrightarrow Me_3COH + R\cdot$$
$$R\cdot + Cu(II) \longrightarrow R^+ + Cu(I)$$
$$R^+ + SOH \longrightarrow ROS + H^+$$

Scheme 4

$$ROOH + Co(II) \longrightarrow RO\cdot + HO^- + Co(III)$$
$$RO\cdot + ROOH \longrightarrow ROH + ROO\cdot$$
$$RO\cdot(ROO\cdot) + \text{[cyclohexene]} \longrightarrow ROH(ROOH) + \text{[cyclohexenyl]}\cdot$$
$$\text{[cyclohexenyl]}\cdot + Co(III) \longrightarrow \text{[cyclohexenyl]}^+ + Co(II)$$
$$\text{[cyclohexenyl]}^+ + ROOH(ROH) \longrightarrow \text{[cyclohexenyl]}\text{—OOR(—OR)} + H^+$$

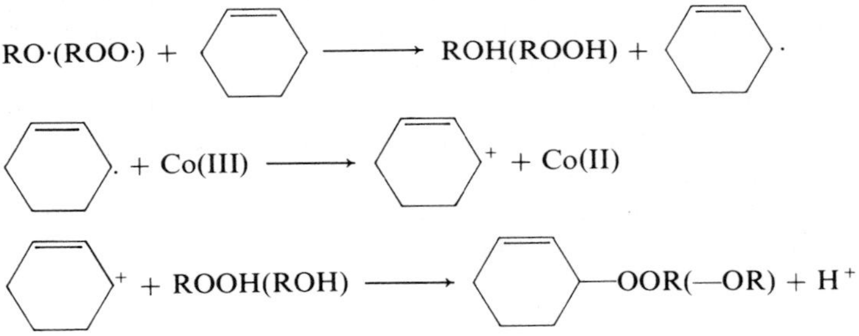

Scheme 5

$$Cu(I) + S_2O_8{}^{2-} \longrightarrow Cu(II) + SO_4{}^{2-} + SO_4^{\cdot-}$$
$$SO_4^{\cdot-} + p\text{-i-PrC}_6H_4CH_3 \longrightarrow SO_4{}^{2-} + p\text{-}i\text{-PrC}_6H_4CH_3^{\cdot+}$$
$$\longrightarrow H^+ + p\text{-i-PrC}_6H_4CH_2\cdot$$
$$p\text{-i-PrC}_6H_4CH_2\cdot + Cu(II) \longrightarrow p\text{-i-PrC}_6H_4CH_2{}^+ + Cu(I)$$
$$p\text{-i-PrC}_6H_4CH_2{}^+ + H_2O \longrightarrow p\text{-i-PrC}_6H_4CH_2OH + H^+$$

Scheme 6

$$RC(O)OOH + Fe(II) \longrightarrow Fe(III) + HO^- + RCO_2\cdot$$
$$RCO_2\cdot \longrightarrow R\cdot + CO_2$$
$$R\cdot + PyH^+ \longrightarrow \text{HN[dihydropyridinyl radical cation]}(H)(R) \rightleftharpoons \text{HN[dihydropyridinyl radical]}\text{—R} + H^+$$
$$\text{HN[dihydropyridinyl radical]}\text{—R} + Fe(III) \longrightarrow \text{HN[pyridinium]}^+\text{—R} + Fe(II)$$

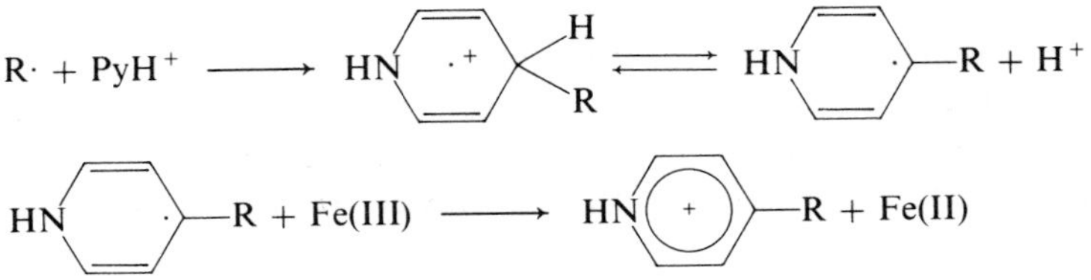

Scheme 7

The processes of Schemes 1–7 can be catalytic in the oxidation-reduction couple. The true mechanisms are often more complicated than the reaction schemes indicate, and the formation of various complexes involving the metal ion must be considered as well as the possible intervention of ligand transfer processes either within a cage or in noncage reactions. In Schemes 1–7 the "initiation" step is also one of the propagation steps and a clearly defined sequence of separate initiation, propagation and termination steps does not exist.

The processes to be considered in this review will be restricted to those where the propagation and initiation reactions are clearly different and where one member of any oxidation-reduction couple has a low persistency and can be considered to be a reactive intermediate. Among the processes to be considered which can be incorporated into chain reactions are:

(i) electron transfer between a radical and an anion
(ii) electron transfer from a radical anion to a neutral reducible substance
(iii) electron transfer from an easily oxidized neutral radical to a neutral reducible substance
(iv) electron transfer to a reducible radical or radical cation from a neutral reducible substance

2 Free radical chain processes involving nucleophiles

AUTOXIDATION OF CARBANIONS

The autoxidations of certain carbanions, such as fluorenide anion, have been demonstrated to proceed via a chain mechanism involving electron transfer (Russell, 1953; Russell *et al.*, 1965, 1968) as shown in Scheme 8. In Me_2SO

$$R\cdot + O_2 \longrightarrow ROO\cdot$$
$$ROO\cdot + R{:}^- \longrightarrow ROO^- + R\cdot$$
$$ROO^- + Me_2SO \longrightarrow RO^- + Me_2SO_2$$
$$Ar_2CHOOH + B^- \longrightarrow Ar_2C{=}O + BH + OH^-$$
$$ROOH + R^- \longrightarrow RO\cdot + R\cdot + HO^-$$

Scheme 8

the intermediate hydroperoxide is cleanly converted into the alkoxide which may be further oxidized via the dianion to the ketone. Carbanion autoxidation can be initiated by electron acceptors such as nitroaromatics as shown in (1) and (2). The rates of electron transfer from fluorenide anions to

$$R{:}^- + ArNO_2 \longrightarrow R\cdot + ArNO_2^{-\cdot} \quad (1)$$
$$ArNO_2^{-\cdot} + O_2 \longrightarrow ArNO_2 + O_2^{-\cdot} \quad (2)$$

nitroaromatics in the absence of oxygen (Russell *et al.*, 1964) parallels the catalysed autoxidation rates (Russell *et al.*, 1962a, 1968). With Ph_2CH^- and Ph_3C^-, the free radical chain is apparently rapidly initiated by electron transfer from the carbanion to molecular oxygen, and the rate of autoxidation of the parent hydrocarbon is equal to the rate of its ionization with Me_3COK in Me_2SO(80%)-Me_3COH(20%) solution (Russell and Bemis, 1965). With fluorene the rate of oxidation becomes equal to the rate of ionization only in the presence of large amounts of the excellent electron acceptor p-$CF_3C_6H_4NO_2$ (Russell and Weiner, 1969). The ease of autoxidation of many carbanions follows the sequence, $\pi C(CH_3)_2^- > \pi CHCH_3^- > \pi CH_2^-$ (π = RCO, RO_2C, O_2N, $ArSO_2$) and reflects the ease of electron transfer from the carbanion to O_2 or ROO· (Russell *et al.*, 1962b).

The autoxidation of Grignard reagents or dialkylmercurials proceeds by a free radical chain mechanism involving the attack of ROO· upon RMgX (Walling and Buckler, 1955; Lamb *et al.*, 1966) or R_2Hg (Razuvaev *et al.*, 1960; Aleksandrov *et al.*, 1964). These reactions may well involve electron transfer or at least a transition state for the formal S_H2 substitution reaction resembling [1]

$$[ROO^- \overset{+}{M}Y\ R\cdot]$$

[1]

The anion of 2-nitropropane in EtOH/EtOLi will undergo a free radical chain autoxidation (Russell, 1953). However, the anion under these conditions does not undergo an electron transfer reaction with molecular O_2. The reaction can be initiated by the presence of a nitroaromatic or by the unionized 2-nitropropane according to (1) and (2) (Russell *et al.*, 1965). The initiation step can be accelerated by photolysis whereby the electron acceptor is excited to the π^* state. The chain reaction of $Me_2C{=}NO_2^-$ with O_2 is autocatalytic suggesting the formation of O_2NCMe_2OOH (Scheme 8). However, the major reaction pathway appears to follow Scheme 9 (Russell,

$$O_2NCMe_2\cdot + O_2 \longrightarrow O_2NCMe_2OO\cdot$$

$$O_2NCMe_2OO\cdot + Me_2C{=}NO_2^- \longrightarrow \begin{cases} O_2NCMe_2OO^- + O_2NCMe_2\cdot \\ O_2NCMe_2OOCMe_2NO_2^{-\cdot} \end{cases}$$

$$O_2NCMe_2OOCMe_2NO_2^{-\cdot} \longrightarrow NO_2^- + O_2NCMe_2OOCMe_2\cdot \longrightarrow O_2NCMe_2O\cdot + Me_2CO$$

$$O_2NCMe_2O\cdot + Me_2C{=}NO_2^- \longrightarrow O_2NCMe_2\cdot + O_2NCMe_2O^- \longrightarrow NO_2^- + Me_2CO$$

Scheme 9

1967) and involves the addition of the peroxy radical to the nitronate anion with the ultimate production of $Me_2C{=}O$ and NO_2^- in nearly quantitative yield. As shown in Scheme 9, an intermediate radical anion of an aliphatic nitro compound is formed which then decomposes to yield NO_2^-, $Me_2C{=}O$ and an alkoxy radical which can continue the chain reaction. Scheme 9 embodies the basic steps involved in many free radical chain processes. The free radical chain propagates not only by an electron transfer step but also by the addition of a radical to an anion to yield a radical anion which subsequently fragments. These steps are of importance in the so-called (Kim and Bunnett, 1970) $S_{RN}1$ sequence of Scheme 10.

$$R\cdot + N^- \longrightarrow RN^{\overline{\cdot}} \tag{3}$$

$$RN^{\overline{\cdot}} + RX \longrightarrow RX^{\overline{\cdot}} + RN \tag{4}$$

$$RX^{\overline{\cdot}} \longrightarrow R\cdot + X^- \tag{5}$$

$$RX + N^- \longrightarrow RN + X^- \tag{6}$$

Scheme 10

SUBSTITUTION BY A MECHANISM INVOLVING RADICALS AND RADICAL ANIONS

The $S_{RN}1$ process of Scheme 10[1] was originally recognized in the reactions of $Me_2C{=}NO_2^-$ with O_2NCMe_2Cl or *p*-nitrobenzyl halides (Kornblum *et al.*, 1966; Russell and Danen, 1966, 1968). In the electron transfer step of Scheme 10, the electron is transferred from an easily oxidized radical anion ($RN^{\overline{\cdot}}$) to an easily reduced substrate (RX). The importance of the nitro group in these substitution reactions must be connected with the low energy of its LUMO which leads to stability of $RN^{\overline{\cdot}}$ whenever R or N contains a nitro group. In Table 1 are collected a number of nitro-containing substrates (RX) which will undergo the $S_{RN}1$ reaction with the anions shown.

The nitro substituent in R can be at the carbon bonded to the leaving group X as in O_2NCR_2X or in a vinylogous position such as *p*-$O_2NC_6H_4CR_2X$. Substrates such as *p*-$O_2NC_6H_4C(O)CR_2X$ are also reactive in the $S_{RN}1$ process, perhaps because of intramolecular electron transfer between the nitro group and the carbonyl group giving rise to the ketyl radical anion which readily fragments. Radicals such as $Me_3C\cdot$ or

[1] The acronym $S_{RN}1$ connotates a free radical chain Substitution reaction in which the new bond is formed by attack of a Radical (R·) upon a Nucleophile (N^-) and in which the bond to the nucleofuge (X) is broken in a unimolecular process. When allylic rearrangement is involved, the process is termed $S_{RN}1'$.

TABLE 1

Examples of aliphatic $S_{RN}1$ substitutions with nitro-containing substrates (RX)

$$(RX + N^- \xrightarrow{h\nu} RN + X^-)$$

R	X	N^-	Reference
Me_2CNO_2	Cl, Br, I, $ArSO_2$, PhS, p-ClC_6H_4S, p-$O_2NC_6H_4S$, o-$O_2NC_6H_4S$, PhSO, p-ClC_6H_4SO, SCN, NO_2	$Me_2C{=}NO_2{}^-$, c-$C_6H_{10}{=}NO_2^-$, $(EtO_2C)_2CR^-$, $(EtO)_2PO^-$, $(EtO)_2PS^-$, N_3^-, $ArSO_2^-$, ArS^- (Ar = 2-pyridyl, 4-pyridyl, pyrimidin-2-yl, 4-5-dihydro-1,3-thiazol-2-yl, 1-methylimidazol-2-yl, 1-3-benzothiazol-2-yl, p-ClC_6H_4 p-$O_2NC_6H_4$, o-$O_2NC_6H_4$	Russell and Danen, 1966, 1968; Russell *et al.*, 1971, 1982d; Russell and Hershberger, 1980a; Kornblum and Boyd, 1970; Kornblum *et al.*, 1971, 1973, 1974; Kornblum, 1975; Zeilstra and Engberts, 1973; Bowman and Richardson, 1980, 1981; Bowman and Symons, 1983; Bowman *et al.*, 1984; Al-Khalil *et al.*, 1986
$R^1R^2CNO_2$ R^1, R^2 = Me; Me, Et; $-(CH_2)_5-$	Cl, Br	$RC(CN)CO_2Et^-$ (R = i-Pr, $PhCH_2$), $(EtO_2C)_2CR^-$ (R = Et, n-Bu, $PhCH_2$), $EtO_2CC(R)COMe^-$ (R = Et, n-Bu, $PhCH_2$), $(MeCO)_2CR^-$ (R = Me, n-Bu) [structures: cyclopentenolate with COMe; cyclopentenolate with CO_2Et; dihydrofuranolate with CO_2Et]	Ono *et al.*, 1983

TABLE 1 (*continued*)

R	X	N^-	Reference
Me_2CNO_2	SAr (Ar = 1-methylimidazol-2-yl, pyrimidin-2-yl, 2-pyridyl, 1,3-benzothiazol-2-yl, 4,5-dihydro-1,3-thiazol-2-yl)	RS^- (R = *p*-chlorophenyl, benzyl, L-cysteine), $Me_2C{=}NO_2^-$	Bowman *et al.*, 1984
$MeC(R)NO_2$ R = CH_2OTHP[a] $CH_2CH_2CO_2Me$	Cl	$(EtO_2C)_2CH^-$, $(EtO_2C)_2CEt^-$, $EtO_2CC(CH_3)COCH_3^-$, $EtO_2CCHCOCH_3^-$, $Me_2C{=}NO_2^-$, $MeC(CH_2CH_2CO_2Me){=}NO_2^-$	Beugelmans *et al.*, 1983
$MeO_2CCH_2CH_2C(CH_2CH_2OTHP)NO_2$	Cl	$(EtO_2C)_2CH^-$, $EtO_2\dot{C}CHCOMe$	Beugelmans *et al.*, 1983
$NCCH_2CH_2C(CH(CH_3)OTHP)NO_2$	Cl	$(EtO_2C)_2CH^-$	Beugelmans *et al.*, 1983
c-$C_6H_{10}NO_2$	NO_2	$CH_3C(CH_2CH_2CN){=}NO_2^-$, $CH_3CH_2C(CH_2OCH_3){=}NO_2^-$, $CH_3C(CH_3CH_2COCH_3){=}NO_2^-$, c-$C_6H_{10}{=}NO_2$	Kornblum and Cheng, 1977
Me_2CNO_2	Cl, NO_2	$R'COCHCO_2Et^-$, $R'COCHCOR'^-$, 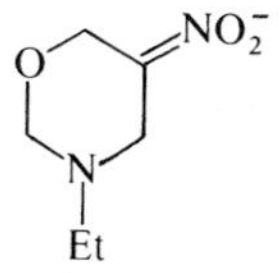	Crozet and Vanelle, 1985; Russell *et al.*, 1981b

[a] THP = tetrahydropyranyl

$(EtO_2C)_2CEt$	NO_2	$Me_2C{=}NO_2^-$	Russell *et al.*, 1971
$Me_2C(Z)$ $Z = COR', CO_2R', CN,$ $p\text{-}O_2NC_6H_4N{=}N$	NO_2	$Me_2C{=}NO_2^-$	Russell *et al.*, 1971; Kornblum and Boyd, 1970
R^1R^2CCN $R^1, R^2 = Me; Me, Et;$ $Me, Me(CH_2)_4; \text{-}(CH_2)_4\text{-},$ $\text{-}(CH_2)_{11}\text{-}$	NO_2	$R^1R^2C{=}NO_2^-$ $R^1, R^2 = H; Me; Me, H; Et, H;$ $Me(CH_2)_5, H; \text{-}(CH_2)_4\text{-},$ $\text{-}(CH_2)_5; \text{-}(CH_2)_{11}\text{-},$ $PhS^-, p\text{-}ClC_6H_4S^-$	Kornblum *et al.*, 1984; Russell *et al.*, 1971; Bowman *et al.*, 1984
Me_2CN_3	NO_2	$PhS^-, p\text{-}ClC_6H_4S^-$ $p\text{-}ClC_6H_4SO_2^-, N_3^-$	Bowman *et al.*, 1984; Al-Khalil *et al.*, 1986
$R^1COCHR^2, R^1O_2CCHR^2,$ $PhCHR^2$ $R^1 = Ph, Me, Et, i\text{-}Pr$ $R^2 = H, Me, Et, n\text{-}Pr$	NO_2	PhS^-	Mayake and Yamamura, 1986
$p\text{-}R^1C_6H_4C(R^2)Me$ $R^1 = H, CN, PhSO_2,$ $PhCO, 3,5\text{-}(CF_3)_2$ $R^2 = Me, Et$	NO_2	$Me_2C{=}NO_2^-, CH_2{=}NO_2^-,$ $Me_3CH{=}NO_2^-, MeC(Et){=}NO_2^-$	Kornblum *et al.*, 1978b Kornblum, 1982; Kornblum and Erickson, 1981
Me_3C	NO_2	$CH_2{=}NO_2^-$	Kornblum and Erickson, 1981
R_fCMe_2 $R_f = n\text{-}C_6F_{13}, n\text{-}C_8F_{17}$	NO_2	$Me_2C{=}NO_2^-$	Feiring, 1983
$p\text{-}YC_6H_4CMe_2$ $Y = PhSO_2, CN$	NO_2	$CH_3S^-, Me_2C{=}NO_2^-, MeCH{=}NO_2^-$	Kornblum, 1982; Kornblum *et al.*, 1979a,b
$p\text{-}YC_6H_4C(R)MeCMe_2$ $Y = PhSO_2, CN, PhCO,$ $R = Me, Et$	NO_2	CH_3S^-	Kornblum *et al.*, 1979a,b
$3,5\text{-}(CF_3)_2C_6H_3CMe(R^1)$ $CMe(R^2)$ $R^1 = Et, R^2 = Me$ $R^1 = Me, R^2 = Et$	NO_2	CH_3S^-	Kornblum *et al.*, 1979b
$PhCOCMe_2$	NO_2	$Me_2C{=}NO_2^-$	Kornblum *et al.*, 1970a

TABLE 1 (*continued*)

R	X	N^-	Reference
$MeCOCH\text{—}C(\text{—}CH_2\text{—})(Me)\text{—}NO_2^-$ (cyclopropane ring: CH_2 bridging CH and C) $\longrightarrow$	$MeCOCHCH_2C(Me){=}NO_2^-$	$Me_2C{=}NO_2^-$	Russell and Dedolph, 1985
p-$O_2NC_6H_4CH_2$	Cl, $C_6Cl_5CO_2$, Me_2S^+, Me_3N^+, c-$C_5H_4N^+$, CH_3SO_2, PhSO	$Me_2C{=}NO_2^-$, PhS^-, $(EtO_2C)_2CMe^-$, $(EtO)_2PS^-$	Kornblum, 1975; Kornblum *et al.*, 1966; Russell and Danen, 1966, 1968; Russell and Pecoraro, 1979; Russell *et al.*, 1982c
p-$O_2NC_6H_4CH_2$	Cl, Me_2S^+	p-$O_2NC_6H_4CHX^-$	Russell and Pecoraro, 1979
p-$O_2NC_6H_4CHZ$ Z = Cl, Br[b]	Cl, Br	$Me_2C{=}NO_2^-$	Freeman and Norris, 1976; Freeman *et al.*, 1978
p-$O_2NC_6H_4CMe_2$	Cl, NO_2, PhSO, $PhSO_2$	$Me_2C{=}NO_2^-$, NO_2^-, PhS^-, $(EtO_2C)_2CR^-$, ArO^-, CN^- N_3^-, $PhSO_2^-$, amines	Kornblum *et al.*, 1967a,b, 1968, 1970b, 1980; Kornblum and Stuchal, 1970; Kornblum, 1975; Russell amd Pecoraro, 1979
p-$O_2NC_6H_4CMe_2$	CMe_2NO_2	MeS^-	Kornblum *et al.*, 1979a
p-$O_2NC_6H_4C(R^1)R^2$ R^1, R^2 = H; Me; H, Me; H, Et; H, i-Pr; H, t-Bu; Me, Et; Me, i-Pr; Me, CH_2CMe_3	Cl, NO_2	$R^1R^2C{=}NO_2^-$ (R^1, R^2 = Me, i-Pr; H, Me; H, i-Pr; Me, Et; Me, i-Pr; Me, t-Bu; Me, CH_2CMe_3	Norris and Randles, 1976, 1982a,b
p-$O_2NC_6H_4CMe_2$	NO_2	9-Ph-fluorenyl$^-$, 9-PhS-fluorenyl$^-$	Bordwell and Clemens, 1982
p-$O_2NC_6H_4CH_2$	Cl	(ring structure: O, N–Et, $=NO_2^-$); c—$C_6H_{10}{=}NO_2^-$	Crozet and Vanelle, 1985

[b] Product of $E_{RC}1$ reaction

o- or *p*-$O_2NC_6H_4CH_2$	Cl	$CH_3C(CH_2OTHP){=}NO_2^-$, $CH_3C(CH_2CH_2CO_2Et){=}NO_2^-$	Beugelmans *et al.*, 1983
p-$O_2NC_6H_4C(CH_3)CH_2Z$ Z = CH_2CO_2Et, OTHP	NO_2	$Me_2C{=}NO_2^-$, $(EtO_2C)_2CH^-$, $MeCOCHCO_2Et^-$, $MeC(CH_2CH_2CO_2Et){=}NO_2^-$	Beugelmans *et al.*, 1984
[5-nitro-2-thienyl]CH(t-Bu)	Cl	$ArSO_2^-$	Norris, 1983
[5-nitro-2-thienyl]CMe_2	Cl	PhS^-	Newcombe and Norris, 1978
[1-methyl-5-nitro-2-imidazolyl]CH_2	Cl	$Me_2C{=}NO_2^-$, c-$C_5H_8{=}NO_2^-$, c-$C_6H_{10}{=}NO_2^-$, c-$C_7H_{12}{=}NO_2^-$, c-$C_{10}H_{18}{=}NO_2^-$, 2-adamantyl$=NO_2^-$, 2-norbornyl$=NO_2^-$, $PhC(Me){=}NO_2^-$, $CH_3C(CH_2OTHP){=}NO_2^-$	Crozet *et al.*, 1985
p-$O_2NC_6H_4CH{=}CHCH$(t-Bu)[c]	Cl	$Me_2C{=}NO_2^-$, $MeC(CO_2Et)_2^-$, *p*-$MeC_6H_4SO_2^-$	Barker and Norris, 1979

[c] Product of $S_{RN}1'$ reaction

$PhCOCMe_2\cdot$ can be formed from the fragmentation of $Me_3CNO_2^{\overline{\cdot}}$ or $PhCOCMe_2NO_2^{\overline{\cdot}}$ and trapped by nitronate anions ($R^1R^2C{=}NO_2^-$) in the $S_{RN}1$ sequence (Kornblum *et al.*, 1970a; Kornblum and Erickson, 1981).

The recognized fragmentation steps of nitro radical anions are summarized in (7)–(11) (Russell and Dedolph, 1985). In the fragmentation processes

$$^{\overline{\cdot}}O_2NCMe_2Cl(Br,I,ArSO_2) \longrightarrow O_2NCMe_2\cdot + Cl^-(Br^-,I^-,ArSO_2^-) \quad (7)$$

$$^{\overline{\cdot}}O_2NCMe_2CN(COR,CO_2Et) \longrightarrow \cdot CMe_2CN(COR,CO_2Et) + NO_2^- \quad (8)$$

$$p\text{-}O_2NC_6H_4CMe_2NO_2^{\overline{\cdot}} \longrightarrow p\text{-}O_2NC_6H_4CMe_2\cdot + NO_2^- \quad (9)$$

$$p\text{-}O_2NC_6H_4CMe_2CMe_2NO_2^{\overline{\cdot}} \longrightarrow p\text{-}O_2NC_6H_4CMe_2\cdot + Me_2C{-}NO_2^- \quad (10)$$

$$(Me)(^{\overline{\cdot}}O_2N)\overset{\triangle}{C{-}CHCOMe} \longrightarrow {}^-O_2N{=}C(Me)CH_2\dot{C}HCOMe \quad (11)$$

of (12) and (13) there is no obvious correlation between the spin delocalization in $RX^{\overline{\cdot}}$ and the structure of the fragmentation products.

$$RX^{\overline{\cdot}} \longrightarrow R\cdot + X^- \quad (12)$$

$$RX^{\overline{\cdot}} \longrightarrow R:^- + X\cdot \quad (13)$$

In the generalized structure $XCR_2NO_2^{\overline{\cdot}}$, fragmentation is apparently controlled by thermodynamic considerations and can lead to any of the four fragmentation routes of (14)–(17). These fragmentation pathways are consis-

$$XCR_2NO_2^{\overline{\cdot}} \longrightarrow NO_2^- + R_2C(X)\cdot \quad (14)$$

$$XCR_2NO_2^{\overline{\cdot}} \longrightarrow NO_2 + R_2C(X):^- \quad (15)$$

$$XCR_2NO_2^{\overline{\cdot}} \longrightarrow X:^- + R_2C(NO_2)\cdot \quad (16)$$

$$XCR_2NO_2^{\overline{\cdot}} \longrightarrow X\cdot + R_2C{=}NO_2^- \quad (17)$$

tent with the observations that in the reaction of $R\cdot$ with N^- the functional group required to stabilize $RN^{\overline{\cdot}}$ can be present in either $R\cdot$ or N^-. Thus, an alkyl radical is readily trapped by $R_2C{=}NO_2^-$ or NO_2^- ions while α-nitroalkyl or *p*-nitrobenzyl radicals can be trapped by a variety of anions including $(RO_2C)_2CH^-$ and $(EtO)_2PO^-$. Scheme 11 illustrates an example where the same radical anion can be formed from two different sets of precursors in the $S_{RN}1$ scheme (Russell *et al.*, 1971).

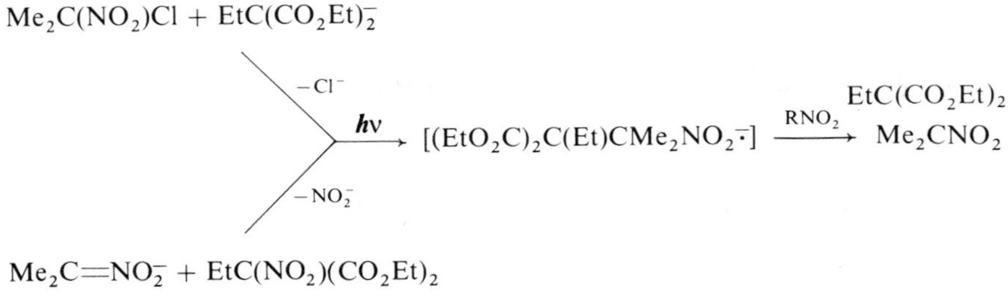

$Me_2C(NO_2)Cl + EtC(CO_2Et)_2^-$

$$\xrightarrow[-NO_2^-]{-Cl^-,\ h\nu} [(EtO_2C)_2C(Et)CMe_2NO_2^{\cdot -}] \xrightarrow{RNO_2} \begin{array}{c} EtC(CO_2Et)_2 \\ Me_2CNO_2 \end{array}$$

$Me_2C{=}NO_2^- + EtC(NO_2)(CO_2Et)_2$

Scheme 11

In the absence of electron transfer from $RN^{\cdot -}$ to RX in Scheme 10, the $S_{RN}1$ process could in principle lead to a polymerization reaction by repetition of the fragmentation and addition steps when $N^- = Me_2C{=}NO_2^-$. In actual fact, this process (18) is prevented by a low

$$RCMe_2NO_2^{\cdot -} + n\text{-}Me_2C{=}NO_2^- \longrightarrow RCMe_2\text{-}(CMe_2)_nNO_2^{\cdot -} + n\text{-}NO_2^- \quad (18)$$

ceiling temperature since the radical anion $RCMe_2CMe_2NO_2^{\cdot -}$ can also fragment as in (17) to $RCMe_2\cdot$ and $Me_2C{=}NO_2^-$ (Kornblum *et al.*, 1979a,b). However, (19) has been observed to occur in competition with the ordinary S_N1 substitution process in the reaction of R_f1 with $Me_2C{=}NO_2^-$ (Feiring, 1983). Reactive olefins can also participate in the $S_{RN}1$ process by competing with N^- for $R\cdot$ as exemplified by (20) (Feiring, 1983, 1985).

$$R_fI + \text{excess}\ \ Me_2C{=}NO_2^- \xrightarrow{h\nu} R_fCMe_2CMe_2NO_2 + I^- + NO_2^- \quad (19)$$

$$R_fI + CH_2{=}CHPh + Me_2C{=}NO_2^- \xrightarrow{h\nu} R_fCH_2CH(Ph)CMe_2NO_2 + I^- \quad (20)$$

The reaction of *p*-nitrobenzylidene dihalides with $Me_2C{=}NO_2^-$ leads to the formation of the $S_{RN}1$ substitution product and the styrene in competitive reactions (21) and (22). The formation of the styrene is favoured by

$$p\text{-}O_2NC_6H_4CHX_2 + Me_2C{=}NO_2^- \xrightarrow{S_{RN}1} p\text{-}O_2NC_6H_4CH(X)CMe_2NO_2 + X^- \quad (21)$$

$$p\text{-}O_2NC_6H_4CHX_2 + Me_2C{=}NO_2^- \xrightarrow{E_{RC}1} p\text{-}O_2NC_6H_4CH{=}CMe_2 + X^- + Me_2C(NO_2)C(NO_2)Me_2 \quad (22)$$

X = Br while the normal process is observed with X = Cl. The styrene formation has been explained by the mechanism of Scheme 12 and termed an $E_{RC}1$ process (Girdler and Norris, 1975; Freeman and Norris, 1976; Freeman *et al.*, 1978; Norris, 1983).[2]

$$\mathrm{Ar} = p\text{-}\mathrm{O_2NC_6H_5}$$
$$\mathrm{ArCHBr_2}^{\overline{\cdot}} \longrightarrow \mathrm{ArCHBr}\cdot + \mathrm{Br}^-$$
$$\mathrm{ArCHBr}\cdot + \mathrm{Me_2C{=}NO_2^-} \longrightarrow \mathrm{ArCH(Br)CMe_2NO_2}^{\overline{\cdot}} \longrightarrow \mathrm{Ar\dot{C}HCMe_2NO_2} + \mathrm{Br}^-$$
$$\mathrm{ArCHCMe_2NO_2} + \mathrm{Me_2C{=}NO_2^-} \longrightarrow \mathrm{ArCH{=}CMe_2} + \mathrm{NO_2^-} + \mathrm{Me_2C(NO_2)}\cdot$$
$$\mathrm{Me_2C(NO_2)}\cdot + \mathrm{Me_2C{=}NO_2^-} \longrightarrow \mathrm{Me_2C(NO_2)C(NO_2)Me_2}^{\overline{\cdot}}$$
$$\mathrm{Me_2C(NO_2)C(NO_2)Me_2}^{\overline{\cdot}} + \mathrm{ArCHBr_2} \longrightarrow \mathrm{Me_2C(NO_2)C(NO_2)Me_2} + \mathrm{ArCHBr_2}^{\overline{\cdot}}$$

Scheme 12

The $S_{RN}1'$ reaction (23) has also been observed with allylic halides containing the *p*-nitrophenyl substituent with $N^- = \mathrm{Me_2C{=}NO_2^-}$, $\mathrm{MeC(CO_2Et)_2^-}$ and $p\text{-}\mathrm{MeC_6H_4SO_2^-}$ (Barker and Norris, 1979).

$$p\text{-}\mathrm{O_2NC_6H_4CH{=}CHCH(Cl)Bu\text{-}t} + \mathrm{N}^- \xrightarrow{h\nu} p\text{-}\mathrm{O_2NC_6H_4CH(N)CH{=}CHBu\text{-}t} \quad (23)$$

Aliphatic substrates other than nitroalkanes which will participate in the S_{RN} chain reaction with anions are summarized in Table 2. All of the substrates (RX) are characterized by low reduction potentials with electron transfer leading to a radical anion of low persistency or occurring in a dissociative manner.

Electron transfer between a nucleophile (N^-) and the substrate (RX) can occur with cage recombination of N· and R· to yield the substitution product. Among the systems where this SET process has been postulated to occur are reactions involving easily oxidized nucleophiles such as RS^- with substrates such as 9-bromofluorene, 2,2-dimethyl-1-iodo-5-hexene, Ph_3CBr (Ashby *et al.*, 1985a) or 2-iodobutane (Bank and Noyd, 1973). Substitution reactions of 1,1-dibromocyclopropanes with PhS^- involves escape from the cage and the $S_{RN}1$ chain reaction occurs in Me_2SO or liquid NH_3 (Meijs, 1984, 1985, 1986). The reaction of nucleophiles with easily reduced substrates incapable of sustaining the $S_{RN}1$ chain is also thought to involve caged radical recombination in cases such as the reaction of $RC{\equiv}C^-$ with O_2NCMe_2X (X = Cl, NO_2) (Russell *et al.*, 1979b; Jawdosiuk *et al.*, 1979) or *m*-$O_2NC_6H_4CMe_2Cl$ with anions such as $\mathrm{Me_2C{=}NO_2^-}$ (Kornblum, 1975).

[2] The acronym $E_{RC}1$ has been defined as an Elimination process occurring by a Radical Chain mechanism involving a unimolecular decomposition of the radical anion derived from the substrate followed by electron transfer with a reductant to yield a carbanionic species capable of undergoing a β-elimination reaction.

TABLE 2

Aliphatic substitutions reported to occur by the S_{RN}-type of reaction where RX does not contain a nitro group

$(RX + N^- \xrightarrow{h\nu} RN + X^-)$

R	X	N^-	Reference
1°, 2°, 3°-Alkyl	HgX (X = Cl, Br, I, O_2CR)	$R^1R^2C{=}NO_2^-$ (R^1, R^2 = H; Me; H, Me; H, Ph; Ph, Me), $R^1R^2C{=}C(O^-)Ph$ (R^1, R^2 = Me; Ph; H, Me; H, Ph), Ph_2CR^- (R = H, CN, Ph), Ph_2P^-, $PhC(CO_2Et)_2^-$, N_3^- NO_2^-, phthalimidyl$^-$	Russell *et al.*, 1979b, 1982b; Russell and Khanna, 1985a,b
p-$NCC_6H_4CR_2$ R = H, Me	$Me_3\overset{+}{N}$, Cl	$Me_2C{=}NO_2^-$, $MeCH{=}NO_2^-$	Kornblum and Fifolt, 1980
p-$YC_6H_4COMe_2$ Y = H, CN	Cl, Br	$Me_2C{=}NO_2^-$	Russell and Ros, 1985
2,4-$(NC)_2C_6H_3CMe_2$	$PhSO_2$	$Me_2C{=}NO_2^-$ + $EtC(NO_2)(CO_2Et)_2$	Kornblum and Fifolt, 1980
$YC_6H_4CX_2$ X = Br, Cl Y = p-CN, m or p-CO_2Et	Cl, Br	$Me_2C{=}NO_2^-$	Norris, 1983
CF_2Cl	Cl	PhS^-	Rico *et al.*, 1983
CCl_3, CBr_3	Cl, Br	$RCOCH_2^-$, RSO_2CHR^-	Meyers *et al.*, 1977; Meyers, 1978
n-C_6F_{13}, n-C_8F_{17}	I	$Me_2C{=}NO_2^-$	Feiring, 1983
CF_3, C_2F_5, i-C_3F_7, $(CF_2)_6$	I	ArS^-, $HO_2CCH_2S^-$, 2-benzothiazolyl-S^-	Boiko *et al.*, 1977; Popov *et al.*, 1977
$PhCH{=}CH$	Br	$CH_3C(O^-){=}CH_2$	Bunnett *et al.*, 1976
3,3-Dimethyl-1-indenyl	I	$CH_3C(O^-){=}CH_2$, PhS^-	Bunnett *et al.*, 1976
1-Cyclopentenyl	I	$CH_3C(O^-){=}CH_2$	Bunnett *et al.*, 1976
2-Norbornenyl	I	$CH_3C(O^-){=}CH_2$, PhS^-	Bunnett *et al.*, 1976
Me_3CCH_2	Br	PhS^-, $PhSe^-$, Ph_2P^-. Ph_2As^-	Pierini *et al.*, 1985

TABLE 2 (*continued*)

R	X	N^-	Reference
1-Adamantyl	Br, I	Ph_2P^-, Ph_2As^-, Te^{-2}, Se^{-2}, $AdTe^-$, $AdSe^-$	Rossi *et al.*, 1982; Palacios *et al.*, 1984, 1985
7-Norcaranyl	Br	Ph_2P^-, Ph_2As^-	Rossi *et al.*, 1984
9-Triptycenyl	Br	Ph_2P^-, Ph_2As^-	Palacios *et al.*, 1984
9,10-Triptycendiyl	I	$PhC(O^-){=}CH_2$	Ashby and Argyropoulos, 1985
Vinyl	Halide	$Co(CO)_4^-$	Brunet *et al.*, 1983
2°, 3°-Alkyl	I	$[\pi\text{-AllylNiBr}]_2$	Hegedus and Miller, 1975[a]
1°-Alkyl	Br	Me_3Sn^-	Ashby *et al.*, 1985b
9-Phenylfluorenyl	Br	PhS^-	Singh and Jayaraman, 1974
1-Bromo-1-cyclopropyl	Br	PhS^-	Meijs, 1984, 1985, 1986
$PhCOCH_2$	HgCl	$R^1R^2{=}NO_2^-$ (R^1, R^2 = H; Me, H, Me), $R^1R^2C{=}C(O^-)Ph$ (R^1, R^2 = Me; Me, H; Ph, H), $Me_3C(O^-){=}CH_2$, $RC(CO_2Et)_2^-$ (R = H, Me, Ph)	Russell and Khanna, 1986

[a] See, however, Hedges and Thompson, 1985

Table 2 contains examples of the $S_{RN}1$ process where neither the substrate (RX) nor the nucleophile N^- contains a nitro group. However, groups must be present which can stabilize the intermediate $RN^{\cdot-}$ if a chain process is to occur. Among the substituents other than nitro which can be present in either R or N, and which will stabilize $RN^{\cdot-}$ by virtue of a low energy π^* MO, are cyanophenyl, benzoyl, thiophenyl, and in some cases simply the phenyl ring itself. Thus, t-Bu· generated by electron transfer to t-BuHgCl will add to carbanions such as $PhC(O^-){=}CR^1R^2$, Ph_2CH^-, Ph_2P^-, fluorenide$^-$, Ph_3C^-, Ph_2CCN^-, $PhC(CO_2Et)_2^-$ or phthalimide but not to anions such as $t\text{-}BuCOCH_2^-$, $HC(CO_2Et)_2^-$ or $(EtO)_2PO^-$ (Russell and Khanna, 1985a,b). On the other hand, electron transfer to $PhCOCH_2HgCl$ forms $PhCOCH_2\cdot$ which now contains π-unsaturation and which will add readily to anions such as $t\text{-}BuCOCH_2^-$ or $RC(CO_2Et)_2^-$ (Russell and Khanna, 1986).

The addition of a radical to a nucleophile to form a radical anion is a key step in the $S_{RN}1$ sequence. The exergonicity of the process must be important in controlling the overall rate of the reaction. However, there is by no means a strict correlation between rate and exergonicity. With electron donor radicals such as t-Bu·, the rates of addition to nucleophiles can actually decrease as the reaction becomes more exergonic, i.e. as the anion becomes more basic when other thermodynamic factors are held constant (Russell and Khanna, 1985a). This suggests that the transition state for the addition

$$R\cdot + \pi^- \longrightarrow \underset{[2]}{R^+\ \pi\cdot^{-2}} \longrightarrow R\pi^{\cdot-} \tag{24}$$

of t-Bu· to an unsaturated anion (π^-) involves the SOMO-LUMO interaction as exemplified in [2]. Towards t-Bu· the reactivity decreases with an increase in exothermicity in the following sequence:

$$PhC(O^-){=}CPh_2 > PhC(O^-){=}CHPh > PhC(O^-){=}C(Me)Ph > PhC(O^-){=}CHMe > PhC(O^-){=}CMe_2$$

With nitronate anions, the reactivity decreases from $H_2C{=}NO_2^-$ to $MeCH{=}CNO_2^-$ to $Me_2C{=}NO_2^-$ while towards phenylacetonitrile anions the reactivity is higher for Ph_2CCN^- than for $PhCHCN^-$. For exergonic reactions, the reactivity towards t-Bu· decreases as the energy of the LUMO of the anion increases by changing the substituent at the carbanionic centre from Ph to H to Me even though the exothermicity of (24) increases.

Towards the electrophilic $PhCOCH_2\cdot$ an opposite trend in relative reactivities is observed, with reactivities of nucleophiles increasing with the basicity of the anion and the exergonicity of (24). Thus, the reactivity sequences observed in addition of $PhCOCH_2\cdot$ are $MeC{=}NO_2^-$ >

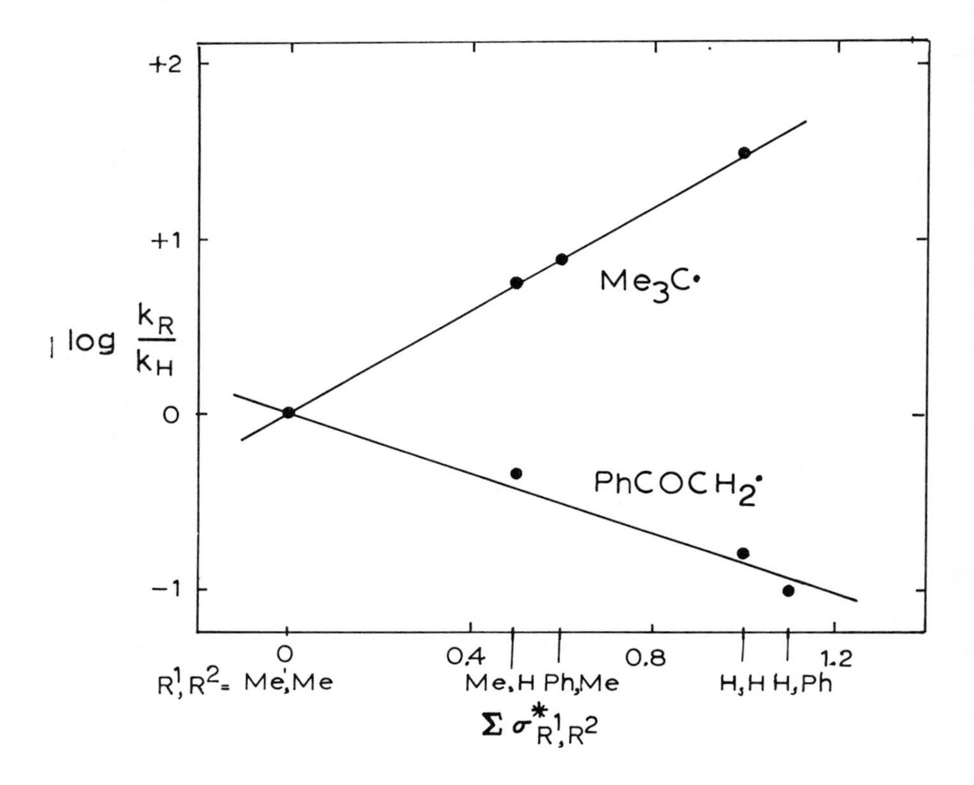

FIG. 1 Relative reactivities of $R^1R^2C{=}NO_2^-$ towards $Me_3C\cdot$ and $PhCOCH_2\cdot$ at 35°C in Me_2SO. (Data are from Russell and Khanna, 1985a, 1986)

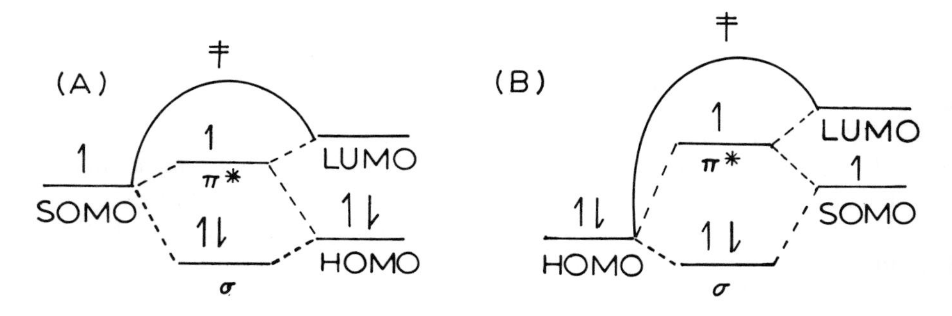

FIG. 2 Reaction of a free radical with a nucleophile. (A) Localized nucleophilic radical (e.g. $Me_3C\cdot$) and delocalized nucleophile (e.g. $Me_2C{=}NO_2^-$); $E^{\ddagger}$ increases with LUMO energy (SOMO constant). (B) Delocalized electrophilic radical (e.g. $PhCOCH_2\cdot$) with localized nucleophile; $E^{\ddagger}$ decreases as HOMO energy increases (SOMO and LUMO constant)

$MeCH{=}NO_2^- > H_2C{=}NO_2^-$; $PhC(O^-){=}CMe_2 > PhC(O^-){=}CHMe > PhC(O^-){=}CHPh > PhC(O^-){=}CPh_2$; $HC(CO_2Et)_2^- > MeC(CO_2Et)_2^- > PhC(CO_2Et)_2^-$. Towards $PhCOCH_2\cdot$, the energy of activation for addition to a nucleophile is apparently controlled by the SOMO–HOMO interaction and reactivity increases by changing the substituent from Ph to H to Me. Figure 1 plots the relative reactivities of $R^1R^2C{=}NO_2^-$ towards $Me_3C\cdot$ and $PhCOCH_2\cdot$ as a function of $\Sigma\sigma^*$ for R^1 and R^2. In Fig. 2, the orbital interaction between a donor radical (e.g. $Me_3C\cdot$) and an acceptor radical (e.g. $PhCOCH_2\cdot$) with an unsaturated nucleophile (π^-) are illustrated to show the dominance of the SOMO–LUMO interaction for the donor radical and the SOMO–HOMO interaction for the acceptor radical.

The reactions of alkylmercury halides with nucleophiles under free radical conditions (i.e. photostimulation, heat, ultrasound) follows the S_{RN} reaction path. However, apparently $RHgX^{-\cdot}$ is not a reaction intermediate, and electron transfer from the radical anion to RHgX is dissociative. Thus, reactions (4) and (5) of Scheme 10 occur in a concerted fashion to yield the two step chain process of Scheme 13. Evidence for the dissociative electron

$$R\cdot + N^- \longrightarrow RN^{-\cdot}$$
$$RN^{-\cdot} + RHgX \longrightarrow RN + R\cdot + Hg^0 + X^-$$
$$\overline{RHgX + N^- \longrightarrow RN + Hg^0 + X^-}$$

Scheme 13

transfer step is that in direct competitive reactions between pairs of alkylmercury chlorides and $Me_2C{=}NO_2^-$, involving electron transfer from $RCMe_2NO_2^{-\cdot}$ to RHgCl, the relative reactivities observed are $PhCH_2 : \text{t-Bu} : i\text{-Pr} : \text{n-Bu} = 4.7 : 1.0 : 0.07 : <0.005$ (Russell and Khanna, 1985b). Thus, the stability of the $R\cdot$ being formed is reflected in the energy of the transition state leading to the formation of $R\cdot$. Since kinetic chains with more than 50 molecules reacting per initiation fragment are involved, essentially every $RCMe_2NO_2^{-\cdot}$ formed must react by electron transfer to yield product and essentially every $R\cdot$ formed must add to N^- to yield $RN^{-\cdot}$. The observed relative reactivities of the two alkylmercury chlorides based on the product yields can thus be assigned to the step in which $R\cdot$ is produced.

$$RHgCl + RCMe_2NO_2^{-\cdot} \rightleftharpoons RHgCl^{-\cdot} + RCMe_2NO_2 \quad (25)$$

$$R^1HgCl^{-\cdot} + R^2HgCl \rightleftharpoons R^2HgCl^{-\cdot} + R^1HgCl \quad (26)$$

$$R^1Hg\cdot + R^2HgX \rightleftharpoons R^1HgX + R^2Hg\cdot \quad (27)$$

Equilibria such as (25)–(27) are excluded since neither $Me_3CCMe_2NO_2$ (or even $PhNO_2$) nor n-BuHgCl retard the rate or shorten the kinetic chain length in the reaction of t-BuHgCl with $Me_2C{=}NO_2^-$. Furthermore, in the reaction of $NaBH_4$ with RHgCl, the chain sequence of Scheme 14 is

$$\begin{gathered}
\mathrm{RHgCl} + \mathrm{NaBH_4} \xrightarrow{\text{fast}} \mathrm{RHgH} + \mathrm{NaBH_3Cl} \\
\mathrm{R\cdot} + \mathrm{RHgH} \longrightarrow \mathrm{RH} + \mathrm{RHg\cdot} \\
\mathrm{RHg\cdot} \longrightarrow \mathrm{R\cdot} + \mathrm{Hg^0} \\
\mathrm{R\cdot} + \mathrm{CH_2{=}C(Cl)CN} \longrightarrow \mathrm{RCH_2\dot{C}(Cl)CN} \\
\mathrm{RCH_2\dot{C}(Cl)CN} + \mathrm{RHgH} \longrightarrow \mathrm{RCH_2CH(Cl)CN} \\
\hline
\mathrm{RHgCl} + \mathrm{NaBH_4} + \mathrm{CH_2{=}C(Cl)CN} \longrightarrow \mathrm{RCH_2CH(Cl)CN} + \mathrm{Hg^0} + \mathrm{NaBH_3Cl}
\end{gathered}$$

Scheme 14

involved (Russell and Guo, 1984). In the presence of an alkene such as $CH_2{=}C(Cl)CN$, a telomerization-type process is involved (Giese, 1983, 1985; Giese and Grönigen, 1984; Giese and Horler, 1985). If an equilibrium between Hg(I) species, e.g. (27) with X = H or Cl, were involved, competition should yield a predominance of the reaction product from the most stable alkyl radical since the stability of the alkyl radical should control the rate of decomposition of RHg·. In actual fact, at high $CH_2{=}C(Cl)CN$ concentrations to insure the trapping of all alkyl radicals, the reaction of an excess of a 1 : 1 mixture of t-BuHgCl and n-BuHgCl yields $t\text{-}BuCH_2CH(Cl)CN$ and $n\text{-}BuCH_2CH(Cl)CN$ in nearly a 1 : 1 ratio (Russell, 1986). Since equilibrium between $R^1Hg\cdot$ and R^2HgH is slow relative to the decomposition of RHg· to yield R·, it seems safe to assume that a similar equilibrium in the RHgCl system is not important. [It is usually observed that such redistribution reactions are faster for H than for Cl (Russell, 1959).] The dissociative electron transfer mechanism of Scheme 13 is also consistent with the lack of reactivity of aryl or vinylmercury halides in this free radical chain process.

Scheme 13 no longer fits the $S_{RN}1$ acronym since the process in which the bond to the leaving group is broken is a dissociative bimolecular electron transfer process. Another type of $S_{RN}2$ process is described in Scheme 15.

$$\begin{gathered}
\mathrm{RN^{\cdot-}} + \mathrm{RX} \longrightarrow \mathrm{RX^{\cdot-}} + \mathrm{RN} \\
\mathrm{RX^{\cdot-}} + \mathrm{N^-} \longrightarrow \mathrm{RN^{\cdot-}} + \mathrm{X^-} \\
\hline
\mathrm{RX} + \mathrm{N^-} \longrightarrow \mathrm{RN} + \mathrm{X^-}
\end{gathered}$$

Scheme 15

Distinction between the $S_{RN}1$ process of Scheme 10 and the $S_{RN}2$ process of Scheme 15 can be made on the basis of evidence for R· as an intermediate, R· being an intermediate in Scheme 10 but not in Scheme 15. Moreover, the nature of the leaving group should have an effect on selectivity in Schemes 13 and 15 but not in Scheme 10 (Russell *et al.*, 1980). In the case of alkylmercurials, evidence for the intermediacy of R· required by Scheme 13 is furnished by the observation of 5-hexenyl radical cyclization in the reaction with $Me_2C{=}NO_2^-$ (Russell and Guo, 1984). For the $S_{RN}1$ process of Scheme 10, it has been observed that the relative reactivity of anions such as $(EtO)_2PO^-$, $Me_2C{=}NO_2^-$ and $MeC(CO_2Et)_2^-$ are independent of the nature of X in $Me_2C(NO_2)X$ with X = Cl, NO_2 or $PhSO_2$ (Russell *et al.*, 1980). Since X does not affect the selectivity, it is concluded that the new bond is formed between $Me_2C(NO_2)\cdot$ and N^- in a reaction not involving X, i.e. (3).

Evidence for a leaving group effect which has been interpreted in terms of Scheme 15 has been found in the reactions of 2-substituted-2-nitropropanes with lithium enolates of monoketones in nonpolar solvents such as THF (Russell *et al.*, 1979c; Russell *et al.*, 1981a, 1982b,c; Russell, 1982). Competing free radical chain processes leading to substitution (28) and to oxidative dimerization (29) are observed with the competition being a function of the

$$Me_2C(NO_2)X + PhC(O^-){=}CHR \longrightarrow PhCOCH(R)CMe_2NO_2 + X^- \quad (28)$$

$$Me_2C(NO_2)X + 2PhC(O^-){=}CHR \longrightarrow PhCOCH(R)CH(R)COPh + Me_2C{=}NO_2^- + X^- \quad (29)$$

nature of X but not of the concentration of the reactants. It is postulated that in nonpolar solvents with lithium as the counterion, a bimolecular reaction between $Me_2C(X)NO_2^{-\cdot}Li^+$ and $PhC(OLi){=}CHR$ can occur in which the enolate anion transfers an electron in a dissociative manner to $Me_2C(X)NO_2^{-\cdot}Li^+$. The resulting caged radical and anion [3] can collapse to

$$[Me_2C{=}NO_2^-\ X^-\ PhCO\dot{C}HR]$$

[3]

form $PhCOCH(R)CMe_2NO_2^{-\cdot}$ or dissociate to yield the free enolate radical which can attack the lithium enolate to yield the product of oxidative dimerization (Section 2, p. 295.). It is pertinent that, in THF with lithium as the counterion, enolate anions have a much higher reactivity towards free radicals than do nitronate anions (Russell *et al.*, 1980). The two competing chain reactions can be summarized as in Scheme 16. The leaving group X influences the competition between cage collapse and escape. Of course, the structure of R in $PhCO\dot{C}HR$ also plays an important role with the percent-

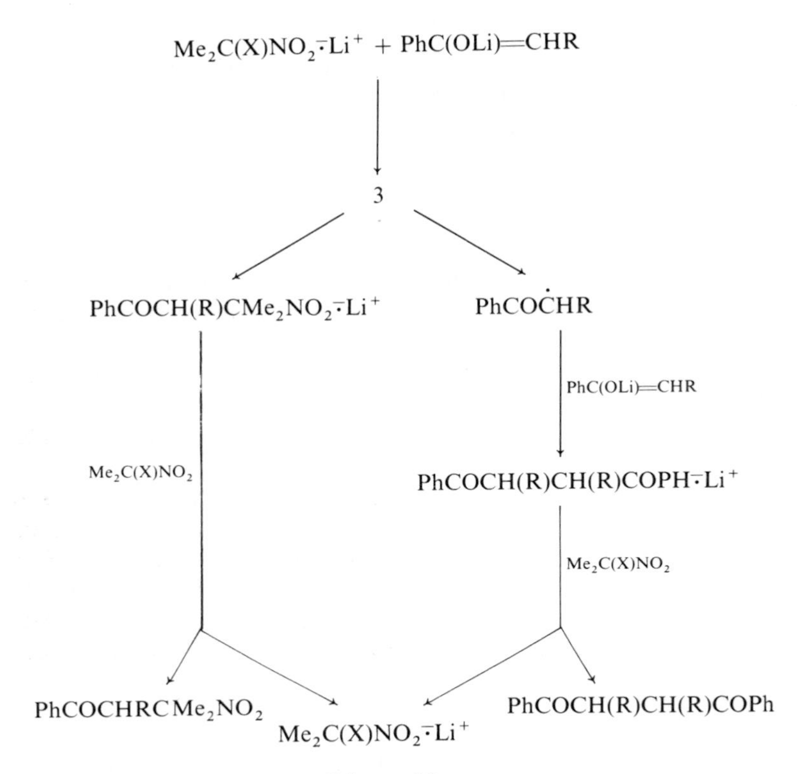

Scheme 16

age of cage escape increasing with steric hindrance to the coupling process or with the reactivity of PhCOĊHR according to the order R = H < Me < Et < i-Pr < Ph (Russell *et al.*, 1981a, 1982b,c). Thus with PhC(OLi)═CHPh a clean chain reaction leading solely to PhCOCH(Ph)CH(Ph)COPh is observed (Section 2, p. 295.).

Spontaneous "dimerization" with elimination occurs for some carbanions by a free radical chain process. The conversion of *p*-nitrobenzyl derivatives to the stilbenes by the action of base occurs by the propagation steps of Scheme 17 (Russell and Pecoraro, 1979). A similar process is apparently observed for 2-(bromomethyl)-5-nitrofuran (Prousek, 1980).

$$p\text{-}O_2NC_6H_4CH_2\cdot + p\text{-}O_2N_6H_4CHX^- \longrightarrow \underset{[4]^{\overline{\cdot}}}{[p\text{-}O_2NC_6H_4CH_2CH(X)C_6H_4NO_2]^{\overline{\cdot}}}$$

$$[4]^{\overline{\cdot}} + p\text{-}O_2NC_6H_4CH_2X \longrightarrow [4] + p\text{-}O_2NC_6H_4CH_2X^{\overline{\cdot}}$$

$$p\text{-}O_2NC_6H_4CH_2X^{\overline{\cdot}} \longrightarrow p\text{-}O_2NC_6H_4CH_2\cdot + X^-$$

$$[4] + B^- \longrightarrow BH + X^- + p\text{-}O_2NC_6H_4CH{=}CHC_6H_4NO_2\text{-}p$$

Scheme 17

The coupling of anions derived from phenyl *p*-toluenesulfonate, N-methyl-*p*-toluenesulfonanilide or their *o*-isomers represents another modification of a chain process involving electron transfer. Scheme 18 has been suggested (Lu and Truce, 1985). The sequence of steps in Scheme 18 is

Z = OPh; MeNPh

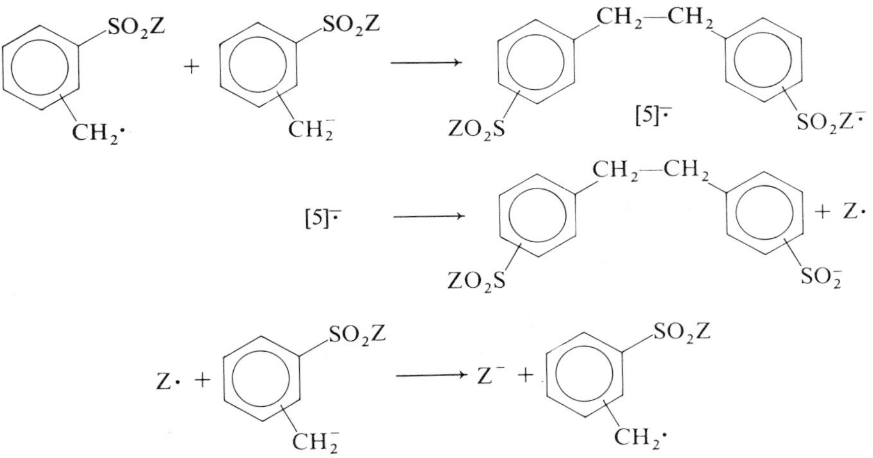

o-$(ZO_2S)C_6H_4CH_2\cdot + o$-$(ZO_2S)C_6H_4CH_2^- \longrightarrow o$-$(ZO_2S)C_6H_4CH_2{-}CH_2C_6H_4(SO_2Z^{\overline{\cdot}})$-$o$ $[5]^{\overline{\cdot}}$

$[5]^{\overline{\cdot}} \longrightarrow o$-$(ZO_2S)C_6H_4CH_2{-}CH_2C_6H_4(SO_2^-)$-$o + Z\cdot$

$Z\cdot + o$-$(ZO_2S)C_6H_4CH_2^- \longrightarrow Z^- + o$-$(ZO_2S)C_6H_4CH_2\cdot$

Scheme 18

addition followed by elimination and then electron transfer. This sequence of steps is involved in the $E_{RC}1$ process of Scheme 12 and the autoxidation of $Me_2C{=}NO_2^-$ (Scheme 9) and is opposite to the sequence of reactions in the $S_{RN}1$ sequence (Scheme 10) where addition is followed by electron transfer and then elimination (Reactions 3—5). Another example of a process involving this sequence of addition-elimination-electron transfer is shown in Scheme 19 for the chain reaction of 1-alkenylmercury halides with anions

$$RCH{=}CHHgCl + A\cdot \longrightarrow R\dot{C}HCH(A)HgCl$$
$$R\dot{C}HCH(A)HgCl \longrightarrow RCH{=}CHA + HgCl$$
$$HgCl + A^- \longrightarrow A\cdot + Hg^0 + Cl^-$$

$$RCH{=}CHHgCl + A^- \longrightarrow RCH{=}CHA + Hg^0 + Cl^-$$

Scheme 19

(A^-) such as $PhSO_2^-$, $(EtO)_2PO^-$, $PhP(OBu)O^-$, or PhS^- (Russell and Hershberger, 1980a,b). Here the reaction has been demonstrated to involve an addition-elimination process followed by electron transfer between monomeric HgCl and A^-.

Another recently discovered chain reaction involving the sequence addition-elimination-electron transfer is summarized in Scheme 20 (Russell and Baik, unpublished). In Scheme 20, the alkoxy group controls the

$$\begin{aligned}
Me_2C(NO_2)\cdot + MeO^- &\longrightarrow MeOCMe_2NO_2^{\overline{\cdot}} \\
MeOCMe_2NO_2^{\overline{\cdot}} &\longrightarrow MeOCMe_2\cdot + NO_2^- \\
MeOCMe_2\cdot + BrCMe_2NO_2 &\longrightarrow Me\overset{+}{O}{=}CMe_2 + BrCMe_2NO_2^{\overline{\cdot}} \\
BrCMe_2NO_2^{\overline{\cdot}} &\longrightarrow Me_2C(NO_2)\cdot + Br^- \\
Me\overset{+}{O}{=}CMe_2 + MeO^- &\longrightarrow (MeO)_2CMe_2
\end{aligned}$$

Scheme 20

fragmentation of $MeOCMe_2NO_2^{\overline{\cdot}}$ to form the most stable radical ($MeOCMe_2\cdot$). [The species $ROCH_2NO_2^{\overline{\cdot}}$ have been formed and studied by esr spectroscopy by the reaction $RO\cdot + CH_2{=}NO_2^-$ (Gilbert *et al.*, 1976, 1977; Gilbert and Norman, 1982).] The resulting $MeOCMe_2\cdot$ is an easily oxidized radical readily converted to the oxonium ion $Me\overset{+}{O}{=}CMe_2$ by electron transfer from the radical to the 2-bromo-2-nitropropane. Other examples of electron transfer from a neutral but easily oxidized radical to a neutral but easily reduced substrate will be considered in Section 3.

Ligand exchange in metal carbonyl complexes can occur by a chain process similar to the $S_{RN}1$ scheme. Exchange of the MeCN for Ph_3P ligand in $(\eta^5\text{-}C_5H_4Me)Mn(CO)_2NCMe$ occurs under anodic catalysis by a process involving the sequence addition-elimination-electron transfer (Hershberger and Kochi, 1982). Under cathodic conditions, substitution of $P(OMe)_3$ for CO occurs in $Fe_3S_2(CO)_8(C_3H_2S_2)$ by an elimination-addition-electron transfer process (Darchen *et al.*, 1982) while exchange of PR_3, $P(OR)_3$ or RNC for CO is observed in a variety of Fe, Ru, Os, Co, Rh clusters in a free radical reaction catalysed by electron transfer from $Ph_2C{=}O^{\overline{\cdot}}$ (Bruce *et al.*, 1982). The sequence of steps in these substitutions can be complicated by molecular reorganizations but involves the addition of one ligand and elimination of CO^- (or vice versa) in the reduced metal cluster followed by electron transfer to the unreacted metal cluster.

PROCESSES LEADING TO OXIDATIVE DIMERIZATION OR DEHYDROGENATION OF THE ANION

Reaction of a free radical with an anion can involve the addition reaction (30) which is fundamental in the $S_{RN}1$ process or an electron transfer from an easily oxidized anion (31). Addition of $A\cdot$ to A^- followed by electron

$$R\cdot + A^- \longrightarrow R{-}A^{\overline{\cdot}} \qquad (30)$$

$$R\cdot + A^- \longrightarrow R{:}^- + A\cdot \qquad (31)$$

transfer from $AA^{\overline{\cdot}}$ to a precursor of R· leads to the oxidative dimerization process of Scheme 21. When the radical precursor (RX) is a 2-substituted

$$R\cdot + A^- \longrightarrow R{:}^- + A\cdot$$
$$A\cdot + A^- \longrightarrow AA^{\overline{\cdot}}$$
$$AA^{\overline{\cdot}} + RX \longrightarrow AA + RX^{\overline{\cdot}}$$
$$RX^{\overline{\cdot}} \longrightarrow R\cdot + X^-$$

$$2\,A^- + RX \longrightarrow AA + R^- + X^-$$

Scheme 21

2-nitropropane, the resulting nitronate anion will react further with RX by the $S_{RN}1$ chain to give the overall stoichiometry of (32). In cases where $AA^{\overline{\cdot}}$ is not stabilized by virtue of a low-lying MO, coupling of A· will lead to the same stoichiometry (32) by a nonchain process (Bordwell and Clemens,

$$2\,A^- + 2\,RX \longrightarrow AA + RR + 2\,X^- \tag{32}$$

1982). In reaction with 2-substituted-2-nitropropanes, Scheme 21 appears to be involved with anions such as PhS^-, $PhC(O^-){=}CHPh$, indoxyl$^-$ or 2,6-di-t-butylphenoxide (Zeldrin and Shechter, 1957; Russell and Kaupp, 1969; Russell *et al.*, 1979b). Reaction of *p*-nitrobenzyl bromide with CN^- or 2-(bromomethyl)-5-nitrofuran with CN^- or RS^- leads to dimerization (33) to the substituted ethane (Prousek, 1980, 1982). Here, R^- formed in Scheme 21 may react in an S_N2 manner with bromo compound to form the dimer.

$$O_2NArCH_2Br \xrightarrow{CN^-/CH_3CN} O_2NArCH_2CH_2ArNO_2 \tag{33}$$

Phenacylmercury chloride will undergo a similar free radical chain process with an easily oxidized anion such as $PhC(O^-){=}CHPh$ as shown in Scheme 22 (Russell and Khanna, 1986). Oxidative dimerization also occurs with $PhC(O^-){=}CPh_2$ and $PhC(CO_2Et)_2^-$ but here coupling of the radicals [$PhCOCPh_2\cdot$, $PhC(CO_2Et)_2\cdot$] in a nonchain process is involved.

$$PhCOCH_2\cdot + Ph(O^-){=}CHPh \longrightarrow PhCOCH_2^- + PhCO\dot{C}HPh$$
$$PhCO\dot{C}HPh + PhC(O^-){=}CHPh \longrightarrow \underset{[6]^{\overline{\cdot}}}{[PhCOCH(Ph)CH(Ph)COPh]^{\overline{\cdot}}}$$
$$[6]^{\overline{\cdot}} + PhCOCH_2HgCl \longrightarrow [6] + PhCOCH_2\cdot + Hg^0 + Cl^-$$

$$2PhC(O^-){=}CHPh + PhCOCH_2HgCl \longrightarrow [PhCOCH(Ph)]_2 + PhCOCH_2^- + Hg^0 + Cl^-$$

Scheme 22

Oxidative dehydrogenation involving electron transfer is observed in systems where the carbanion is easily oxidized to a radical which can lose a proton to form a radical anion (34). Thus, dihydroindigo is oxidized to

$$>\bar{C}-CH< \xrightarrow{-e} >\dot{C}-CH< \rightleftharpoons >\dot{C}-\bar{C}< \xrightarrow{-e} >C=C< \quad (34)$$

indigo by 2-chloro-2-nitropropane in basic solution, and the oxidative dimerization product of 2,6-di-t-butylphenoxide is dehydrogenated to the diphenoquinone (35). In a similar fashion, an excess of $Me_2C(NO_2)_2$ will

$$\text{[dimer]} + 2B^- + ClCMe_2NO_2 \longrightarrow \text{[diphenoquinone]} + Me_2C{=}NO_2^- + Cl^- + 2BH \quad (35)$$

convert fluorenide anion first to 9,9′-bifluorene and then to $\Delta^{9,9'}$-bifluorene (Russell *et al.*, 1979b).

PROCESSES INVOLVING REDUCTIVE ELIMINATION

Elimination of the elements N_2O_4 from a vicinal aliphatic dinitro compound occurs by a free radical mechanism in the presence of PhS^- or Na_2S (Kornblum *et al.*, 1971). Reducing agents such as calcium amalgam are also effective for this process and can be applied to the synthesis of tetra-substituted alkenes with functional groups (Kornblum and Cheng, 1977). The mechanism of the chain reaction with PhS^- or Na_2S is not completely known. Certainly, electron transfer to a nitro group followed by fragmentation of the radical anion must be a key step (36). One possibility for the

$$>C(NO_2)-C(NO_2)< + e^- \longrightarrow [>C(NO_2)-C(NO_2)<]^{\cdot-} \longrightarrow >C(NO_2)-\dot{C}< + NO_2^- \quad (36)$$

chain reaction is shown in Scheme 23 where the α-nitro radical reacts with the sulfur nucleophile to form an intermediate α-nitro sulfide radical anion which can under fragmentation be followed by a β-elimination of a sulfur radical. Alternatively, the elimination of NO_2 from a β-nitro radical may be involved. Similar reductive processes are known in the reaction of naphthalene radical anion with vicinal dibromides or dichlorides or in the reaction of vicinal dibromides with the anion of diphenylacetonitrile (Garst and Barton, 1969; Korzan *et al.*, 1971).

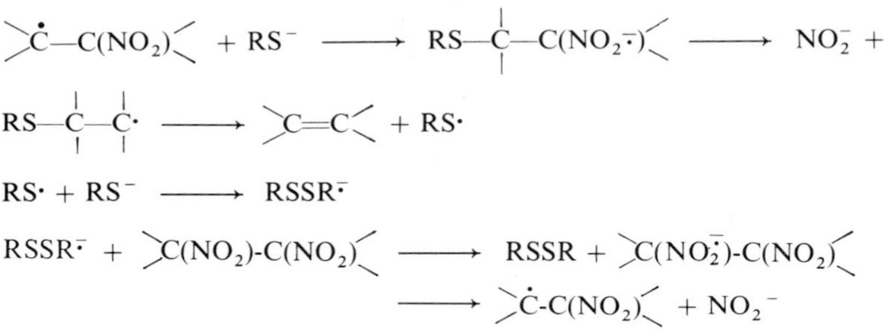

$$>\dot{C}{-}C(NO_2)< \; + RS^- \longrightarrow RS{-}\overset{|}{\underset{|}{C}}{-}C(NO_2^{\cdot -})< \longrightarrow NO_2^- +$$

$$RS{-}\overset{|}{\underset{|}{C}}{-}\overset{|}{\underset{|}{C}}\cdot \longrightarrow >C{=}C< \; + RS\cdot$$

$$RS\cdot + RS^- \longrightarrow RSSR^{\cdot -}$$

$$RSSR^{\cdot -} + >C(NO_2){-}C(NO_2)< \longrightarrow RSSR + >C(NO_2^{\cdot -}){-}C(NO_2)<$$

$$\longrightarrow >\dot{C}{-}C(NO_2)< \; + NO_2^-$$

Scheme 23

PROCESSES INVOLVING HYDROGEN ATOM TRANSFER FROM AN ANION

A variety of free radical chain processes are known in which reactions (37)–(39) are coupled (Scheme 24). Among the negatively charged hydrogen

$$R\cdot + HY^- \longrightarrow RH + Y^{\cdot -} \quad (37)$$

$$Y^{\cdot -} + RX \longrightarrow Y + RX^{\cdot -} \quad (38)$$

$$R{-}X^{\cdot -} \longrightarrow R\cdot + X^- \quad (39)$$

Scheme 24

donors (HY^-) recognized to react in this manner are BH_4^- (Groves and Ma, 1974), AlH_4^- (Hatem and Waegell, 1973; Chung and Chung, 1979; Chung, 1980; Singh *et al.*, 1980, 1981; Singh and Khanna, 1983; Ashby *et al.*, 1984), cyclopentadienyl-$V(CO)_3H^-$ (Kinney *et al.*, 1978). CH_3O^- (Simig and Lempert, 1961; Simig *et al.*, 1977, 1978), CH_3S^- (Kornblum *et al.*, 1978a, 1979a,b), and $PhSiMe_2HF^-$ (Yang and Tanner, 1986). Table 3 lists some typical aliphatic substrates which participate in the chain process of Scheme 24. The chain reactions involving $LiAlH_4$ in THF are presumed to occur with a variety of vinyl, cyclopropyl and bridgehead halides as well as aryl

TABLE 3

Free radical chain reactions involving Scheme 24
($RX + HY^- \longrightarrow RH + XY^-$)

R	X	HY^-	Reference
$Ph_2C(CONMe_2)$	Br	CH_3O^-	Simig and Lempert, 1961; Simig *et al.*, 1977, 1978
Me_3CCH_2	I	CH_3O^-	Boyle and Bunnett, 1974
3°-Alkyl, 3°-benzylic	NO_2	CH_3S^-, $CH_3SCH_2S^-$	Kornblum *et al.*, 1978a, 1979a,b
p-$XC_6H_4CMe_2$ X = $PhSO_2$, CN, PhCO	SMe	CH_3S^-, $CH_3SCH_2S^-$	Kornblum *et al.*, 1979a,b
(*E*) and (*Z*) PhCH=CH	Br	AlH_4^-	Chung, 1980
7,7-Norcarane	Br	BH_4^-	Groves and Ma, 1974
Benzyl, diphenylmethyl, 9-fluorenyl, anthracene-carbinyl	Cl, Br	AlH_4^-	Singh *et al.*, 1981
Alkyl, acyl	Br, Cl	$HCpV(CO)_3^-$	Kinney *et al.*, 1978
Alkyl	HgCl	AlH_4^-	Singh and Khanna, 1983
$PhCOCH_2$	F	$PhSiMe_2H$- - - F^-	Tanner *et al.*, 1986a
2-Octyl, 2-hepten-6-yl, 1-hexen-5-yl; 2,2-dimethyl-1-hexen-5-yl	I	AlH_4^-, AlH_3	Ashby *et al.*, 1984

halides (Barltrop and Bradbury, 1973; Chung and Chung, 1979; Singh *et al.*, 1980; Chung and Filmore, 1983). Free radical chain reactions initiated by AlH_4^- or AlH_3 are recognized to occur in the reduction of alkyl iodides such as 2-iodooctane, 6-iodo-1-heptene, 6-iodo-1-hexene, 2,2-dimethyl-1-iodo-5-hexene (Ashby *et al.*, 1984). Similar reactions of CH_3O^- are also recognized in the reductive substitution reaction of aromatic systems, particularly p-$NO_2C_6H_4I$ or p-$NO_2C_6H_4N{=}NOMe$, p-$NO_2C_6H_4N_2^+$ (Detar and Turetzky, 1955, 1956; Bunnett and Takayama, 1968a,b; Boyle and Bunnett, 1974).

When the reduced product (RH) of Scheme 24 can be deprotonated by the basic HY^- to give R^-, the $S_{RN}1$ chain reaction between R^- and RX (Scheme 10) may occur as exemplified by (40) (Zorin *et al.*, 1983).

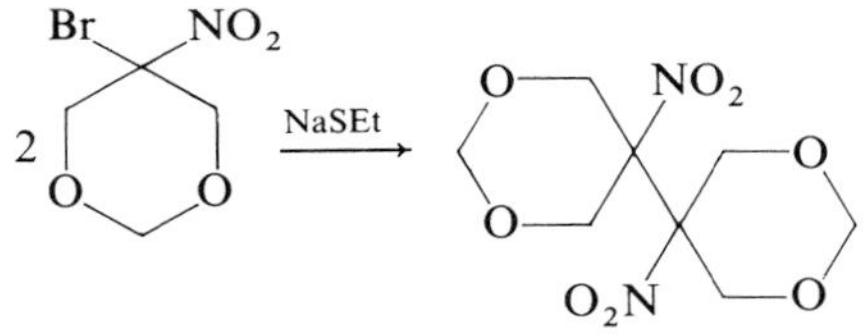

2 [5-bromo-5-nitro-1,3-dioxane] $\xrightarrow{\text{NaSEt}}$ [5,5'-dinitro-5,5'-bi(1,3-dioxane)] (40)

3 Free radical chain processes involving electron transfer between neutral substances

REACTIONS OF 1,4-DIHYDROPYRIDINES, DIALKYLANILINES, HYDRAZINES, ENAMINES AND, PYRIDINES

Easily oxidized neutral radicals can be one-electron donors to easily reduced neutral molecules as in Scheme 20. The reverse process involving neutral acceptor radicals and neutral oxidizable substrates is of course possible but is not a commonly recognized chain propagation reaction in free radical

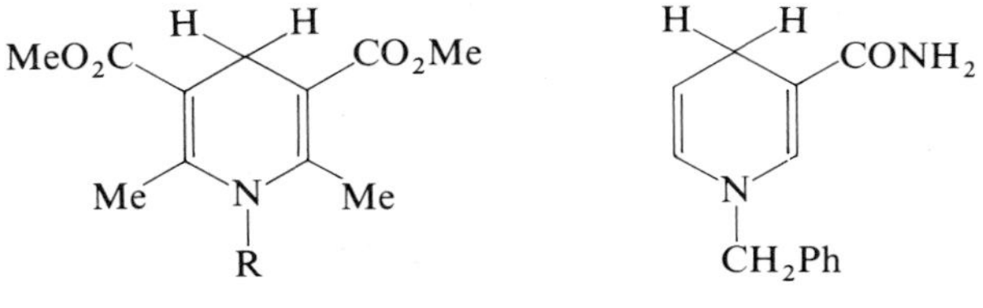

[7] (R = Me, H) [8]

processes. 1,4-Dihydropyridines, such as [7] and [8], participate in a variety of chain processes by virtue of the ease with which reactions (41) and (42) occur where πH_2 is a dihydropyridine and A is a neutral acceptor molecule.

$$R\cdot + \pi H_2 \longrightarrow RH + \cdot\pi H \tag{41}$$

$$\cdot\pi H + A \longrightarrow \pi H^+ + A^{\cdot-} \tag{42}$$

When $A^{\cdot-}$ can regenerate R· by a fragmentation process, a chain reaction will ensue. Thus [7] is oxidized to the pyridinium cation $[7]^+$ in chain reactions by electron acceptor molecules such as MeC(O)OOC(O)Me, PhC(O)OO-C(O)Ph, t-$BuO_3C(CH_3)_3$, t-BuOOBu-t, $BrCCl_3$, I_2, HOOH or $S_2O_8^{-2}$ (Huyser *et al.*, 1964, 1972; Huyser and Kahl, 1970; Huyser and Harmony, 1974). Among the oxidants which are effective in the conversion of [8] to the cation are tetramethylthiuram disulfide (Huyser and Harmony, 1974), alkylmercury acetates or alkylthallium dichloride (Kurosawa *et al.*, 1980, 1981), nitroalkanes such as $R_2C(NO_2)Br$, $R_2C(NO_2)CN$, $R_2C(NO_2)CO_2R$, $R_2C(NO_2)COAr$, $R_2C(NO_2)SO_2Ar$ (Kill and Widdowson, 1976; Ono *et al.*, 1980, 1985), and $PhCOCH_2X$ with X = F, Cl, Br (Tanner *et al.*, 1986). Electron transfer to these reagents forms, after fragmentation, the species $MeCO_2$· (or Me·), $PhCO_2$·, CCl_3·, I·, Me_2N·, alkyl·, $R_2C(NO_2)$·, or $PhCOCH_2$· which serve as R· in Scheme 25. In the case of RHgCl or $BrCCl_3$, the electron transfer from [7]· or [8]· is probably dissociative while for the substituted nitroalkanes and probably for the peresters the radical anion ($RX^{\cdot-}$) is a true intermediate.

$$R\cdot + [7] \text{ or } [8] \longrightarrow [7]\cdot \text{ or } [8]\cdot + RH$$

$$[7]\cdot \text{ or } [8]\cdot + RX \longrightarrow [7]^+ \text{ or } [8]^+ + RX^{\cdot-} \text{ or } R\cdot + X^-$$

$$RX^{\cdot-} \longrightarrow R\cdot + X^-$$

Scheme 25

Reduction of $PhCOCH_2Br$ to $PhCOCH_3$ by NADH occurs by the mechanism of Scheme 25 (Tanner *et al.*, 1986). The reaction is not catalysed by horse liver alcohol dehydrogenase, and the free radical chain dehydrogenations of 1,4-dihydropyridines seem not to be involved in enzymatic reactions involving NADH as a hydrogen donor.

The formal hydride transfer between NAD^+ analogues has been interpreted as a multistep but nonchain process involving electron and proton transfers (Ohno *et al.*, 1981) or alternatively as a one-step hydride transfer with an appreciable primary isotope effect (Ostovic *et al.*, 1983). Although not as yet recognized, a free radical chain mechanism should be possible in certain cases where [8]· transfers an electron to an easily reducible NAD^+ analogue such as N-methylacridinium ion. The chain could continue by hydrogen atom abstraction from [8] by the N-methylacridinyl radical.

The reaction of substituted aminomethyl radicals with benzoyl, acetyl or cyclohexanesulfonylacetyl peroxides involves at least partially the free radical process of Scheme 26 (Horner and Schwenk, 1944; Horner and Betzel, 1953; Horner and Anders, 1962; Walling and Indictor, 1958; Hrabak and Lokaj, 1970). The same products are also formed by the ionic Polonovski-type reaction involving the intermediate $PhN(Me)_2O_2CPh^+$ $PhCO_2^-$ which decomposes to $PhN(Me)CH_2O_2CPh$ *via* $Ph\overset{+}{N}(Me)(O_2CPh)CH_2^-$.

$$PhN(Me)CH_2\cdot + PhC(O)OOC(O)Ph \longrightarrow$$
$$Ph\overset{+}{N}(CH_3){=}CH_2 + PhCO_2\cdot + PhCO_2^-$$
$$PhCO_2\cdot \longrightarrow Ph\cdot + CO_2$$
$$Ph\cdot(PhCO_2\cdot) + PhNMe_2 \longrightarrow PhH(PhCO_2H) + PhN(CH_3)CH_2\cdot$$
$$Ph\overset{+}{N}(CH_3)CH_2 + PhCO_2^- \longrightarrow PhN(CH_3)CH_2O_2CPh$$

Scheme 26

Oxidation of phenylhydrazine to phenyldiimide by $BrCCl_3$, halogens or $S_2O_8{}^{-2}$ appears to involve similar processes (Scheme 27).

$$PhNHNH_2 + Cl_3C\cdot \longrightarrow Ph\dot{N}NH_2 + HCCl_3$$
$$Ph\dot{N}NH_2 + BrCCl_3 \longrightarrow PhN{=}NH_2^+ + Br^- + Cl_3C\cdot$$

Scheme 27

The electron donor radicals in Schemes 26 and 27 are generated by hydrogen abstraction reactions by radicals which themselves are either acceptors ($PhCO_2\cdot$, $Cl_3C\cdot$, $I\cdot$, $SO_4{}^{\overline{\cdot}}$) or have little donor ability such as Me· or Ph·. Another route to a donor radical is the addition of a radical to an unsaturated system such as an enamine or ynamine. Thus, the addition of $CF_3\cdot$, $\cdot CF_2Cl$, $\cdot CF_2Br$ or $\cdot CF_2CF_2Br$ to an enamine converts an acceptor radical into a donor radical (e.g. $R_2N\dot{C}HC(CF_3)R_2$) and the free radical process of Scheme 28 occurs (Cantacuzène and Dorme, 1975; Rico *et al.*, 1981, 1983).

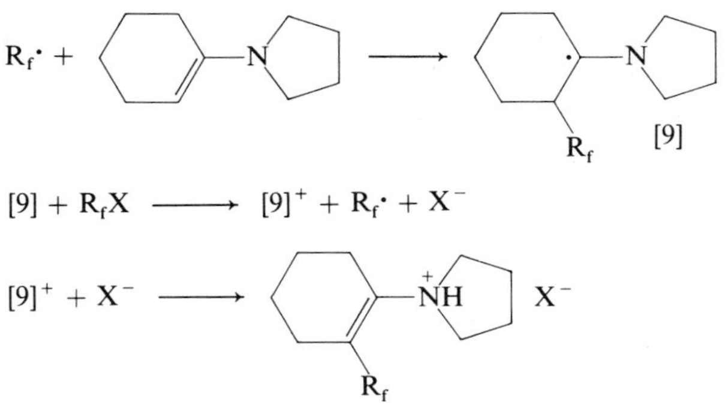

$R_f\cdot$ + [1-(cyclohex-1-enyl)pyrrolidine] ⟶ [9]

[9] + R_fX ⟶ $[9]^+ + R_f\cdot + X^-$

$[9]^+ + X^-$ ⟶ [2-R_f-cyclohexenyl pyrrolidinium] X^-

Scheme 28

Addition of *p*-nitro-α,α-dimethylbenzyl radical to a variety of coordinatively unsaturated primary, secondary or tertiary amines or ammonia produces the radical zwitterion [10] (Kornblum and Stuchal, 1970). The radical zwitterion will transfer an electron to *p*-nitrocumyl chloride in the $S_{RN}1$ process.

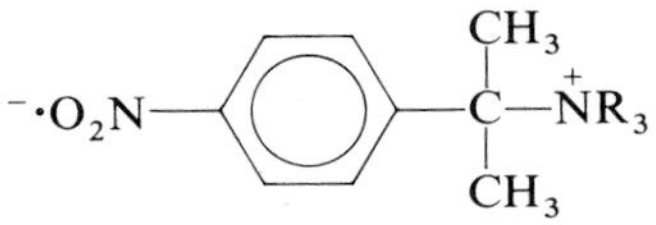

$$^{-}\cdot O_2N-C_6H_4-C(CH_3)_2-\overset{+}{N}R_3$$

[10]

Addition of alkyl radicals to the pyridine (43) or pyridinium ring (44) will also lead to our easily oxidized pyridinyl radical. Radical substitution in pyridinium ions occurs readily in a variety of reactions involving H_2O_2,

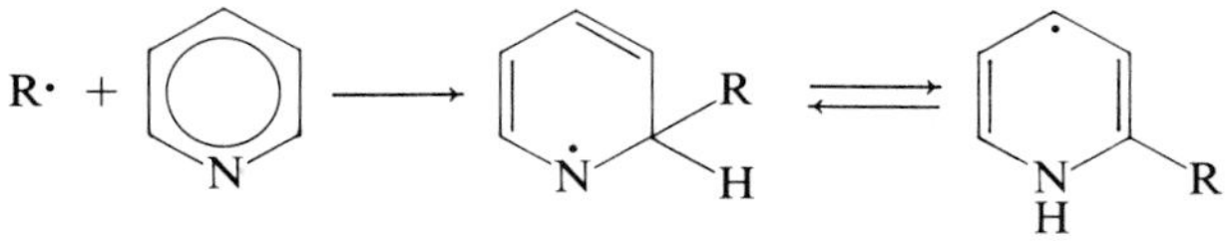

$$R\cdot + \text{pyridine} \longrightarrow \text{2-R-2-H-dihydropyridin-1-yl} \rightleftharpoons \text{2-R-1H-pyridin-4-yl} \tag{43}$$

$$R\cdot + \text{pyridinium (N}^+\text{–H)} \longrightarrow \text{4-H-4-R-dihydropyridinium radical cation (N}\cdot^+\text{–H)} \rightleftharpoons \text{4-R-1H-pyridin-4-yl} + H^+ \tag{44}$$

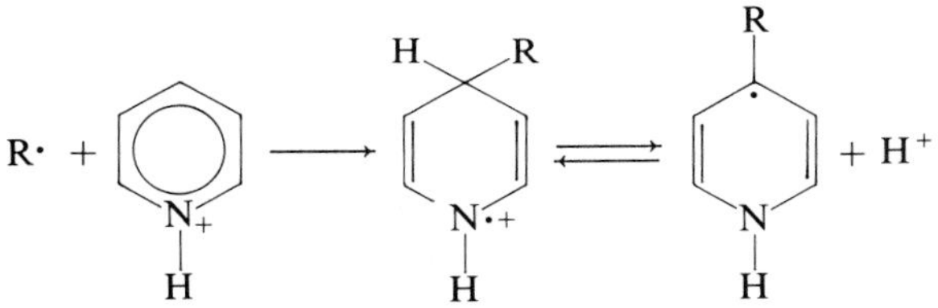

NH_2OH, t-BuOOH or PhC(O)OOC(O)Ph and metal ions such as Ti(III) or Fe(II), e.g. Scheme 7 (Minisci *et al.*, 1985). A reaction (Scheme 29) can be observed in the absence of metal ion where the solvent (SH) is a primary alcohol and S· an α-hydroxyalkyl radical (Minisci *et al.*, 1985).

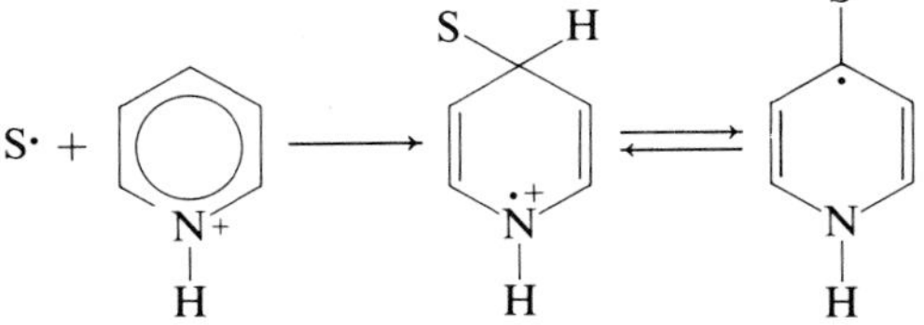

$$S\cdot + \text{pyridinium (N}^+\text{–H)} \longrightarrow \text{4-S-4-H-dihydropyridinium radical cation (N}\cdot^+\text{–H)} \rightleftharpoons \text{4-S-1H-pyridin-4-yl [11]}$$

$$[11] + PhC(O)OOC(O)Ph \longrightarrow [11]^+ + PhCO_2^- + PhCO_2\cdot$$

$$PhCO_2\cdot \longrightarrow Ph\cdot + CO_2$$

$$Ph\cdot(PhCO_2\cdot) + SH \longrightarrow PhH(PhCO_2H) + S\cdot$$

Scheme 29

Alkylmercury halides or carboxylates undergo a free radical aromatic substitution reaction with pyridine, pyridinium ions or dialkylanilines. Addition of R· to these substrates produces the easily oxidized radicals

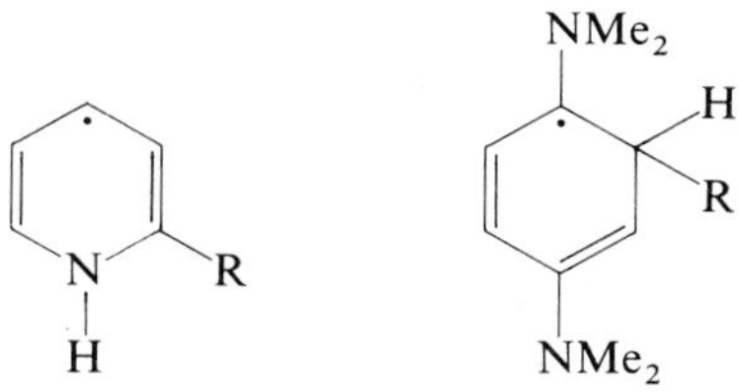

NMe_2 H R N R H NMe_2

[12] [13]

[11]–[13] which undergo dissociative electron transfer with RHgX to regenerate R· as shown in Scheme 30 (Russell *et al.*, 1985). Since the stability of R· formed in the dissociative electron transfer step controls the rate of this

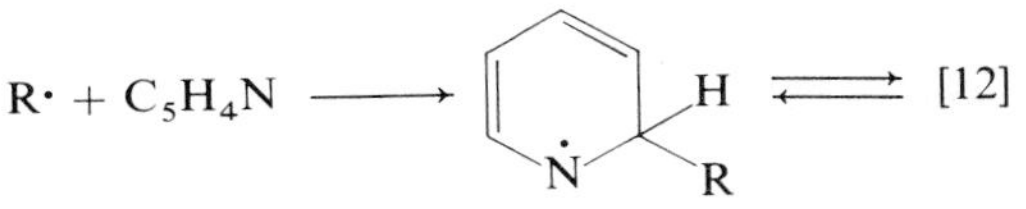

$$R\cdot + C_5H_4N \longrightarrow [\text{H, R-dihydropyridinyl}] \rightleftharpoons [12]$$

$$[12] + RHgCl \longrightarrow [12]^+ + R\cdot + Hg^0 + Cl^-$$

Scheme 30

process, it follows that the overall rate of reaction of pyridine with RHgCl increases according to the order R = n-Bu < i-Pr < t-Bu. In reaction with N,N,N′,N′-tetramethyl-*p*-phenylenediamine, it is also observed that $PhCH_2HgCl$ is more reactive than n-BuHgCl. The preferred position of attack of an alkyl radical upon pyridine is at the α-position to give [12] after tautomerization. The ratio of *o*/*p* attack by R increases with the donor ability of R· according to the order R = n-Bu < i-Pr < t-Bu. Attack of R· upon pyridinium ion occurs with a slight preference at the *p*-position to yield [11]; again the *p*/*o* ratio is a function of the donor ability of R· and increases from R = n-Bu to i-Pr to t-Bu.

REACTIONS OF THE TRIBUTYLTIN RADICAL

The tributyltin radical is easily oxidized to the cation. Therefore, it seems reasonable that $Bu_3Sn\cdot$ should transfer an electron to an easily reduced substrate such as R_3CNO_2, $ArCH_2I$ or RHgX as in (45). This process has

$$Bu_3Sn\cdot + RX \longrightarrow Bu_3Sn^+ + X^- + R\cdot \qquad (45)$$

been reported to occur in the reaction of nitroalkanes (Tanner *et al.*, 1981) and benzylic iodides (Blackburn and Tanner, 1980). In reaction of $Bu_3Sn\cdot$ with an aliphatic nitro group, an alternative chain reaction (46) not involving electron transfer must be considered (Ono *et al.*, 1985). The observation

$$R{-}NO_2 + Bu_3Sn\cdot \longrightarrow R{-}\underset{\displaystyle\cdot O}{\underset{|}{N}}{-}O{-}SnBu_3 \longrightarrow R\cdot + Bu_3SnONO \quad (46)$$

that $Me_2C(SO_2C_6H_4CH_3\text{-}p)NO_2$ undergoes reaction with [8] according to Scheme 30 but fails to undergo any reaction with Bu_3SnH has been interpreted as evidence that (46) is often the preferred route of reaction for $Bu_3Sn\cdot$. The reaction of Bu_3SnH with alkylmercury chlorides does appear to involve the electron transfer of (45). When Bu_3SnH is utilized in the Giese reaction (Giese and Horler, 1985) with $CH_2{=}CH(Cl)CN$ (Scheme 31), it is

$$R\cdot + CH_2{=}C(Cl)CN \longrightarrow RCH_2\dot{C}(Cl)CN$$
$$RCH_2\dot{C}(Cl)CN + Bu_3SnH \longrightarrow RCH_2CH(Cl)CN + Bu_3Sn\cdot$$
$$Bu_3Sn\cdot + RHgCl \longrightarrow Bu_3Sn^+ + Cl^- + Hg^0 + R\cdot$$

Scheme 31

observed that in a direct competition, t-BuHgCl is more than 100 times as reactive as n-BuHgCl (Russell, 1986). In another reaction involving $Bu_3Sn\cdot$, it is observed that the reaction of (*E*)-$PhCH{=}CHSnBu_3$ with RHgX forms (*E*)-$PhCH{=}CHR$ with a relative reactivity of R = t-Bu : c-C_6H_{11} : n-Bu of 1.0 : 0.011 : 0.001. The substitution process of Scheme 32 occurs in the

$$R\cdot + PhCH{=}CHSnBu_3 \longrightarrow Ph\dot{C}HCH(R)SnBu_3$$
$$Ph\dot{C}HCH(R)SnBu_3 \longrightarrow PhCH{=}CHR + Bu_3Sn\cdot$$
$$Bu_3Sn\cdot + RHgCl \longrightarrow Bu_3Sn^+ + Cl^- + Hg^0 + R\cdot$$

Scheme 32

reaction of vinyl or acetylenic tin, compounds with RHgCl (Russell *et al.*, 1984; Russell and Ngoviwatchai, 1985) and involves the sequence of addition-elimination-electron transfer. Figure 3 illustrates the relative reactivity observed in the electron transfer reaction of $Bu_3Sn\cdot$ with a 1 : 1 mixture of t-BuHgCl and n-BuHgCl (> 100 : 1) as well as the unselective reaction observed in the BH_4^- reduction (~1 : 1) which involves Hg(I) intermediates (i.e. Scheme 14). Data for the reaction of $LiAlH_4$ in THF with RHgCl is included in Fig. 3 which indicates that ~80% of the reaction occurs by the $AlH_3^{\cdot-}$ electron transfer mechanism (Scheme 24) and 20% by the mechanism involving RHgH.

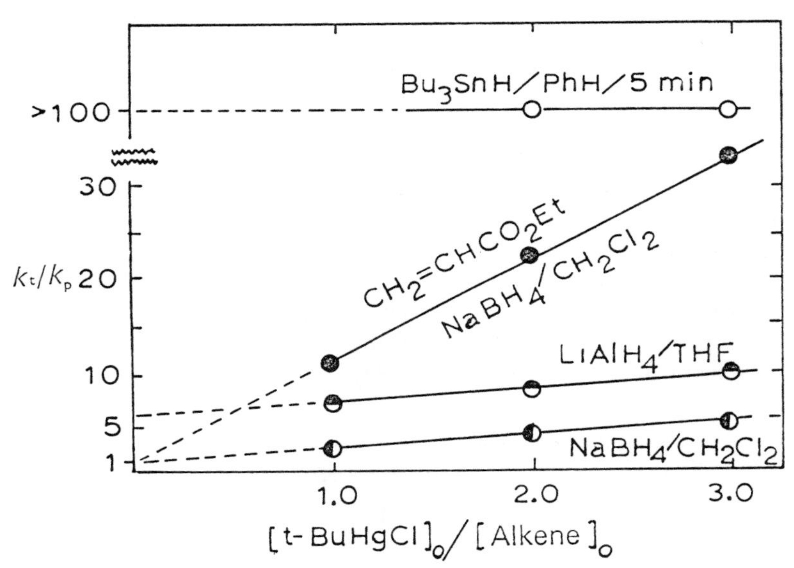

FIG. 3 Relative reactivities (k_t/k_p) of t-BuHgCl(H) and n-BuHgCl(H) towards $RCH_3\dot{C}(Cl)CN$ or $RCH_2\dot{C}HCO_2Et$ in competitive reactions with $CH_2{=}C(Cl)CN$ and $CH_2{=}CHCO_2Et$ at 25°C. (Data are from Russell, 1986.) Reactions in the presence of BH_4^- involve hydrogen atom abstraction from RHgH, while reactions in the presence of Bu_3SnH involve electron transfer from $Bu_3Sn\cdot$ to RHgCl

A reduction mechanism analogous to Scheme 24 has been observed with the easily oxidized $Ph_3Sn\cdot$. The reaction of Ph_3SnH with $PhCOCH_2F$ (Tanner *et al.*, 1985b), $PhCOCH_2Cl$ or $PhCOCH_2Br$ (Tanner *et al.*, 1986b) proceeds *via* the loss of F^-, Cl^- or Br^- from $PhCOCH_2X^{\cdot -}$ to yield $PhCOCH_2\cdot$ which reacts with Ph_3SnH to form $PhCOCH_3$ and $Ph_3Sn\cdot$. The chain is continued by electron transfer from $Ph_3Sn\cdot$ to $PhCOCH_2X$. Reduction of phenyl cyclopropyl ketone by Ph_3SnH but not by $(n\text{-}Bu)_3SnH$ also proceeds via electron transfer from the tin radical to form the ketyl which undergoes β-elimination and hydrogen atom abstraction to yield n-butyrophenone in methanol solution (Tanner *et al.*, 1985b).

REACTIONS OF ELECTRON ACCEPTOR RADICALS WITH ALKYLMERCURY HALIDES

Alkylmercurials can serve as electron donors to an electron acceptor radical. It is, however, difficult to distinguish the electron transfer (47) from a S_H2 reaction occurring in one step (48). Electron accepting radicals such as

$$A\cdot + RHgX \longrightarrow [A{:}^- \; {}^+HgX + R\cdot] \longrightarrow AHgX + R\cdot \quad (47)$$

$$A\cdot + RHgX \longrightarrow AHgX + R\cdot \quad (48)$$

$PhS\cdot$, $PhSe\cdot$, $PhTe\cdot$, $PhSO_2\cdot$ or t-$BuCH_2\dot{C}HE$ with E = $(EtO)_2PO$, $PhSO_2$ or p-$NO_2C_6H_4$ participate in such reactions (Russell and Tashtoush, 1983; Russell *et al.*, 1986a). The relative reactivities of t-BuHgCl, i-PrHgCl, and n-BuHgCl are shown in Table 4 (Russell, 1986). It is also reported that $PhCO_2\cdot$ reacts with R_2Hg with a rate increasing according to the sequence R = Me < 1°-alkyl < 2°-alkyl < 3°-alkyl (Nugent and Kochi, 1977). The reactions of Scheme 33 occur with A· = $PhSO_2\cdot$, $PhSO\cdot$, $PhS\cdot$, $I\cdot$ (Russell *et al.*, 1984). With A = HgCl, the reaction also occurs, but now the electron transfer process occurs as shown in (49).

$$R\cdot + R'CH{=}CHA \longrightarrow R'\dot{C}HCH(R)A$$
$$R'\dot{C}HCH(R)A \longrightarrow R'CH{=}CHR + A\cdot$$
$$A\cdot + RHgCl \longrightarrow AHgCl + R\cdot$$

Scheme 33

$$HgCl + RHgCl \longrightarrow HgCl_2 + Hg^0 + R\cdot \quad (49)$$

TABLE 4

Relative reactivities of RHgCl at 35–40°C in reactions (47)/(48), (49)[a]

Attacking radical	Relative reactivity of RHgCl t-Bu : i-Pr : n-Bu
PhS·	1 : 0.08 : <0.003
PhSe·	1 : 0.3 : <0.004
$RCH_2\dot{C}HP(O)(OEt)_2$	1 : 0.2 : <0.003
HgCl	1 : 0.011[b] : 0.0001

[a] Russell, 1986
[b] cC_6H_{11}

Reactions of 1,1-diarylethylenes illustrate the importance of the acceptor or donor abilities of the diarylmethyl radical on the course of the photostimulated reaction with t-BuHgCl (Russell *et al.*, 1986b). With 1,1-diphenylethylene, a chain reaction does not occur and the t-$BuCH_2CPh_2\cdot$ radicals mainly disproponate to form t-$BuCH{=}CPh_2$ and t-$BuCH_2CHPh_2$ in a 1 : 1 ratio (50). When the structure of the diarylethylene is modified so that t-$BuCH_2CAr_2\cdot$ becomes an acceptor or donor radical, a chain reaction with

$$\text{t-BuHgCl} + CH_2{=}CPh_2 \xrightarrow{h\nu} \text{t-BuCH}{=}CPh_2 + \text{t-BuCH}_2CHPh_2 + HgCl_2 + Hg^0 \quad (50)$$

$$\text{t-BuHg} + CH_2{=}C(C_6H_4OMe\text{-}p)_2 \xrightarrow{h\nu} \text{t-BuCH}{=}C(C_6H_4OMe\text{-}p)_2 + HCl + Hg^0 \quad (51)$$

$$\text{t-BuHgCl} + CH_2{=}C(C_6H_4NO_2\text{-}p)_2 \xrightarrow{h\nu} \text{t-BuCH}_2C(C_6H_4NO_2\text{-}p)_2^- + ClHg^+ \quad (52)$$

t-BuHgCl occurs. With Ar = p-$MeOC_6H_4$, the diarylmethyl radical is a donor species; electron transfer to RHgCl now occurs and the overall process (51) is observed. On the other hand, with Ar = p-$O_2NC_6H_4$, the diarylmethyl radical is an acceptor radical and electron transfer from RHgCl results in process (52). Since anions such as $RC(C_6H_4NO_2\text{-}p)_2^-$ are unreactive towards t-Bu·, the dominant reaction product isolated from (52) is t-$BuCH_2CH(C_6H_4NO_2\text{-}p)_2$. Surprisingly, a free radical chain process (53) between t-BuHgCl and $CH_2{=}CPh_2$ does occur in the presence of an excess of $MeC(CO_2Et)_2^-$ (Russell *et al.*, 1986b).

$$\text{t-BuHgCl (excess)} + Ph_2C{=}CH_2 + MeC(CO_2Et)_2^- \xrightarrow[h\nu,\ 2\ h]{DMSO} \text{t-BuCH}_2C(Ph)_2\text{Bu-t} \ (65\%) \quad (53)$$

In (53), $Ph_2C{=}CH_2$ is an effective trap for t-Bu· while $MeC(CO_2Et)_2^-$ has no measurable reactivity towards t-Bu·. In the presence of the malonate anion, electron transfer (54) may occur to yield t-$BuCH_2C(Ph)_2^-$. The

$$\text{t-BuCH}_2CPh_2\cdot + MeC(CO_2Et)_2^- \rightleftharpoons Me\dot{C}(CO_2Et)_2 + \text{t-BuCH}_2CPh_2^- \quad (54)$$

t-$BuCH_2CPh_2^-$ efficiently traps t-Bu· in Scheme 13 to form the di-t-butylated product. Apparently $MeC(CO_2Et)_2\cdot$ acts as A· in reactions (47)/(48) to regenerate t-Bu·. The photostimulated reaction (53) is drastically retarded by 10 mol % of $(t\text{-}Bu)_2NO\cdot$ whereas in the absence of $MeC(CO_2Et)_2^-$ the nonchain reaction (50) is not appreciably affected by the nitroxide. Reaction (53) does not occur when the less easily oxidized $HC(CO_2Et)_2^-$ is employed. Furthermore, the oxidative dimerization of $MeC(CO_2Et)_2^-$ does not occur when excess t-BuHgCl is employed but $(EtO_2C)_2C(Me)C(Me)(CO_2Et)_2$ becomes a significant product when the concentration of t-BuHgCl is low.

A further example of electron transfer in the reactions of RHgCl is the allylic substitution of Scheme 34 which can be observed with a donor Q· (e.g. $Bu_3Sn\cdot$) or an acceptor Q· (e.g. $PhSO_2\cdot$) (Russell and Ngoviwatchai, unpublished).

$$R\cdot + CH_2{=}CHCH_2Q \longrightarrow RCH_2\dot{C}HCH_2Q$$
$$RCH_2\dot{C}HCH_2Q \longrightarrow RCH_2CH{=}CH_2 + Q\cdot$$
$$D\cdot + RHgCl \longrightarrow D^+ + Cl^- + Hg^0 + R\cdot$$
$$A\cdot + RHgCl \longrightarrow AHgCl + R\cdot$$

Scheme 34

4 Free radical chain reactions involving radical cations

ALIPHATIC SUBSTITUTION PROCESSES

Intra- and inter-molecular substitution reactions (55) and (56) involving the dialkylamino radical cations formed from protonated N-haloamines are well known and are based upon the initial work of Hofmann (1883) and of

$$R\cdot + R_2'NHX^+ \longrightarrow RX + R_2'NH^{\ddagger} \quad (55)$$
$$R_2'NH^{\ddagger} + RH \longrightarrow R_2'NH_2^+ + R\cdot \quad (56)$$

Löffler and Freytag (1909). The free radical nature of these processes was firmly established by Wawzonek and Thelan (1950) and the intramolecular process of the Hofmann–Löffler–Freytag reaction has been extended to intermolecular halogenation by Minisci *et al.* (1965), Deno *et al.* (1971), Spanswick and Ingold (1979), Tanner and Mosher (1978) and by Tanner *et al.* (1985a). The Hofmann–Löffler–Freytag reaction does not involve electron transfer although strong polar effects are operative in the attack of $R_2'NH^{\ddagger}$ upon a system of carbon—hydrogen bonds.

Reactions analogous to the $S_{RN}1$ process of Scheme 10 are not important for cations in aliphatic substitution processes. There are a number of chain reactions leading to aromatic substitution *via* radical cation intermediates involving addition of a nucleophile to an aromatic radical cation (Eberson and Jönsson, 1980; Eberson, 1983). Benzylic substitution involving Co(III) as a reactant occurs readily (Heiba *et al.*, 1969) and involves the fragmentation processes (57) and (58). Although there are many examples of the

$$[C_6H_5CH_3]^{\ddagger} \longrightarrow C_6H_5CH_2\cdot + H^+ \quad (57)$$
$$[p\text{-}MeOC_6H_4CH_2SPh]^{\ddagger} \longrightarrow p\text{-}MeOC_6H_4CH_2{}^+ + PhS\cdot \quad (58)$$

formation of radical anions by the addition of a radical to an anion, there are few examples of the addition of a radical to a cation to form a radical cation capable of continuing a chain reaction.

The electrochemical oxidation of N-benzylaziridine to yield the macrocycle [14] illustrates a chain process involving both fragmentation of a radical cation and electron transfer (Scheme 35) (Kossi *et al.*, 1980).

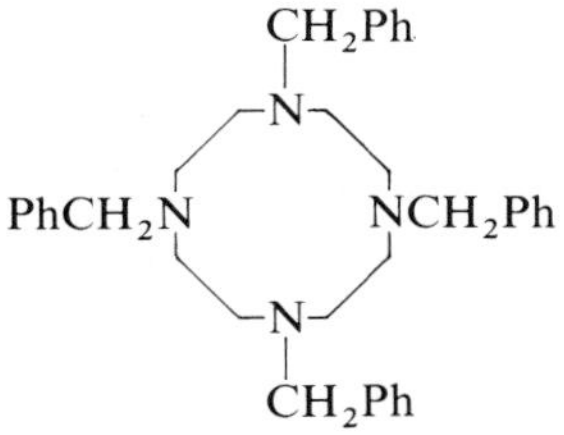

$$\text{cyclo-}[N(CH_2Ph)CH_2CH_2]_4$$

[14]

$$PhCH_2N\triangleleft \xrightarrow{-e} PhCH_2\dot{N}^{+}\triangleleft$$

$$PhCH_2\dot{N}^{+}\triangleleft + PhCH_2N\triangleleft \longrightarrow PhCH_2\dot{N}CH_2CH_2\overset{+}{N}(CH_2Ph)\triangleleft$$

$$\longrightarrow \longrightarrow PhCH_2\dot{N}CH_2CH_2N(CH_2Ph)CH_2CH_2N(CH_2Ph)CH_2CH_2\overset{+}{N}(CH_2Ph)\triangleleft$$

$$\xrightarrow{PhCH_2N\triangleleft} [14] + PhCH_2\dot{N}^{+}\triangleleft$$

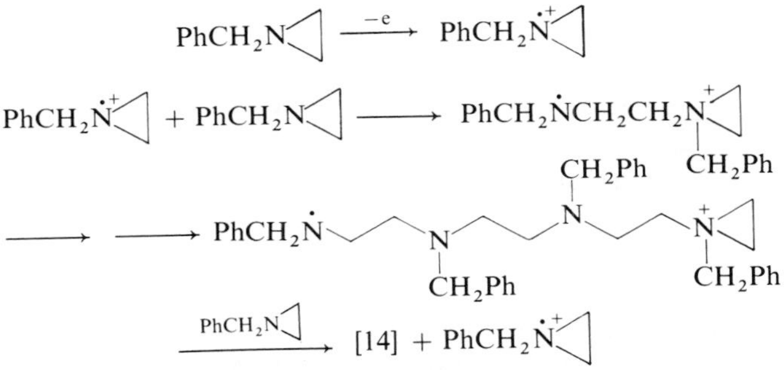

Scheme 35

The isomerization of oxiranes to ketones occurs under anodic oxidation with the consumption of less than 0.1 faraday per mole (Delaunay *et al.*, 1982). The chain sequence of Scheme 36 has been suggested.

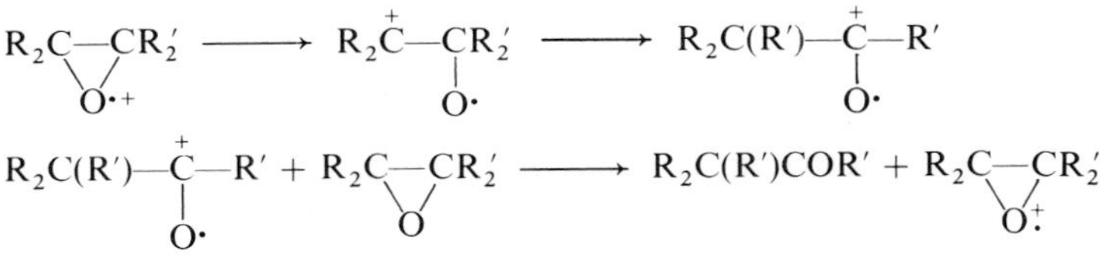

$$R_2C\underset{O^{\cdot+}}{—}CR'_2 \longrightarrow R_2\overset{+}{C}—CR'_2(O^{\cdot}) \longrightarrow R_2C(R')—\overset{+}{C}(O^{\cdot})—R'$$

$$R_2C(R')—\overset{+}{C}(O^{\cdot})—R' + R_2C\underset{O}{—}CR'_2 \longrightarrow R_2C(R')COR' + R_2C\underset{O^{\dot{+}}}{—}CR'_2$$

Scheme 36

The conversion of diazoalkanes to tetrasubstituted ethylenes by electrochemical (Jugelt and Pragst, 1968) or chemical oxidants (Bethell *et al.*, 1979) is believed to follow a chain pathway involving addition, elimination and

$$R_2\dot{C}\text{-}\overset{+}{N}\equiv N + R_2C{-}N_2 \longrightarrow \underset{\displaystyle R_2C{-}N{=}N\cdot}{R_2C\text{-}N_2^{+}} \xrightarrow{-2N_2} R_2C \overset{\cdot+}{-} CR_2$$

$$R_2\dot{C}^{+}{-}CR_2 + R_2C{=}N_2 \longrightarrow R_2C{=}CR_2 + R_2\dot{C}{-}\overset{+}{N}\equiv N$$

Scheme 37

electron transfer steps as shown in Scheme 37. It has been suggested by Alder (1980) that hydrogen–deuterium exchange at the α-position of tertiary amines (Gardini and Bargon, 1980) can involve the chain sequence of Scheme 38.

$$R_2N\dot{C}R_2' + BD^{+} \rightleftharpoons R_2\dot{N}\overset{+}{C}DR_2' + B$$

$$R_2\dot{N}\overset{+}{C}DR_2' + R_2NCHR_2' \rightleftharpoons R_2NCDR_2' + R_2\dot{N}\overset{+}{C}HR_2'$$

$$R_2\dot{N}\overset{+}{C}HR_2' + B \rightleftharpoons R_2N\dot{C}R_2' + BH^{+}$$

$$R_2NCHR_2' + BD^{+} \longrightarrow R_2NCDR_2' + BH^{+}$$

Scheme 38

The mechanism of Scheme 38 explains the unusual exchange process (59) with inside protonation of a bicyclic diamine (Alder *et al.*, 1979).

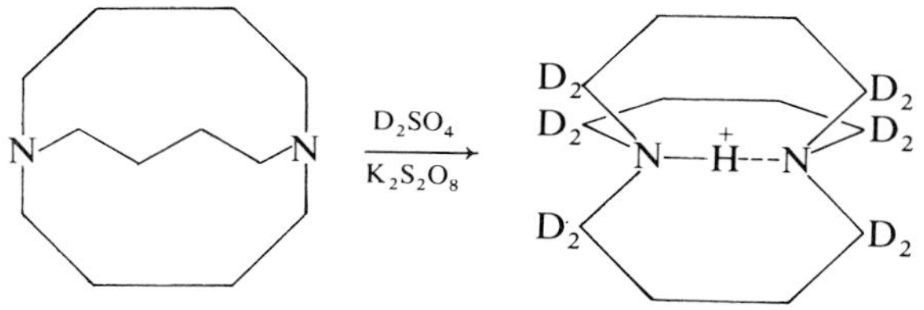

$$\text{N}\quad\text{N} \xrightarrow[K_2S_2O_8]{D_2SO_4} \text{N}{-}\overset{+}{H}{-}{-}\text{N} \; (D_2 \times 6) \qquad (59)$$

ALKENE DIMERIZATION AND AUTOXIDATION

Alkenes can be converted into radical cations by a variety of oxidizing agents such as Fe(III), Ce(IV), Cu(II), $Ar_3N^{\dot{+}}$, $NO^{+}PF_6^{-}$ or a mixture of diphenylpicrylhydrazyl and Ph_3C^{+} (Ledwith, 1972; Tang *et al.*, 1978; Nelsen *et al.*, 1984). Electrochemical oxidation or photochemical electron transfer to an excited sensitizer molecule also efficiently leads to radical cations.

Alkene dimerizations were originally described by Ledwith (Ledwith, 1972; Carruthers *et al.*, 1969) for the 2 + 1 cycloaddition reactions of

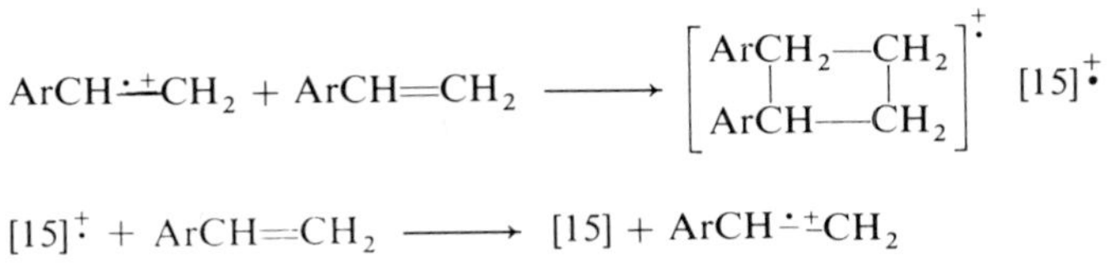

$$ArCH\overset{\cdot+}{-}CH_2 + ArCH{=}CH_2 \longrightarrow \left[\begin{array}{c} ArCH_2{-}CH_2 \\ | \quad\quad | \\ ArCH{-}CH_2 \end{array}\right]^{\cdot+} \quad [15]^{\cdot+}$$

$$[15]^{\cdot+} + ArCH{=}CH_2 \longrightarrow [15] + ArCH\overset{\cdot+}{-}CH_2$$

Scheme 39

enamines such as N-vinylcarbazole or 1,1-bis(*p*-dimethylaminophenyl)ethylene. The process of Scheme 39 also occurs with phenyl vinyl ether and dimethylindene (Mattes and Farid, 1982) with initiation by various metal ions, by electrochemical oxidation (Cedheim and Eberson, 1976) or by photosensitizers such as acetone or fluorenone, particularly in the presence of O_2. In the presence of the photosensitizer (S) and oxygen, the initiation

$$ArCH{=}CH_2 + S^* \rightleftharpoons ArCH\overset{\cdot+}{-}CH_2\ S^{\cdot-} \xrightarrow{O_2} ArCH\overset{\cdot+}{-}CH_2 + S + O_2^{\cdot-} \quad (60)$$

step can be formulated as shown in (60). Triarylamines promote the metal-catalysed reactions presumably by the added initiation process (61).

$$M^{+(n)} + Ar_3'N \longrightarrow M^{+(n-1)} + Ar_3'N^{\cdot+} \xrightarrow{ArCH{=}CH_2} Ar_3'N + ArCH\overset{\cdot+}{-}CH_2 \quad (61)$$

The stereochemistry of the cyclization process suggests that stepwise addition of $ArCH\overset{\cdot+}{-}CH_2$ to $ArCH{=}CH_2$ followed by cyclization is less favourable than a concerted reaction leading directly to [15] (Pabon and Bauld, 1983, 1984). Mixed dimers are formed in the 2 + 1 cyclization reaction with high chemo- and regio-selectivity (Pabon *et al.*, 1984). The reaction of the radical cation of a dienophile with a diene can also lead to a concerted 4 + 1 cyclization (Bellville *et al.*, 1981; Bauld, *et al.*, 1983b) with high chemo- and regio-selectivity as well as *endo* stereoselectivity (Pabon *et al.*, 1983; Bellville and Bauld, 1982; Bellville, *et al.*, 1983), but in certain instances the 2 + 1 cyclization to a cyclobutane predominates over the Diels–Alder route (Pabon *et al.*, 1984). The 2 + 1 or 4 + 1 cyclization can be initiated by $Ar_3N^{\cdot+}$ or by photochemical electron transfer to a molecule such as 1,4-dicyanobenzene (Bauld *et al.*, 1983a,b; Neunteufel and Arnold, 1973; Farid and Shealer, 1973).

The 2 + 1 and 4 + 1 cyclizations are considered to involve a concerted but nonsynchronous process in solution (Bellville *et al.*, 1983). However, in the gas phase the addition of butadiene radical cation to butadiene proceeds *via* a stepwise mechanism involving [16] as an intermediate (Groenewald and Gross, 1984a,b). The addition of $CH_2{=}CH{-}CH{=}CH_2^{\cdot+}$ to

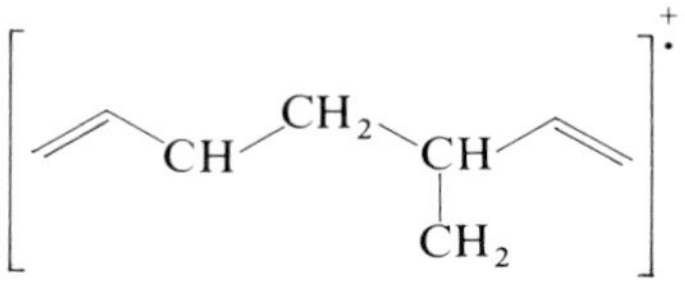

$$\left[CH_2{=}CH{-}CH{-}CH_2{-}CH(CH_2){-}CH{=}CH_2 \right]^{+\cdot}$$

[16]

$CH_2{=}CH{-}CH{=}CH_2$ has been studied by tandem mass spectroscopy (Dass and Gross, 1984), metastable ion analysis (Groenewold *et al.*, 1984) and by FT–MS (Groenewold and Gross, 1984a,b).

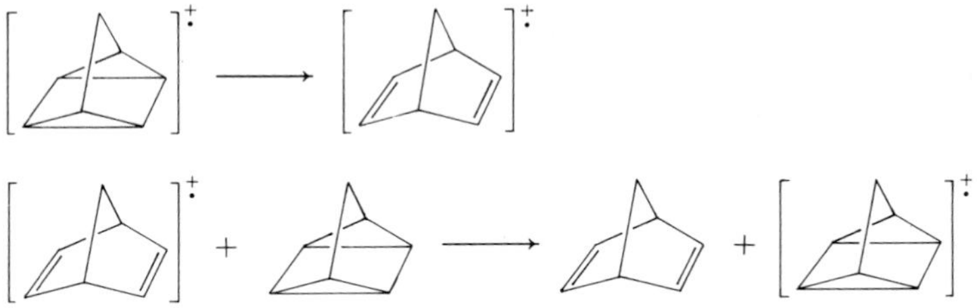

Scheme 40

The reversal of the 2 + 1 cyclization is recognized to proceed by a chain process, particularly in the quadricyclene–norbornadiene interconversion (Scheme 40) (Borsub and Kutal, 1984; Yasufuka and Kutal, 1984).

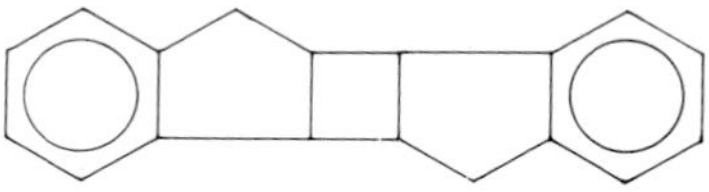

[17]

A variety of 2 + 2 cyclodimers will undergo reversion in a photosensitized redox reaction (Pac *et al.*, 1977; Majima *et al.*, 1978, 1980). In the presence of a sensitizer (S) such as anthracene and *p*-NCC_6H_4CN, indene dimers [17] and other cyclobutane derivatives participate in Scheme 41.

$$S^* + p\text{-}NCC_6H_4CN \longrightarrow S^{+\cdot} + p\text{-}NCC_6H_4CN^{-\cdot}$$

$$S^{+\cdot} + [17] \longrightarrow S^{+\cdot} \text{----} [17]$$

$$S^{+\cdot} \text{----} [17] \longrightarrow 2\ \text{(indene)} + S^{+\cdot}$$

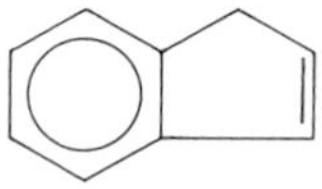

Scheme 41

Reaction of molecular oxygen with enamines has been proposed to proceed *via* the chain mechanism of Scheme 42 (Malhotra *et al.*, 1968).

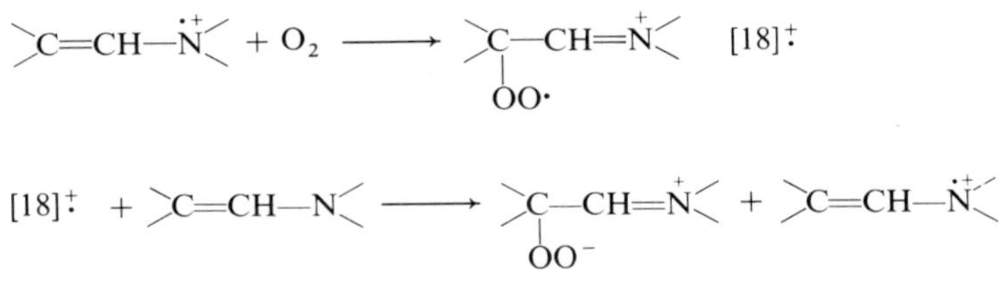

$$\gt C{=}CH{-}\dot{N}^{+}\lt \; + \; O_2 \longrightarrow \gt C(OO\cdot){-}CH{=}\overset{+}{N}\lt \quad [18]^{\ddagger}$$

$$[18]^{\ddagger} + \gt C{=}CH{-}N\lt \longrightarrow \gt C(OO^-){-}CH{=}\overset{+}{N}\lt \; + \; \gt C{=}CH{-}\dot{N}^{+}\lt$$

Scheme 42

Alkenes and 1,3-dienes also react with triplet oxygen in the presence of $Ar_3N^{\ddagger}$, Lewis Acids or Ph_3C^+ with ultraviolet irradiation (Barton *et al.*, 1972, 1974, 1975; Hayes *et al.*, 1978) to form cyclic peroxides such as [19]

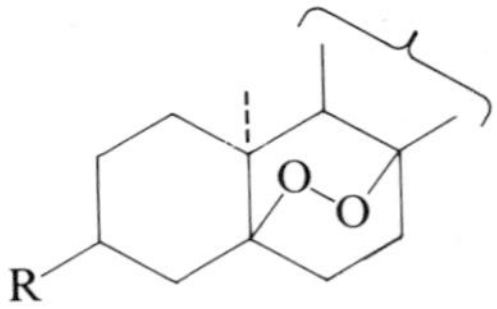

R

[19]

$H_2C{-}CH_2$

$Ar_2C \qquad CAr_2$

$O{-}O$

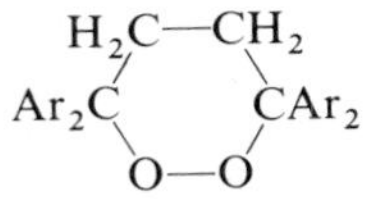

[20]

and [20]. It was reported by Tang *et al.* (1978) that these processes could involve a chain reaction according to Scheme 43. Clearly a chain process is involved in the conversion of alkenes such as adamantylideneadamantane to

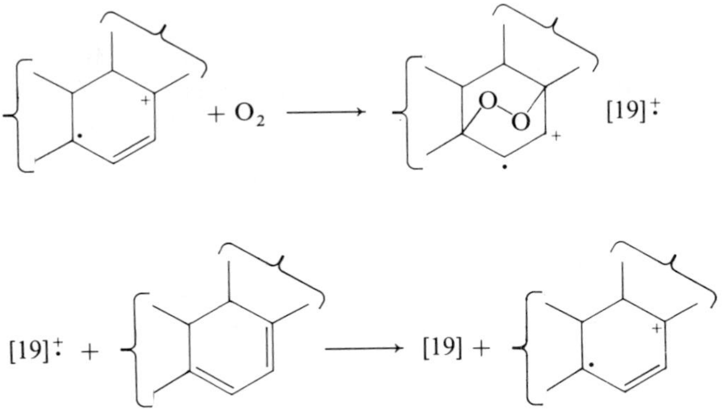

$$+ \; O_2 \longrightarrow [19]^{\ddagger}$$

$$[19]^{\ddagger} + \longrightarrow [19] +$$

Scheme 43

the dioxetane either anodically or catalysed by electron transfer to $Ar_3N^{\dot{+}}$ in CH_2Cl_2 (Clennan *et al.*, 1981; Nelsen *et al.*, 1984). Chain lengths up to 600 have been reported for the process of Scheme 44, and at low tempera-

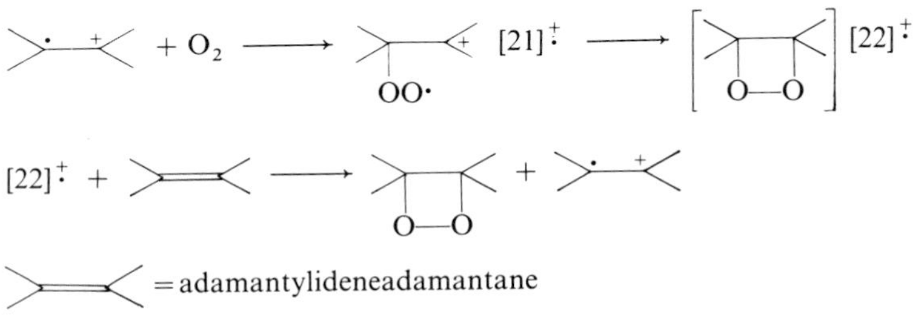

$$\text{Ad}\dot{-}\overset{+}{\text{Ad}} + O_2 \longrightarrow [21]^{\dot{+}} \longrightarrow [22]^{\dot{+}}$$

$$[22]^{\dot{+}} + \text{Ad}{=}\text{Ad} \longrightarrow \text{dioxetane} + \text{Ad}\dot{-}\overset{+}{\text{Ad}}$$

Ad=Ad = adamantylideneadamantane

Scheme 44

tures the dioxetane radical cation $[22]^{\dot{+}}$ has been observed by esr spectroscopy (Nelsen *et al.*, 1986). Unsymmetrically substituted alkenes produce the dioxetanes nonstereospecifically as expected for a mechanism involving $[21]^{\dot{+}}$ as an intermediate (Ando *et al.*, 1982). Reactions involving singlet oxygen can be excluded since $^1O_2$ fails to form the dioxetane from olefins with unprotected alkyl groups. Thus, isopropylideneadamantane reacts with $^3O_2$ by the chain mechanism of Scheme 44 to form the dioxetane but reacts with $^1O_2$ to yield only the allylic hydroperoxide (Nelsen *et al.*, 1984; Nelsen and Teasley, 1986).

Under proper conditions, chain reactions involving radical cations should be important in *cis-trans* isomerizations of alkenes and 1,2-disubstituted cyclopropanes, particularly with aryl substituents (Roth and Schilling, 1979, 1980a). There is abundant evidence that electron transfer is involved in the *cis-trans* isomerizations of 1,2-diarylcyclopropanes photosensitized by naphthalenes and cyano-substituted aromatics (Hixson *et al.*, 1974; Hixson, 1979; Wong and Arnold, 1979) and in the isomerization of 1,1,2,2-tetraphenylcyclopropane to 1,1,3,3-tetraphenylpropene (Arnold and Humphreys, 1979; Wayner and Arnold, 1985). However, on the basis of CIDNP signals observed in the presence of chloranil, it appears that the isomerization between the radical cations of *cis*- and *trans*-1,2-diphenylcyclopropane occurs slowly (or not at all) in competition with the back electron transfer from the chloranil radical anion (Roth and Schilling, 1980b, 1981). Processes initiated by electron transfer to radical cations such as $Ar_3N^{\dot{+}}$, or more powerful oxidizing agents which can act in an irreversible manner, would have a better chance to occur by a chain process in which the rearranged radical cation continues the chain by electron transfer with the unreacted alkene or cyclopropane.

5 Concluding remarks

The occurrence of electron transfer reactions greatly expands the scope of free radical chain processes and provides a crossing of the boundary between polar and homolytic processes. The formation of a radical ion from a neutral substrate or a radical from an ion can be considered to be a type of activation (Chanon, 1982, 1985). Electron transfer between radical ions and easily oxidized or reduced substrates is a common process. Somewhat less common in the absence of metal ions are electron transfers between neutral radicals and easily oxidized or reduced diamagnetic substances. The mechanistic possibilities for free radical chain processes are greatly expanded by the recognition that there is a bridge between radical ion chemistry and the chemistry of separated radicals and ions since in appropriate cases radical ions can dissociate to form a radical and an ion, or conversely, radicals may add to ions to form radical ions. It is indeed this bridging of the chemistry of radical ions and of free radicals which is the basis of numerous thermal or photostimulated substitution processes in organic, organometallic and inorganic chemistry of which the $S_{RN}1$ process is perhaps the most notable example.

The occurrence of free radical chain processes involving electron transfer is, of course, most likely to occur when easily oxidized or reduced substrates are involved, with radical anion formation being favoured by strongly basic or reductive conditions and radical cations by acidic or oxidative environments.

The experimental proof of a free radical chain mechanism usually requires relatively simple kinetic analysis although questions inevitably remain concerning the exact timing and nature of some of the individual steps which contribute to the overall chain reaction. Electron transfer can, of course, also be involved in multistep but nonchain processes which interconvert diamagnetic reactants and products (Russell *et al.*, 1964; Eberson, 1982). In general, in these cases one expects a spectrum of substitution mechanisms with the radical chain processes representing one extreme and reaction proceeding *via* transition states involving an electron shift process (Pross, 1985) as the other extreme. In between these extremes are multistep but nonchain processes involving intermediates, either free or in a cage, resulting from electron transfer processes. However, reaction by the radical chain mechanism is not in all cases simply an alternative to a one-step polar process since the product and the selectivity observed can be quite different in the two processes. This is particularly well illustrated in the reactions of $Me_2C{=}NO_2^-$ which undergo alkylation at oxygen in typical S_N2 processes but which undergo carbon alkylation in S_{RN} substitution reactions.

References

Alder, R. W. (1980). *J. Chem. Soc., Chem. Commun.* 1184

Alder, R. W., Casson, A. and Sessions, R. B. (1979). *J. Am. Chem. Soc.* **101**, 3652

Aleksandrov, Yu. A., Druzhkov, O. N. Zhil'tzov, S. F. and Razuvaev, G. A. (1964). *Dokl. Akad. SSSR* **157**, 1395

Al-Khalil, S. I., Bowman, W. R. and Symons, M. C. R. (1986). *J. Chem. Soc., Perkin Trans 1* 555

Ando, W., Kake, Y. and Takata, R. (1982). *J. Am. Chem. Soc.* **104**, 7314

Arnold, D. R. and Humphreys, D. W. R. (1979). *J. Am. Chem. Soc.* **101**, 2743

Ashby, E. C. and Argyropolous, J. N. (1985). *J. Org. Chem.* **50**, 3274

Ashby, E. C., DePriest, R. N., Goel, A. B., Wenderoth, B. and Pham, T. N. (1984). *J. Org. Chem.* **49**, 3545

Ashby, E. C., Park, W. S., Goel, A. B. and Su, W.-Y. (1985a). *J. Org. Chem.* **50**, 5184

Ashby, E. C., Su. W.-Y. and Dham, T. N. (1985b). *Organometallics* **4**, 1493

Bank, S. and Noyd, D. A. (1973). *J. Am. Chem. Soc.* **95**, 8203

Barker, S. D. and Norris, R. K. (1979). *Tetrahedron Lett.* 1973

Barltrop, J. A. and Bradbury, D. (1973). *J. Am. Chem. Soc.* **95**, 5085

Barton, D. H. R., LeClerc, G., Magnus, P. D. and Menzies, I. D. (1972). *J. Chem. Soc., Chem. Commun.* 447

Barton, D. H. R., Hayes, R. K., Magnus, P. D. and Menzies, I. D. (1974). *J. Chem. Soc., Chem. Commun.* 511

Barton, D. H. R., Hayes, R. K., LeClerc, G., Magnus, P. D. and Menzies, I. D. (1975). *J. Chem. Soc., Perkin Trans. 1* 2055

Bauld, N. L., Bellville, D. J. and Pabon, R. A. (1983a) *J. Am. Chem. Soc.* **105**, 5158

Bauld, N. L., Bellville, D. J., Pabon, R. A., Chelsky, R. and Green, G. J. (1983b). *J. Am. Chem. Soc.* **105**, 2378

Bellville, D. J. and Bauld, N. L. (1982) *J. Am. Chem. Soc.* **104**, 2655

Bellville, D. J., Wirth, D. D. and Bauld, N. L. (1981) *J. Am. Chem. Soc.* **103**, 718

Bellville, D. J., Bauld, N. L., Pabon, R. A. and Gardner, S. (1983) *J. Am. Chem. Soc.* **105**, 3584

Bethell, D., Handoo, K. L., Fairhurst, S. A. and Sutcliffe, L. H. (1979). *J. Chem. Soc., Perkin Trans 2* 707

Beugelmans, R., Lechevallier, A. and Rosseau, H. (1983). *Tetrahedron Lett.* 1187

Beugelmans, R., Lechevallier, A. and Rosseau, H. (1984). *Tetrahedron Lett.* 2347

Blackburn, E. V. and Tanner, D. D. (1980). *J. Am. Chem. Soc.* **102**, 692

Boiko, V. N., Shchupak, G. N. and Yagupol'skii, L. M. (1977). *J. Org. Chem. USSR, [Engl. Trans.]* **13**, 1972

Bordwell, F. G. and Clemens, A. H. (1982). *J. Org. Chem.* **47**, 2510

Borsub, N. and Kutal, C. (1984). *J. Am. Chem. Soc.* **106**, 4826

Bowman, W. R. and Richardson, G. D. (1980). *J. Chem. Soc., Perkin Trans 1* 1407

Bowman, W. R. and Richardson, G. D. (1981). *Tetrahedron Lett.* 1551

Bowman, W. R. and Symons, M. C. R. (1983). *J. Chem. Soc., Perkin Trans 2* 25

Bowman, W. R., Rakshit, D. and Valmas, M. D. (1984). *J. Chem. Soc., Perkin Trans 1* 2327

Boyle, W. J. and Bunnett, J. F. (1974). *J. Am. Chem. Soc.* **96**, 1418

Bruce, M. I., Kehoe, D. C., Matisons, J. G., Nicholson, B. K., Rieger, P. H. and Williams, M. L. (1982). *J. Chem. Soc. Chem. Commun.* 443

Brunet, J.-J., Sidot, C. and Caubere, P. (1983). *J. Org. Chem.* **48**, 1166

Bunnett, J. F. and Takayama, H. (1968a). *J. Org. Chem.* **33**, 1924

Bunnett, J. F. and Takayama, H. (1968b). *J. Am. Chem. Soc.* **90**, 5173

Bunnett, J. F., Creary, X. and Sundberg, J. E. (1976). *J. Org. Chem.* **41**, 1707
Cantacuzène, D. and Dorme, R. (1975). *Tetrahedron Lett.* 2037
Carruthers, R. A., Crellin, R. A. and Ledwith, A. (1969). *J. Chem. Soc., Chem. Commun.* 252
Cedheim, L. and Eberson, D. (1976). *Acta Chem. Scand.* **B30**, 527
Chanon, M. (1982). *Bull. Soc. Chim. Fr.* 197
Chanon, M. (1985). *Bull. Soc. Chim. Fr.* 209
Chung, S.-K. (1980). *J. Org. Chem.* **45**, 3513
Chung, S.-K. and Chung, F.-F. (1979). *Tetrahedron Lett.* 2473
Chung, S.-K. and Filmore, K. L. (1983). *J. Chem. Soc., Chem. Commun.* 358
Citterio, A., Minisci, F., Porta, O. and Sesana, G. (1977). *J. Am. Chem. Soc.* **99**, 7960
Clennan, E. L., Simmons, W. and Almgren, C. W. (1981). *J. Am. Chem. Soc.* **103**, 2098
Cohen, T., McMullen, C. H. and Smith, K. (1968). *J. Am. Chem. Soc.* **90**, 6866
Crozet, M. P. and Vanelle, P. (1985). *Tetrahedron Lett.* 323
Crozet, M. P., Surzur, J.-M., Vanelle, P., Ghiglione, C. and Maldonada, J. (1985). *Tetrahedron Lett.* 1023
Darchen, A., Mahé, C. and Patin, H. (1982). *J. Chem. Soc., Chem. Commun.* 243
Dass, C. and Gross, M. L. (1984). *J. Am. Chem. Soc.* **106**, 5775
Delaunay, J., Lebouc, A., Tallec, A. and Simonet, J. (1982). *J. Chem. Soc., Chem. Commun.* 387
Deno, N. C., Billups, W. E., Fishbein, R., Pierson, C., Whalen, R. and Wyckoff, J. C. (1971). *J. Am. Chem. Soc.* **93**, 438
Detar, D. F. and Turetzky, M. N. (1955). *J. Am. Chem. Soc.* **77**, 1745
Detar, D. F. and Turetzky, M. N. (1956). *J. Am. Chem. Soc.* **78**, 3925, 3928
Dickerman, S. C. and Weiss, K. (1957). *J. Org. Chem.* **22**, 1020
Dickerman, S. C., Weiss, K. and Ingberman, A. K. (1956). *J. Org. Chem.* **21**, 380
Dickerman, S. C., Weiss, K. and Ingberman, A. K. (1958). *J. Am. Chem. Soc.* **80**, 1904
Eberson, L. (1982). *Adv. Phys. Org. Chem.* **18**, 79
Eberson, L. (1983). *J. Mol. Cat.* **20**, 27
Eberson, L. and Jönsson, L. (1980). *J. Chem. Soc., Chem. Commun.* 1187
Farid, S. and Shealer, S. E. (1973). *J. Chem. Soc., Chem. Commun.* 677
Feiring, A. E. (1983). *J. Org. Chem.* **48**, 347
Feiring, A. E. (1985). *J. Org. Chem.* **50**, 3269
Freeman, D. J. and Norris, R. K. (1976). *Austral. J. Chem.* **29**, 2631
Freeman, D. J., Norris, R. K. and Woolfenden, S. K. (1978). *Austral. J. Chem.* **31**, 2477
Gardini, G. P. and Bargon, J. (1980). *J. Chem. Soc., Chem. Commun.* 757
Garst, J. F. and Barton, F. E. (1969). *Tetrahedron Lett.* 587
Giese, B. (1983). *Angew. Chem. Int. Ed. Engl.* **22**, 753
Giese, B. (1985). *Angew. Chem. Int. Ed. Engl.* **24**, 553
Giese, B. and Gröningen, K. (1984). *Tetrahedron Lett.* 2473
Giese, B. and Horler, H. (1985). *Tetrahedron* **41**, 4025
Gilbert, B. C. and Norman, R. O. C. (1982). *Can. J. Chem.* **60**, 1379
Gilbert, B. C., Holmes, R. G. G., Laue, H. A. H. and Norman, R. O. C. (1976). *J. Chem. Soc., Perkin. Trans. 2* 1047
Gilbert, B. C., Holmes, R. G. G. and Norman, R. O. C. (1977). *J. Chem. Res. (M)* 1001
Girdler, D. J. and Norris, R. K. (1975). *Tetrahedron Lett.* 2375

Groenewold, G. S. and Gross, M. L. (1984a). *J. Am. Chem. Soc.* **106**, 6575
Groenewold, G. S. and Gross, M. L. (1984b). *J. Am. Chem. Soc.* **106**, 6569
Groenewold, G. S., Chess, E. K. and Gross, M. L. (1984). *J. Am. Chem. Soc.* **106**, 537
Groves, J. T. and Ma, K. W. (1974). *J. Am. Chem. Soc.* **96**, 6527
Hatem, J. and Waegell, B. (1973). *Tetrahedron Lett.* 2023
Hayes, R. K., Roberts, M. K. S. and Wilmot, I. D. (1978). *Austral. J. Chem.* **31**, 1737
Hegedus, L. S. and Miller, L. L. (1975). *J. Am. Chem. Soc.* **97**, 459
Hegedus, L. S. and Thompson, D. H. P. (1985) *J. Am. Chem. Soc.* **107**, 5663
Heiba, E. I., Dessau, R. M. and Koehl, W. J. (1969). *J. Am. Chem. Soc.* **91**, 6830
Hershberger, J. W. and Kochi, J. K. (1982). *J. Chem. Soc., Chem. Commun.* 212
Hixson, S. S. (1979). *Org. Photochem.* **4**, 191
Hixson, S. S., Boyer, J. and Gallucci, C. (1974). *J. Chem. Soc., Chem. Commun.* 540
Hofmann, A. W. (1883). *Chem. Ber.* **16**, 558, 586
Horner, L. and Anders, B. (1962). *Chem. Ber.* **95**, 2470
Horner, L. and Betzel, C. (1953). *Justus Liebigs Ann. Chem.* **175**, 579
Horner, L. and Schwenk, E. (1944). *Angew Chem.* **61**, 411
Hrabák, F. and Lokaj, J. (1970). *Collect. Czech. Chem. Commun.* **35**, 1081
Huyser, E. S. and Harmony, J. A. K. (1974). *In* "Methods in Free-Radical Chemistry" (ed. E. S. Huyser), Vol 5, p. 101. Marcel Dekker, Inc., New York
Huyser, E. S. and Kahl, A. A. (1970). *J. Org. Chem.* **35**, 3742
Huyser, E. S., Bredeweg, C. J. and VanScoy, R. M. (1964). *J. Am. Chem. Soc.* **86**, 4148
Huyser, E. S., Harmony, J. A. K. and McMillian, F. L. (1972). *J. Am. Chem. Soc.* **94**, 3176
Jawdosiuk, M., Makosza, M., Mudryk, B. and Russell, G. A. (1979). *J. Chem. Soc., Chem. Commun.* 488
Jugelt, W. and Pragst, F. (1968). *Angew Chem. Int. Ed.* **7**, 290
Kharasch, M. S. and Sosnovsky, G. (1958). *J. Am. Chem. Soc.* **80**, 765
Kharasch, M. S., Pauson, P. and Nudenberg, W. (1953). *J. Org. Chem.* **18**, 332
Kill, R. G. and Widdowson, D. A. (1976). *J. Chem. Soc., Chem. Commun.* 755
Kim, J. K. and Bunnett, J. F. (1970). *J. Am. Chem. Soc.* **92**, 7463, 7464
Kinney, R. J., Jones, W. D. and Bergman, R. G. (1978). *J. Am. Chem. Soc.* **100**, 635
Kochi, J. K. (1957). *J. Am. Chem. Soc.* **79**, 2942
Kochi, J. K. (1967). *Science* **155**, 415
Koelsch, C. F. and Boekelheide, V. (1944). *J. Am. Chem. Soc.* **66**, 412
Kornblum, N. (1975). *Angew. Chem. [Int. Ed. Engl.]* **14**, 734
Kornblum, N. (1982). *In* "The Chemistry of Functional Groups. Supplement F" (ed. S. Patai), Ch. 10, p. 361. John Wiley and Sons, New York
Kornblum, N. and Boyd, S. D. (1970). *J. Am. Chem. Soc.* **92**, 5784
Kornblum, N. and Cheng, L. (1977). *J. Am. Chem. Soc.* **42**, 2944
Kornblum, N. and Erickson, A. S. (1981). *J. Org. Chem.* **46**, 1037
Kornblum, N. and Fifolt, M. J. (1980). *J. Org. Chem.* **45**, 360
Kornblum, N. and Stuchal, F. W. (1970). *J. Am. Chem. Soc.* **92**, 1804
Kornblum, N., Michael, R. E. and Kerber, R. C. (1966). *J. Am. Chem. Soc.* **80**, 5660, 5662
Kornblum, N., Davies, T. M., Earl, G. W., Green, G. S., Holy, N. L., Kerber, R. C., Manthey, J. W., Musser, M. T. and Snow, D. H. (1967a). *J. Am. Chem. Soc.* **89**, 725
Kornblum, N., Davies, T. M., Earl, G. W., Holy, N. L., Kerber, R. C., Musser, M. T. and Snow, D. H. (1967b). *J. Am. Chem. Soc.* **89**, 5714
Kornblum, N., Davies, T. M., Earl, G. W., Pinnick, H. W. and Stuchal, F. W. (1968). *J. Am. Chem. Soc.* **90**, 6219

Kornblum, N., Boyd, S. D. and Stuchal, F. W. (1970a). *J. Am. Chem. Soc.* **92**, 5783
Kornblum, N., Swiger, R. T., Earl, G. W., Pinnick, H. W. and Stuchal, F. W. (1970b). *J. Am. Chem. Soc.* **92**, 5513
Kornblum, N., Boyd, S. D., Pinnick, H. W. and Smith, R. G. (1971). *J. Am. Chem. Soc.* **93**, 4316
Kornblum, N., Kestner, M. M. Boyd, S. C., and Cattran, L. C. (1973). *J. Am. Chem. Soc.* **95**, 3356
Kornblum, N., Boyd, S. D. and Ono, N. (1974). *J. Am. Chem. Soc.* **96**, 2580
Kornblum, N., Carlson, S. C. and Smith, R. G. (1978a). *J. Am. Chem. Soc.* **100**, 289
Kornblum, N., Carlson, S. C., Widner, J., Fifolt, M. J., Newton, B. N. and Smith, R. G. (1978b). *J. Org. Chem.* **43**, 1394
Kornblum, N., Carlson, S. C. and Smith, R. G. (1979a). *J. Am. Chem. Soc.* **101**, 647
Kornblum, N., Widmer, J. and Carlson, S. C. (1979b). *J. Am. Chem. Soc.* **101**, 658
Kornblum, N., Ackerman, P. and Swiger, R. T. (1980). *J. Org. Chem.* **45**, 5294
Kornblum, N., Singh, H. K. and Boyd, S. D. (1984). *J. Org. Chem.* **49**, 358
Korzan, D. G., Chen, F. and Ainsworth, C. (1971). *J. Chem. Soc. Chem. Commun.* 1053
Kossi, R., Simonet, J. and Dauphin, G. (1980). *Tetrahedron Lett.* 3575
Kurosawa, H., Okada, H. and Yasuda, M. (1980). *Tetrahedron Lett.* 959
Kurosawa, H., Okada, H. and Hattori, T. (1981). *Tetrahedron Lett.* 4495
Lamb, R. C., Ayers, P. W., Toney, M. K. and Garst, J. F. (1966). *J. Am. Chem. Soc.* **88**, 4261
Ledwith, A. (1972). *Acc. Chem. Res.* **5**, 133
Löffler, K. and Freytag, C. (1909). *Chem. Ber.* **42**, 3427
Lu, J.-J. and Truce, W. E. (1985). *J. Org. Chem.* **50**, 4885
Majima, T., Pac, C., Nakasone, A. and Sakurai, H. (1978). *J. Chem. Soc., Chem. Commun.* 490
Majima, T., Pac, C. and Sakurai, H. (1980). *J. Am. Chem. Soc.* **102**, 5265
Malhotra, S. K., Hostynek, J. J. and Lundin, A. F. (1968). *J. Am. Chem. Soc.* **90**, 6565
Mattes, S. L. and Farid, S. (1982). *Acc. Chem. Res.* **15**, 80
Mayake, H. and Yamamura, K. (1986). *Bull. Chem. Soc. Jpn.* **59**, 89
Meijs, G. F. (1984). *J. Org. Chem.* **49**, 3863
Meijs, G. F. (1985). *Tetrahedron Lett.* 105
Meijs, G. F. (1986). *J. Org. Chem.* **51**, 606
Meyers, C. Y. (1978). "Topics in Organic Sulfur Chemistry", p. 207. University Press, Ljubljana, Yugoslavia
Meyers, C. Y., Matthews, W. S., Ho, L. L., Kolb, V. M. and Parady, T. E. (1977). *In* "Catalysis in Organic Synthesis", p. 197. Academic Press, New York
Minisci, F., Galli, R. and Cecere, M. (1965). *Tetrahedron Lett.* 4663
Minisci, F., Citterio, A., Vismara, E. and Giordano, C. (1985). *Tetrahedron* **41**, 4170
Nelsen, S. F. and Teasley, M. F. (1986). *J. Org. Chem.*, **51**, 3221
Nelsen, S. F., Kapp, D. L. and Teasley, M. F. (1984). *J. Org. Chem.* **49**, 579
Nelsen, S. F., Kapp, D. L., Gerson, F. and Lopez, J. (1986). *J. Am. Chem. Soc.* **108**, 1027
Neunteufel, R. A. and Arnold, D. R. (1973). *J. Am. Chem. Soc.* **95**, 4080
Newcombe, P. J. and Norris, R. K. (1978). *Austral. J. Chem.* **31**, 2463
Nonhebel, D. C. and Waters, W. A. (1957). *Proc. Roy. Soc., A* **242**, 16
Norris, R. K. (1983). *In* "The Chemistry of Functional Groups, Supplement D", (eds S. Patai and Z. Rappoport), Ch. 16, p. 681. John Wiley and Sons, New York
Norris, R. K. and Randles, D. (1976). *Austral. J. Chem.* **29**, 2621
Norris, R. K. and Randles, D. (1982a). *Austral. J. Chem.* **35**, 1621
Norris, R. K. and Randles, D. (1982b). *J. Org. Chem.* **47**, 1047

Nugent, W. A. and Kochi, J. K. (1977). *J. Organometal. Chem.* **124**, 327, 349, 371
Ohno, A., Shio, T., Yamamoto, H. and Oka, S. (1981). *J. Am. Chem. Soc.* **103**, 2045
Ono, N., Tamura, R. and Kaji, A. (1980). *J. Am. Chem. Soc.* **102**, 2851
Ono, N., Tamura, R., Eto, H., Hamamoto, I., Nakatuska, T., Hayami, J.-I. and Kaji, A. (1983). *J. Org. Chem.* **48**, 3678
Ono, N., Miyake, H., Kamimura, A., Hamamoto, I., Tamura, R. and Kaji, A. (1985). *Tetrahedron* **41**, 4013
Ostovic, D., Roberts, R. M. G. and Kreevoy, M. M. (1983). *J. Am. Chem. Soc.* **105**, 7629
Pabon, R. A. and Bauld, N. L. (1983). *J. Am. Chem. Soc.* **105**, 633
Pabon, R. A. and Bauld, N. L. (1984). *J. Am. Chem. Soc.* **106**, 1145
Pabon, R. A., Bellville, D. J. and Bauld, N. L. (1983). *J. Am. Chem. Soc.* **105**, 5158
Pabon, R. A., Bellville, D. J. and Bauld, N. L. (1984). *J. Am. Chem. Soc.* **106**, 2730
Pac, C., Nakasone, A. and Sakurai, H. (1977). *J. Am. Chem. Soc.* **99**, 5866
Palacios, S. M., Santiago, A. N. and Rossi, R. A. (1984). *J. Org. Chem.* **49**, 4609
Palacios, S. M., Alonso, R. A. and Rossi, R. A. (1985). *Tetrahedron* **41**, 4147
Pierini, A. B., Penenory, A. B. and Rossi, R. A. (1985). *J. Org. Chem.* **50**, 2739
Popov, V. I., Boiko, V. N. and Yagupol'skii, L. M. (1977). *J. Org. Chem. USSR, [Engl. Trans.]* **13**, 1985
Pross, A. (1985). *Acc. Chem. Res.* **19**, 212
Prousek, J. (1980). *Czech. Chem. Commun.* **45**, 3347
Prousek, J. (1982). *Czech. Chem. Commun.* **47**, 1334
Razuvaev, R. A., Petukhov, G. G., Zhil'tsov, S. F. and Kudryavtsev, L. F. (1960). *Dokl, Akad. Sci.* **135**, 87
Rico, I., Cantacuzène, D. and Wakselman, C. (1981). *Tetrahedron Lett.* 3405
Rico, I., Cantacuzène, D. and Wakselman, C. (1983). *J. Org. Chem.* **48**, 1979
Rossi, R. A., Palacios, S. M. and Santiago, A. N. (1982). *J. Org. Chem.* **47**, 4654
Rossi, R. A., Santiago, A. N. and Palacios, S. M. (1984). *J. Org. Chem.* **49**, 3387
Roth, H. D. and Schilling, M. L. M. (1979). *J. Am. Chem. Soc.* **101**, 1898
Roth, H. D. and Schilling, M. L. M. (1980a). *J. Am. Chem. Soc.* **102**, 4303
Roth, H. D. and Schilling, M. L. M. (1980b). *J. Am. Chem. Soc.* **102**, 7956
Roth, H. D. and Schilling, M. L. M. (1981). *J. Am. Chem. Soc.* **103**, 7210
Russell, G. A. (1953). *J. Am. Chem. Soc.* **75**, 5100
Russell, G. A. (1959). *J. Am. Chem. Soc.* **81**, 4815, 4825
Russell, G. A. (1967). *Pure and Appl. Chem.* **15**, 185
Russell, G. A. (1982). *Chemia Stosowana* **26**, 317
Russell, G. A. (1986). *Preprints of Papers, Div. of Petroleum Chem.*, p. 891. 192nd National Meeting of the Am. Chem. Soc., Anaheim, California
Russell, G. A. and Bemis, A. G. (1965). *Chem. and Ind. (London)* 1262
Russell, G. A. and Danen, W. C. (1966). *J. Am. Chem. Soc.* **88**, 5663
Russell, G. A. and Danen, W. C. (1968). *J. Am. Chem. Soc.* **90**, 347
Russell, G. A. and Dedolph, D. F. (1985). *J. Org. Chem.* **50**, 2378
Russell, G. A. and Guo, D. (1984). *Tetrahedron Lett* 5239
Russell, G. A. and Hershberger, J. (1980a). *J. Chem. Soc., Chem Commun.* 216
Russell, G. A. and Hershberger, J. (1980b). *J. Am. Chem. Soc.* **102**, 7603
Russell, G. A. and Kaupp, G. (1969). *J. Am. Chem. Soc.* **91**, 3851
Russell, G. A. and Khanna, R. K. (1985a). *J. Am. Chem. Soc.* **107**, 1450
Russell, G. A. and Khanna, R. K. (1985b). *Tetrahedron* **41**, 4133
Russell, G. A. and Khanna, R. K. (1987). *Adv. in Chem. Ser., Am. Chem. Soc.* **215**, in press

Russell, G. A. and Ngoviwatchai, P. (1985). *Tetrahedron Lett.* 4975
Russell, G. A. and Pecoraro, J. M. (1979). *J. Am. Chem. Soc.* **101**, 3331
Russell, G. A. and Ros, F. (1985). *J. Am. Chem. Soc.* **107**, 2506
Russell, G. A. and Tashtoush, H. (1983). *J. Am. Chem. Soc.* **105**, 1398
Russell, G. A. and Weiner, S. A. (1969). *Chem. and Ind. (London)* 659
Russell, G. A., Janzen, E. G., Becker, H.-D. and Smentowski, F. (1962a). *J. Am. Chem. Soc.* **84**, 2652
Russell, G. A., Moye, A. J. and Nagpal, K. (1962b). *J. Am. Chem. Soc.* **84**, 4154
Russell, G. A., Janzen, E. G. and Strom, E. T. (1964). *J. Am. Chem. Soc.* **86**, 1807
Russell, G. A., Geels, E. J., Moye, A. J., Janzen, E. G., Bemis, A. G., Mak, S. and Strom, E. T. (1965). *Adv. in Chem. Ser., Am. Chem. Soc.* **51**, 112
Russell, G. A., Bemis, A. G., Geels, E. J., Janzen, E. G., and Moye, A. J. (1968). *Adv. in Chem. Ser., Am. Chem. Soc.* **75**, 174
Russell, G. A., Norris, R. K. and Panek, E. J. (1971). *J. Am. Chem. Soc.* **93**, 5839
Russell, G. A., Hershberger, J. and Owens, K. (1979a). *J. Am. Chem. Soc.* **101**, 1312
Russell, G. A., Jawdosiuk, M. and Makosza, M. (1979b). *J. Am. Chem. Soc.* **101**, 2355
Russell, G. A., Jawdosiuk, M. and Ros, F. (1979c). *J. Am. Chem. Soc.* **101**, 3378
Russell, G. A., Ros, F. and Mudryk, B. (1980). *J. Am. Chem. Soc.* **101**, 7601
Russell, G. A., Mudryk, B. and Jawdosiuk, M. (1981a). *J. Am. Chem. Soc.* **103**, 4610
Russell, G. A., Mudryk, B. and Jawdosiuk, M. (1981b). *Synthesis* 62
Russell, G. A., Hershberger, J. and Owens, K. (1982a). *J. Organomet. Chem.* **225**, 43
Russell, G. A., Mudryk, B., Jawdosiuk, M. and Wrobel, Z. (1982b). *J. Org. Chem.* **47**, 1879
Russell, G. A., Mudryk, B., Ros, F. and Jawdosiuk, M. (1982c). *Tetrahedron* **38**, 1059
Russell, G. A., Ros, F., Hershberger, J. and Tashtoush, H. (1982d). *J. Org. Chem.* **47**, 4316
Russell, G. A., Tashtoush, H. and Ngoviwatchai, P. (1984). *J. Am. Chem. Soc.* **106**, 4622
Russell, G. A., Khanna, R. K. and Guo, D. (1985). *J. Org. Chem.* **50**, 3423
Russell, G. A., Jiang, W. and Hu, S. S. (1986a). *J. Org. Chem.* **51**, 5499
Russell, G. A., Khanna, R. K. and Guo, D. (1986b). *J. Chem. Soc., Chem. Commun.* 632
Simig, G. and Lempert, K. (1961). *Chem. Ber.* **61**, 607
Simig, G., Lempert, K., Tamas, J. and Szepsey, P. (1977). *Tetrahedron Lett.* 1151
Simig, G., Lempert, K., Vali, Z., Toth, G. and Tamas, J. (1978). *Tetrahedron* **34**, 2371
Singh, P. R. and Jayaraman, B. (1974). *Indian J. Chem.* **12**, 1306
Singh, P. R. and Khanna, R. K. (1983). *Tetrahedron Lett.* 1411
Singh, P. R., Nigham, A. and Khurana, J. M. (1980). *Tetrahedron Lett.* 4753
Singh, P. R., Khurana, J. M. and Nigham, A. (1981). *Tetrahedron Lett.* 2901
Spanswick, J. and Ingold, K. U. (1970). *Can. J. Chem.* **48**, 546
Tang, R., Yue, H. J., Wolf, J. F. and Mares, F. (1978). *J. Am. Chem. Soc.* **100**, 5248
Tanner, D. D. and Mosher, M. W. (1978). *Can J. Chem.* **47**, 715
Tanner, D. D., Blackburn, E. V. and Diaz, G. E. (1981). *J. Am. Chem. Soc.* **103**, 1557
Tanner, D. D., Arhart, R. and Meintzer, C. (1985a). *Tetrahedron* **41**, 4261
Tanner, D. D., Diaz, G. E. and Potter, A. (1985b). *J. Org. Chem.* **50**, 2149
Tanner, D. D., Singh, H. K. and Yang, D. (1986a) *Rev. Chem. Intermed.* **7**, 13

Tanner, D. D., Singh, H. K., Yang, D. and Stein, A. R. (1986b). *Preprints of Papers, Div. of Petroleum Chem.*, p. 897. 192nd National Meeting of the Am. Chem. Soc., Anaheim, California
Walling, C. and Buckler, S. A. (1955). *J. Am. Chem. Soc.* **77**, 6032
Walling, C. and Indictor, N. (1958). *J. Am. Chem. Soc.* **80**, 5814
Waters, W. A. (1942). *J. Chem. Soc.* 266
Wawzonek, S. and Thelan, P. J. (1950). *J. Am. Chem. Soc.* **72**, 2118
Wayner, D. D. M. and Arnold, D. R. (1985). *Can. J. Chem.* **63**, 871
Wong, P. C. and Arnold, D. R. (1979). *Tetrahedron Lett.* 2101
Yasufuka, K. and Kutal, C. (1984). *Tetrahedron Lett.* 4893
Zeilstra, J. J. and Engberts, J. B. F. N. (1973). *Recl. Trav. Chim. Pays-Bas* **92**, 954
Zeldrin, L. and Shechter, H. (1957). *J. Am. Chem. Soc.* **79**, 4708
Zorin, V. V., Kukovitskii, D. M., Zlotskii, S. S., Todres, Z. V. and Rakhomankulov, D. L. (1983). *J. Org. Chem. SSSR* **53**, 906

Author Index

Numbers in italic refer to the pages on which references are listed at the end of each article

Cumulative Index of Authors

Cumulative Index of Titles